Here Be Dragons

The Rise of SpaceX and the Journey to Mars

by Stewart Money

APOGEE PRIME

To Laura,
and the discovery of one particular harbour

Contents

Acknowledgements

The hardest part is sometimes simply getting started, and on that note I would like to thank Dr. John Logsdon for an email out of the blue which had the unintended effect of providing motivation as I was deciding whether to undertake this endeavor. Also a thank you to Jeff Foust both for publishing a series of articles in The Space Review, and for being an endless source of information regarding the political aspects of the American space program.

Also, many thanks to my cousin Joe Moorman, site architect for Innerspace.net, but even more so for many years of dreaming, almost as many scheming, and in the end, the enthusiastic encouragement to just make it so.

And of course to my editor Rob Godwin with Apogee Books. Thank you for sound advice, and for taking a chance on a first time author at a time when the publishing industry is undergoing tremendous upheaval.

Many thanks to the people with SpaceX, both past and present, who have helped along the way.

While *Here Be Dragons* is an independent account of the first era of Space Exploration Technologies Corp. and was neither authorized nor reviewed by SpaceX, it would still not have been possible without their gracious help. Any mistakes or omissions are solely my own.

In particular I would like to thank Emily Shanklin for answering questions, arranging visits and in general being of invaluable assistance to this project. I hope you like it.

And to Elon, thanks for getting the ball rolling. It's a great big solar system out there, and it's time we started to check it out.

To my wife Laura, who made this effort possible, providing both encouragement and motivation, thank you for having the patience to see this through to conclusion. I love you.

Finally, for a certain Black Lab, who oversaw every keystroke, every correction and every spilled cup of coffee along the way, what can I say? Thanks Einie.

Prologue

There was a time, in the early 1960's, when the future of the United States, and of the human race itself, seemed inextricably intertwined with the fate of a regularly growing stream of rockets and converted missiles roaring off the small reinforced concrete pads planted amongst the scrub and dunes of Florida's Cape Canaveral. What had only a few years before been a testing ground for missile systems designed to deliver death and destruction on a regional and even planetary scale, became instead a port of departure for voyages not of death; but to expand life to new worlds. In the process, names such as Redstone, Atlas and Titan became for a time nearly as familiar as Ford, Chevrolet and Chrysler.

It was an incredible, heady time to be alive, and even though the decade would come to be considered as one of the most turbulent, tragic, and ultimately formative in the American experience, Cape Canaveral was a time and a place where the sense of progress and of the unbounded potential of the human experience was literally in the air. By the end of the decade, another name, Saturn, supplanted them all, an American flag was on the Moon, and Neil Armstrong and Buzz Aldrin were back on Earth.

It is often said that things are rarely as good, or as bad, as they seem to be at the time. Funding for the Apollo program peaked in 1966, and monetarily at least, the nation's commitment to space exploration began a long, slow decline beginning with the termination of the Saturn production line and the cancellation of the final three lunar missions; Apollo 18, 19 and 20.

Although the nation continued to spend money on NASA, on average 3/4 of one percent of the Federal budget during the intervening years, human exploration of new horizons took a back seat with the cancellation of the Saturn V launch vehicle. So when President Barack Obama breezily asserted in 2010 that his administration was cancelling NASA's ill-fated attempt to recreate Apollo, Project Constellation, because "we've been there before," he at least had the benefit of precedent.

We may have been there, but as it turned out, we apparently missed a couple of attractions which should have been on the itinerary, in particular the apparent deposits of water ice in permanently shadowed craters at the lunar poles. Sadly in 1972, the nation still had the means to go to the Moon, but not the knowledge of where to look. By 2010, we knew where to look, but now lacked the means.

The decision to terminate production of the Saturn V was in some ways a practical response to an approach to space exploration which although successful, was prohibitively expensive to maintain on a sustained basis. The problem was not just the rocket itself, but the massive infrastructure behind the entire project, one which included a small army of support personnel as well as some very large and unique pieces of support hardware such as the Vertical Assembly Building, the Crawler-Transporter and a test stand in Mississippi which has yet to see another engine powerful enough to fully challenge its capacity.

Following three missions to Skylab, and the Apollo Soyuz test project in 1975, each of which was boosted to orbit by the Saturn IB launch vehicle, the nation said goodbye to the last of the Saturn family of rockets, and waited through the first significant "gap" in access to space for the debut of an entirely new space transportation system which offered the promise of lowering the cost to orbit well below the level of the operationally magnificent but extremely expensive and completely expendable moon rockets.

When Space Shuttle Columbia lifted off the pad for the first time in April 1981, the nation was for a while, once more captivated by a sense of what was possible in outer space. Nothing holds the attention span for long however, and so it was not too long before the Shuttle assumed its 30 year role as fact of American life, brought into the spotlight primarily by disaster, but occasionally for better reasons such as missions to repair the Hubble telescope. As a spacecraft which never left low Earth orbit, and flew on a semi regular basis, there was no chance it could command attention from a nation which had seemingly grown bored with even moon landings as early as Apollo 15.

And thus was set up the dichotomy that while the nation spent a fairly steady sum on manned spaceflight for nearly thirty years of the Shuttle program, conducting 135 flights, and launching more than 3.5 million lbs. into orbit, the state of human space exploration in fact offered little advance and even less promise. What had once captured the imagination of a nation, and a planet, became yet another government program, which although offering endless platitudes about learning to live and work in space, seemed to have little to show for it other than an overriding concern with self-perpetuation.

For those who grew up feeding on a steady diet of science fiction served by early masters of the genre, and those who followed in their path, including a former Pan Am Pilot named Gene Roddenberry, the seemingly open frontier of space, first reached in rapid succession by the dizzying advances of Mercury, Gemini and Apollo in a magical decade at Cape Canaveral, receded into an increasingly distant future. Even science fiction seemed to wane. As Mars visionary Robert Zubrin observed in 1996, a closer look at the science fiction section of a local bookstore, once the driving inspiration behind more pioneers than one could count, would reveal that many of the books were in fact not science fiction at all, but instead works of fantasy more concerned with swords and sorcerers. It is a trend which has accelerated to previously inconceivable levels, with vampires, werewolves and zombies now commanding far more attention than colonies on Mars or voyages of exploration. Where once we considered the range of futures a space faring civilization might make manifest, we are now offered reflections of the darker side of the human experience instead.

Of course, bookstores themselves have become increasingly scarce as well, pushed to the limits by one futuristic invention many never saw coming, the internet and with it global connectivity, massive multi player on-line gaming, social media and smart phones. For the time being, technology has driven us inward rather than upward. At least wireless technology has made it possible for us to take the gear outside, where even if by chance, we might occasionally point it towards the Moon or stars and allow it to label the craters and constellations, at least until the next text message.

It is not necessary to bemoan one advance in order to advocate for another, particularly when the relationship is so wonderfully interrelated. After all, the same internet which offers a million ways to divert, also offers Google Earth, Moon, and Mars, and while the printed word still holds appeal, allows Amazon to offer a stunning variety of books inconceivable for any one store, including accounts of space that was, with astronaut memoirs and histories of the Shuttle program, as well as the future in space that might be, science fiction. This book explores through the history of SpaceX, the relationship between the two, which is bound in the present by the simple fiction that the United States is either willing or able to repeat the Apollo experience. And yet, as of 2014, that is precisely what NASA, at the command of the U.S. Congress is attempting to do in developing the Space Launch System, a launch vehicle so very much meant to remind us of the Saturn V, that it even sports the same paint job.

The problem was and still is, that space exploration begins with launching spacecraft, supplies, astronauts and most of all fuel; hundreds and hundreds of tons of fuel, into Earth orbit, and the cost to accomplish that task with hardware and far flung infrastructure has not gotten any cheaper since the Saturn V production line shut down, despite the promises and best intentions of those who originally conceived the Shuttle program.

The Apollo and Shuttle era approaches are not the only possible paths to space. Engineers, business leaders and space enthusiasts have long argued that there are other ways, although with little substantive proof until now. Whether by fate or happenstance, a handful of entrepreneurs who built their businesses and their fortunes in advancing the digital economy, are now applying their efforts and their money to solving the problem of affordable access to space. The effort, with enormous assistance from NASA, is beginning to pay off, as evidenced by the first ever private flight to the International Space Station in May of 2012.

This book, though principally concerned with the rise of Space Exploration Technologies Corporation and its Falcon 9 rocket and Dragon spacecraft, also attempts to tell the story about what has gone wrong, particularly in recent years, in the context of the challenges which SpaceX has embraced. It is a story told in several sections beginning with the company's first rocket, the Falcon 1 which encountered a difficult journey on its way to becoming the first privately developed liquid fueled launch vehicle in history to reach Earth orbit. In this section, intended for either the casual observer of space exploits, who perhaps lost interest years ago, or the new reader who is far more familiar with the elements of social media than launch vehicles, the challenges encountered in the development of the Falcon 1 offer an opportunity to re-introduce the basic elements of launch vehicles and the mechanics of reaching orbit.

The second section focuses on the evolution of SpaceX and the development of the much larger and more capable Falcon 9; a vehicle whose capacity puts it in the very center of the world wide stable of launch vehicles. For SpaceX, and indeed for any new launch vehicle, establishing a thriving transport business to Earth orbit is a necessary and unavoidable precursor for any plan which would once more address the space beyond.

The third section features the development of the Dragon capsule and seeks to introduce some of the basic elements of spacecraft as opposed to launch vehicle operations. In doing so it focuses on the unique program which is the Dragon's first destination, the International Space Station, an orbiting complex only now coming into its own as a true research facility, even as it serves a vital role as the sole anchor tenant for private manned spaceflight.

The fourth and final section, addressing SpaceX's program to develop both a heavy lift as well as a reusable booster, examines the stunning potential for both systems to bring about a new era of deep space exploration at a fraction of the cost of legacy systems.

If there is one thing that the reader should take away from the story of SpaceX, it is that considerably more affordable access to space is, and always has been, readily achievable without long shot investments in break-through technologies. For reasons both malignant and benign, the space infrastructure first established during the Apollo program, and perpetuated throughout the Shuttle era, dominated by politically connected large aerospace contractors and a top down, monotheistic NASA bureaucracy has consistently failed to deliver on the promise of an expanding space frontier. It has, quite simply, kept the future at bay.

Now, however, the future of space exploration and commerce, and mankind's place in the solar system and ultimately the universe, is being driven from multiple directions, but first among them is one company, SpaceX, led by one key individual, its founder Elon Musk, and with the surprising assistance of elements within America's space agency, has already clearly and undeniably provided existence proof that another way to space exists, and that the future, constrained for a time, is once again open to become whatever we choose to make of it.

Much of that future is inextricably intertwined with the planet Mars. The vast expanse of our own home solar system is proving an increasingly interesting place, but if there is one location which seems readily adaptable to human life, it is the Red Planet, and it is a fascination with the potential of Mars to offer a new home for humanity, a second branch of civilization which provides the motivating incentive for Elon Musk and SpaceX, and the story does not make sense except in that light. While NASA, free of the political constraints of the Shuttle Program, is once again at least able to openly discuss going to Mars, that is about all the agency can do, offering only a nebulous timeframe, and no credible plan for actually going there. Meanwhile, SpaceX is focused on building step by efficient step, the infrastructure for allowing not only the first human missions, but the beginning of a large scale human emigration to Mars within the next 20 years.

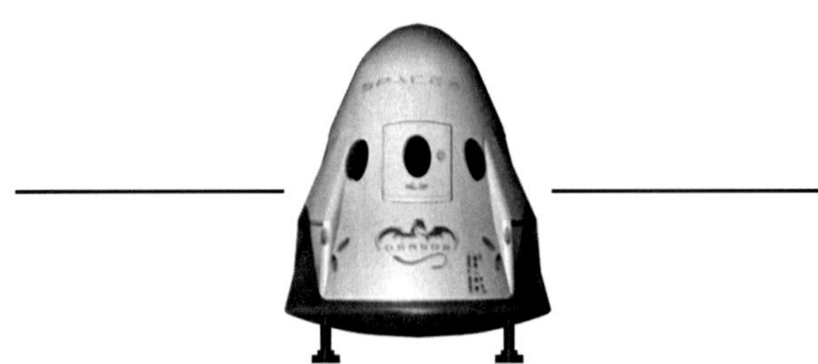

Chapter I: Into the Unknown

Two hundred and fifty years before Genoese captain Christopher Columbus departed Spain in search of a westerly route to India, two predecessors left on a similar quest, one which came to a very different and still mysterious ending. In the spring of 1291, Vandino and Ugolino Vivaldi left the Maritime state of Genoa in an expedition consisting of two galleys, the Sanctus Antonius and the Alegranzia, accompanied by a pair of Franciscan friars, a crew of sixty, and a full stock of supplies. Their goal was to bypass the dangerous and expensive overland trade passage to the Far East and discover a sea route to India. Although the names and dates are known to history, the outcome is shrouded in mystery, and we do not know for certain whether the brothers intended to sail south around the foreboding African continent, or whether more improbably, they sought to sail west into an open and hostile ocean bereft of even the occasional comforting site of terra firma, regardless of how incognita it might be.

Like many lost expeditions which would follow that of the Vivaldi brothers during the age of sail, this one gave rise to a number of stories and myths, many of which were contradictory, but some contained just enough hint of authenticity to be intriguing. One hundred sixty four years later, as the Portuguese were first inching their way around the African continent, an expedition sailed up the Gambia River, reportedly encountering a man who spoke Italian in the Genoese dialect, who claimed to be the sole remaining descendant of survivors of that first expedition.

Whether lost at sea, or grounding somewhere around the vast and untamed shore of 16,000 miles of African coastline, the Vivaldi expedition was almost certainly doomed from the start. As Genoan merchants in a city-state which was rapidly rising on the pillars of an international trade originating in the far east, the Vivaldi brothers were likely in a position to know as much or more about the eastern end of that trade route than other expeditions which would set off in later years. The crew would have been accomplished mariners, well versed in sailing techniques necessary to traverse the Mediterranean in comparative safety. What they lacked however, was both a vessel capable of operating in the open ocean beyond Gibraltar, as well as the critical knowledge and skills for navigating along an unknown and implacably hostile coast. In the end, irrespective of which route they ultimately chose, lacking the right equipment, the right ship, and above all the right knowledge, the final destination was obscurity.

On another spring day, this time in 1433, a small Portuguese sailing vessel cast off its lines and caught a rising wind to carry it out of the harbor of Lagos, slowly turned port, and headed south into the unknown. The voyage itself would be brief, and considered a failure by the man who had commissioned the trip. A year later, Gil Eanes was sent out again, this time with orders not to return until he successfully rounded Cape Bojador, a semi-mystical line of demarcation along the African Coast which had thwarted previous attempts, growing steadily in the imaginations of even experienced mariners to become an insurmountable barrier, the ocean's own forbidden zone. Not a lot is known about Eanes, or the more than a dozen failed voyages which preceded this one, but we do know a great deal about the class of ship, the Portuguese Caravel, which ultimately grew out of the effort. And we know this, the commitment to exploration begun with this first, unhappy journey would be the opening act in a new age of discovery, one which would ultimately fill in all the blank spaces on the map, and create the first renditions of the global trading economy which drives the world to this day.

Five hundred and seventy three years later, on the far side of the planet, on an island very near the destination the Portuguese traders ultimately sought, a small rocket lifted off on the first flight of a new family of launch vehicles, and it too would be a failure, and the first of several. Like its predecessor in the age of sail, the SpaceX Falcon family of boosters incorporated a number of features, both new and old, in a surprisingly modest package which offered one simple advantage, it was an affordable craft designed to carry out a sustained program of exploration and commerce. To begin to understand the potential inherent in the latter, it is useful to look back across the centuries at the record setting performance of the former as the modest vessel which tamed a planet.

Throughout the rise of the early Mediterranean cultures, trade was conducted along coastal routes, never far from the sight of shore, and the design of vessels whether intended for trade or war reflected local sea, current and wind conditions as well as the limitations of local building materials. As empires arose and wealth was accumulated, we witnessed the development of specifically designed, purpose built ships, from ubiquitous merchant galleys to oar-driven triremes powered by three rows of hapless slaves, to opulent vessels of state represented in popular culture as Cleopatra's barge. What was entirely missing however, were vessels designed for exploration, because there was so very little of it.

Not much had changed more than 1,000 years later when the Vivaldi brothers were preparing to make their voyage into the unknown. Consequently they set out in what was available, a pair of coastal trading galleys equipped with oars and a single square sail. Although there is some evidence to suggest the expedition may have made it as far as the Canary Islands, they were already past the point of no return. The particular combination of wind and cyclical ocean current, the turn of the sea, meant that in traveling any further along the African coast, an expedition would encounter treacherous conditions which would have driven them to destruction onto the barren shore. The only safe alternative, learned after trial and error, was an improbable diversion far out into the Atlantic, followed by a turn south into an unknown and unknowable fate.

Even as the Renaissance began to unfold, fueled by the wealth of the eastern trade the Vivaldi brothers sought to capture, all overland routes to the riches of the east ran through Arab lands which brooked no intervention. Thus it was that a hundred seventy years later, an entirely different group of explorers prepared to test the sea route. This time it was not the trade-rich Italian states, but relatively poor Portugal, which considered exploration as a means to develop trade. Led by Infante Henry, Duke of Viseu, better known to history as Prince Henry the Navigator, the small nation embarked on an incremental and sustained era of exploration, the sheer duration of which is almost unimaginable in the modern era. The secret of its success lay in both the method of exploration, a series of incremental voyages stretching out over a century, and in the means, the use of a new type of ship, the Caravel. Among its many virtues, including lateen sails which allowed it to sail into the wind, was the fact that it was just large enough to brave the open ocean, but small enough to be operated efficiently and sailed swiftly by a minimal crew for months at a time to destinations halfway around the planet. Of shallow draft, it could push up coastal rivers, chart reefs and shorelines, sound smaller harbors and be careened on distant island beaches for repair.

Although the Caravel would evolve over time, and ultimately be replaced by larger ships of trade and war, for a period of three hundred years, these remarkable little vessels allowed nations, corporations and even motivated individuals, to explore the expanse of a planet which is 70% water with an efficiency and cost effectiveness unattainable by the much larger vessels which kings and queens regularly commissioned to enshrine their own sense of wealth and power. Centuries later, not much has changed, and if we are to embark on the next great era of human exploration, we must sail in the wake of the Caravels.

Officially, the launch of Sputnik on October 4, 1957 ushered in the space age. For some however, the era was already alive and well in the growing genre of science fiction; led by the trio of great authors, Asimov, Clarke and Heinlein, and the many others who published stories both long and short in Amazing and Astounding magazines. At the same time, the general public was treated to the expansive visions of Wernher von Braun through Life and Colliers magazines in the years leading up to America's first forays into space, energized by the latter's bold declaration "Man Will Conquer Space Soon!" Such exhilarating images of the future would quickly come to odds with the stark picture of the solar system which would soon begin to emerge as the first fleets of American and Soviet spacecraft set out into deeper space.

To the newly evolving field of planetary scientists, the results which came dribbling back in via weak radio signals were no doubt exciting. After all, the ability to see the first images and data streaming in from a new world were a testament to the incredible times we were living in. To everybody else though, those eagerly anticipating the first close up images of worlds which might one day harbor human life, the results were disappointing to say the very least. The Moon of course, we knew well, but Mars? The first successful flyby of the Red Planet, conducted by Mariner 4 in 1965, produced 22 grainy black and white photographs which revealed a dead and dry world. In retrospect it is amazing to consider the fact that even as the U.S. was building and testing the hardware which would place men on the Moon only four years later, relative ignorance of our planetary neighbor was so great, that many would have been hardly more surprised if those first photographs revealed instead abundant evidence of life. Later probes subsequently detailed interesting terrain at least, with Olympus Mons rising as the tallest mountain in the solar system, and Valles Marineris stretching across nearly 25% of the planet's circumference, but when the twin Viking Landers touched down in 1976, what they showed in striking colors amounted to a somewhat boring, flat plain, strewn with small rocks.

The Viking missions marked an end to a remarkable initial program of Mars exploration which even if the results were less than exhilarating, provided a wealth of information about the planet. What began next was the first great "gap" in Mars planetary exploration, caused not coincidentally by the budget demands of the massively complicated Space Shuttle System, and it would not be until 1992 that the United States would attempt to resume robotic exploration of the Red Planet with the ill-fated Mars Observer spacecraft.

If the immediate environs of the solar system were not what we hoped them to be, that was not a problem. We could always build our own worlds, large stations in orbit. For a while perhaps it seemed possible. With the Shuttle set to debut a few years in the future, offering the possibility of high launch rates and low launch prices, the construction of large scale, permanent stations, and eventually even cities in space seemed for a time, not all that remote.

A funny thing happened on the way to the space station though. Even as the Shuttle revealed itself to be a far cry from the cost efficient system promised, a second wave of planetary probes, some launched by the Shuttle, began to peel back the imperfect layer of first impressions, and reveal the solar system to be a much stranger and infinitely more interesting place than we had thought it to be. At the same time, a series of smaller Mars probes commissioned under NASA's Faster, Better, Cheaper mantra, began to suggest that Mars is a more interesting, and plainly more livable world than we thought it to be from the sterile images arriving in 1965. Beginning with Mars Global Surveyor launched in 1996, and the Mars Odyssey orbiter launched in 2001, NASA built the case for the long held suspicion that in addition to the frozen carbon dioxide which makes Mars' polar caps visible from high end telescopes on Earth, the planet also harbors a significant quantity of water ice. Seven years later, NASA announced on July 31st, 2008 that its Phoenix lander had indeed confirmed the presence of water ice on Mars. The implications were clear. Where there is water, light and material for shelter from cold and radiation, there exist the constituent ingredients to support human life and civilization.

Only a few years before the U.S. probes returned to Mars after a 15 year absence, President George H. W. Bush stood on the steps of the Smithsonian Air and Space Museum on July 20th, 1989, the twentieth anniversary of the first lunar landing. Flanked by the crew of Apollo 11, Michael Collins, Buzz Aldrin and Neil Armstrong, the President called on the nation to begin a new era of exploration aimed at mounting the first manned (now crewed) mission to Mars. Unfortunately, no one at NASA picked up the phone. The response was a case study in everything which was wrong with the government's approach to manned space exploration. Rather than present a clear, focused plan for reaching Mars, the agency's various departments responded instead with an all you can eat buffet, endless salad bar wish list of "must do" projects, many of which were only loosely related to Mars to begin with. A price tag was never officially announced, but once the press picked up on one ungrounded and wildly inflated estimate of $500 billion and began repeating it endlessly, the proposal was DOA. For many in the agency it was just as well. NASA was more concerned with flying the Shuttle than actually exploring space and seemed happy to avoid the rude interruption and go back to doing just that. A decade later, with another largely forgotten Apollo anniversary out of the way in 1999, NASA still appeared content to go on flying the Shuttle and building the ISS for at least the next 20 years, with possible extensions into the 2030's on the table for discussion. The latter was possible ironically enough, because due to its incredibly high cost, technical complexity and corresponding low flight rate, as the third decade of Shuttle operations approached, the orbiters were nowhere near exceeding their airframe's designed lifespan, a figure expressed in the total number of flights performed, until many years in the future.

That all came crashing down with the debris of Space Shuttle Columbia on February 1st, 2003. Though the Shuttle program would go through another multi-year stand down and subsequent return to flight, this time it was to finish construction of ISS, and that was all. Despite protests from some quarters which suggested that partially as a result of the learning experience of two tragic accidents, and backed by some empirical data to prove it, the Shuttle program was now actually safer than it ever had been and could continue to operate, another Bush administration introduced a new space policy a little over a year after the accident. The policy, the Vision for Space Exploration called for a renewed program of voyages beyond low Earth orbit, and a new family of launch vehicles to make it happen.

Six years and two blue ribbon Presidential commissions later, it would be readily apparent that after a series of embarrassing development problems and skyrocketing costs, the United States could not afford the new program, Project Constellation, without substantial and highly improbable boosts to NASA's budget. With the Shuttle program approaching its final three flights, and beyond the practical point of return, the nation's space agency found itself in one hell of a mess. With an obligation to launch both supplies and astronauts to the International Space Station until at least 2015, and an ever widening gap in the ability to accomplish either goal, the agency's problems in affordably accessing space had finally grown to the point that they were too big to ignore, and too expensive to fix. The only way out of the quandary was a program of commercial re-supply and astronaut taxi service which was potentially more dangerous to the status quo if it succeeded, than if it failed.

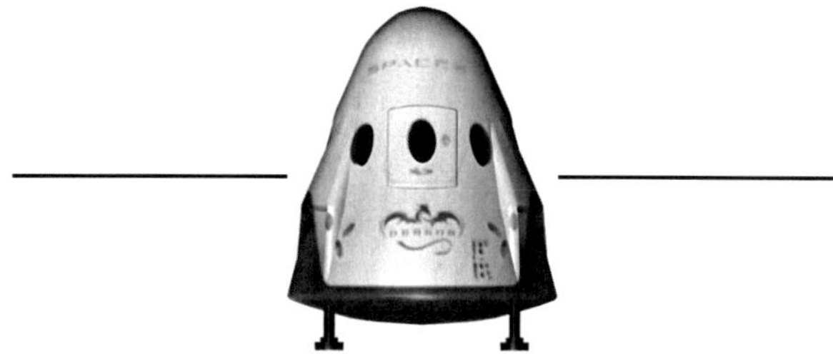

Chapter II: The Mars Oasis Project

SpaceX, like many successful ideas, began as a response to a problem. The specific problem in this case was that in 2002, even for a freshly minted multi-millionaire like Elon Musk, obtaining access to space through the purchase of commercial launch service was a prohibitively expensive affair. American launch vehicles offered by legacy aerospace giants such as Lockheed Martin and Boeing were outrageously expensive even by the standards of the day. The international market offered more affordable, albeit considerably riskier alternatives in the form of space adapted Russian intercontinental ballistic missiles, perhaps the ultimate manifestation of the conversion of swords into plowshares, but there were drawbacks. First, dating back to the Soviet era, the Russians had a lousy delivery history for the destination Elon Musk had in mind, and more importantly, they did not come across as the most reliable business partners one could find.

Beginning with Dennis Tito's historic flight to the International Space Station in 2001, Russia secured its place as the world's first and thus far only country willing to use its considerable space assets to accommodate a handful of ultra-rich space tourists eager to exchange a portion of their wealth for an experience which literally could not be had on Earth. Flush with cash from the recent sale of online payment service provider PayPal to eBay, his second successful internet startup, Musk also wanted to purchase a ride to space, but his ambition was at the same time more humble, and infinitely more grandiose than that of the handful of tourists which would ride the Soyuz rocket to orbit. For Musk, who could have easily followed Tito's path, the ride was not for himself, but rather for something much smaller, a small container of seeds. Admittedly by 2002 there was nothing particularly original about flying seeds into space. After all, seeds comprised one of the first "get away specials" on the Space Shuttle, lofted aboard Challenger in 1983, courtesy of the Park Seed Company. While the Park seeds were primarily fruit and vegetable, their chief accomplishment was to demonstrate that much to the disappointment of school children across the country, seeds flown in space experienced no noticeable bizarre effects and worse yet, still yielded questionable benefits such as abundant spinach and Brussels sprouts, Musk however, had something much more quixotic in mind. He wanted to fly and land his seeds not in Ms. Grundy's 8th grade science class, but on Mars.

Imbued with a passion for "technology which will change the world," which extended well beyond the realms of internet programming and had made him a rich man at a young age, Elon Musk had for some time been contemplating what struck him as one of the greatest technological ironies of all time, the post Apollo lull in space exploration. As an immigrant and a new citizen of the United States, and more particularly one who had made the most of the opportunity which the country offered, it seemed contradictory that what appeared to be the quintessential American enterprise, space exploration, had progressed very little since the heyday of Apollo. The Shuttle program, though impressive in terms of its technological complexity, and the sheer scale of its operations, had long since ceased to inspire excitement or elicit a sense of progress.

Musk's plan, and the reason he wanted to purchase a commercial space launch, was to attempt to inject a new sense of adventure and opportunity into space exploration by providing an indelible symbol of the promise of a new era of exploration. To Elon Musk, like so many others, there was no greater siren call than that sounded by the small, ruddy planet which had fixated mankind's imagination since the dawn of western civilization, Mars.

For those who had been paying attention, the steady program of Mars exploration begun in the mid 1990's was revealing the Red Planet to be a more interesting destination than what it initially appeared to be in humanity's first, brief glance provided in black and white photos from early flyby spacecraft. Following the failed Mars Observer mission launched in 1992, NASA scored a major success and a media home run in 1997 with the arrival in orbit of the Mars Global Surveyor followed by the Mars Pathfinder Lander and its tiny Sojourner rover. Coming only a year after the announcement on August 16th, 1996 that possible signs of microbial life were detected in a Mars meteorite, Allan Hills 84001, which had been collected from Antarctica in 1984, interest in Mars briefly peaked, and then inevitably faded into the background. Except for periodic incidents generated by events both good and bad, such as the twin failures of Mars Climate Orbiter and Mars Polar Lander in 1999, and the success of the Mars Odyssey orbiter in 2002, in a 24 hour news cycle, neither space exploration in general nor Mars itself, offered much to hold attention spans already under regular assault from video games, and emerging technologies such as smart phones and social networking sites as well as the endless other distractions and difficulties of everyday life. After all, how could the remote challenges of exploring a cold, dry and dusty new world compare to the more earthly excitement of faux life entertainment offering a steady stream of melodramas accented with ripped shirts and string bikinis on the next episode of Survivor, the Bachelor or Jersey Shore?

As unlikely as it might seem, igniting a new era of space exploration required somehow rising through the noise and clatter and finding a way to excite the general public about the possibilities it contained. Elon Musk reasoned that one path to accomplish this was to provide an actual physical example of an alien world coming to life, specifically, flowers and vegetables grown on Mars.

That image, if made manifest, could perhaps serve as a single concise point to illustrate to the general public that even today, we have the potential to spread life outwards from Earth to other parts of the solar system. In some ways it would be the counterpoint to the iconic Earthrise photo taken by the crew of Apollo 8 on Christmas Eve, 1968 as they rounded the dark side of the Moon to see our home planet rising, blue, green and white, and so incredibly alive, over the cold desolate lunar landscape. Perhaps the reverse could be just as powerful, the image of verdant, green life from Earth, growing defiantly and magnified against a cold alien world, as opposed to an entire planet viewed from an immense distance. [1]

In order to avoid the issue of planetary contamination, the flower would need to be grown in a nutrient gel inside a thoroughly sealed environmental chamber housed aboard the Martian lander which the rocket he sought to purchase would lift into space. Though isolated from the actual, external Martian environment, it would be a flower grown in a chamber flooded with the thin Martian atmosphere and perhaps most symbolically, under the pink, wan sunlight of the Martian sky. Musk's hope was that the image and the surrounding publicity would be the catalyst to spur a NASA manned mission to Mars and reawakening of the space age. If the idea itself was a stretch, and surely it was, hoping to inspire NASA to carry out a new program of exploration may have been the most unlikely element of the entire scheme.

Elon Musk's initial approach to addressing the problem, though curious in hindsight, reflected the degree to which both outsiders and those within the aerospace establishment had come to believe, or at least to profess, that the reason the U.S. space program remained stagnant after Apollo and through Shuttle program was the lack of a single coalescing principle such as that which guided Apollo. The years since the last Apollo mission have seen a stream of editorials, advisory groups and blue ribbon commissions all expounding on the theme that NASA is "lost in space" and "needs a new direction," but one has never been forthcoming, at least not for long. Forty years of meandering policy have led to the more recent claim that what the agency really needs, even more than a destination, is instead a guiding principle, what is now sometimes referred to as an "enduring value proposition." In truth, both viewpoints, which could be boiled down to a debate between a destination versus a mission driven agenda, overlooked the fact that the cost basis of existing space infrastructure was so out of touch with economic reality that it would rebuff any renewed exploration program, even one initiated at a Presidential level, as President George W. Bush would soon discover.

So how naïve was Elon Musk's original concept? It is only natural that a person's perception is guided by their own experience in life, and for Musk that experience was strongly influenced by the growing phenomenon of peer to peer marketing clearly demonstrated by his experience in leading his two former companies, Zip2 and PayPal to succeed as well as they did. While today the term "going viral" is an everyday figure of speech, its origins lay with the rapid growth of internet companies such as PayPal which often spread by word of mouth or person to person, rather than from direct advertising or marketing. In fact, under Musk's leadership, PayPal went from zero to over a million users with no sales force and no marketing. Instead, word spread from one customer to another, as individual users sought to pay each other with the new medium. With this kind of peer to peer marketing as a background, Musk viewed the challenge of motivating space exploration from a point of view of providing an example which excited individual citizens as opposed to a top down approach of pushing a particular space agenda, observing that the general public responded to "precedents and superlatives." As the world has grown vastly more interconnected in the brief time since Elon Musk originally pursued the Mars Oasis project, it is not altogether out of the question that such an image, presented then, now, or even in the future, might have a more profound effect than would be otherwise possible. [2]

We are all dreamers. As engaged members in a constantly changing industrial society, any healthy human might from time to time have a creative or unusual thought or two on how to improve that society and the trajectory of change. The difference is that most people have neither the inclination, nor the means to follow up on those thoughts. Elon Musk had both. Specifically, he was willing to invest $20 million to bring the Mars Oasis project to life. Unfortunately, that figure roughly approximated the cost estimate for building the lander, which was not all that large. The real sticker shock came when he discovered that the price tag for the only appropriately sized U.S. launch vehicle available at the time, the Boeing Delta II, would start at $50 million.

Realizing that a U.S. launch provider was a financial non-starter, Musk looked at the international market and at Russia in particular. Russia at that point was actively marketing several series of refurbished ballistic missiles as commercial launch vehicles, and compared to everyone else, the price was right. Musk traveled to Moscow three times seeking to secure a rocket for his project. It was not until the third trip however, that he began to question the overall situation which seemed to be leading him towards a Russian rocket in the first place. The question arose, why? After all, while Americans imported an increasing volume from China, they bought very little in the way of industrial products from Russia, with the notable exception of vodka, and the preferred mixer of both gun collectors and revolutionaries around the world, the Kalashnikov AK-47.

That was really the extent of it, among the vast array of high tech or manufactured products which serve the needs of everyday life, a good many are built in Germany, Japan, China, Korea, at home in the U.S. and increasingly even in Mexico, but Russia, almost nothing, so why rockets? To Musk, this suggested that the problem was not in Russia, but instead back in the U.S. with the unreasonably high price of American launch vehicles. For a natural entrepreneur, the apparent paradox, and the discrepancy between Russian and U.S. pricing for what should be an equivalent service, meant that an opportunity waited. Perhaps the better answer was to forgo the Russian option, set aside the immediate goal of the Mars Oasis for a time, and consider the possibility of building his own rocket.

It was not a decision to be taken lightly. In pursuing his somewhat quixotic attempt to jumpstart the U.S. civil space exploration program, Musk ran not only into a personal roadblock, the $50 million dollar cost of the Delta II launch vehicle in 2002, but also smack into the massive obstruction which had kept the United States from accomplishing virtually anything new in terms of human exploration for several decades. It was not just the cost of a single modestly sized launch vehicle, it was in fact an entire system; a network of government agencies, contractors and commercial vendors which seemed perpetually locked in an endless loop of unreasonably high costs which was proving remarkably resistant to change either from within or from without. For someone coming from the outside, and in particular from the constantly evolving and expanding world of computers and internet business, the stagnation was appalling. After all, as early as 1965 the rapid progression in computer processing capability gave rise to Moore's law, which observed that the approximate state of computing power roughly doubles every 18 months. With that as a point of comparison, and driven by a deep seated conviction that technology should progress, and that the future should be better than the past, the relative paralysis in advancing space technology was not only profoundly disappointing, it also seemed wholly unnatural.

What Musk encountered was an environment which although like computer processing power and seemingly technological in nature, was in reality also something very different. The space industry owed as much for its present condition to political ties and a geographically dispersed contractor network, cemented during the buildup of the defense establishment throughout the Cold War, and its quasi civilian cousin, the Apollo program, as it did to the actual hard work of successfully escaping Earth's gravity well. Nowhere was this more obvious than in the two large programs which had dominated the American perception of what it meant to be a space faring nation for most of the last 40 years, and in the process built an air of inevitability around the proposition that getting to space would always be both difficult and expensive.

The stunning growth in capability during the early years of the space race and America's sprint to the Moon tapped into the far flung network of defense contractors, sub-contractors and sub-sub-contractors to produce a remarkable technological success in the Saturn V launch vehicle and the hardware of the Apollo program. Driven by the constraints of President Kennedy's "before the decade is out" timetable, Apollo was the successful result of a national mobilization of people and materials on a virtual war footing, a scale which the nation was unable and unwilling to sustain beyond the first flights. The limited flight history, massive infrastructure costs, and rapid development schedule combined to produce three launch vehicles, the Saturn 1, Saturn 1B and Saturn V which still remain as some of the most expensive launch vehicles in history in terms of a per flight and per pound basis.

It is difficult, if not impossible to objectively determine what it cost to launch either the Saturn V or the 1B on a per flight basis, because the figure varies widely depending on which costs are taken in to account. In round numbers however, the cost per flight of the Saturn V in 2012 dollars amounts to approximately $3 billion per flight. [3] In one sense, it is not as bad as it seems, because with a lift capacity of 260,000 lb. to LEO, the cost per pound comes in to about $11,500 which is less than some current launch vehicles. But for most purposes it is an almost meaningless number, as there is no practical way to take advantage of all that lift capacity, short of packing satellites like olives in a jar.

The Space Shuttle program was billed as an attempt to change the unsustainable cost basis of Project Apollo, which had seen the cancellation of the final three lunar missions originally planned and the termination of the Saturn V production line. Though partially reusable, the Shuttle was still a product of that same system, but with a development budget strung out over a decade instead of condensed into a few short years. It was also hampered from the outset by a series of design considerations made to accommodate divergent commercial, NASA and military roles in order to gain the political support for a development decision. Attempting to be all things to all people, and ordered on the premise of a wildly optimistic flight schedule, the Shuttle program was sold on the promise of a 10 fold reduction in the costs to orbit, but there was little chance it would ever come close to achieving those goals. Obtaining a true assessment of Shuttle flight costs are if anything more difficult than it is for the Saturn series, for many of the same reasons. The Shuttle never lifted off with anything approaching its original estimated payload capacity of 63,500 lbs., a burden which would have made normal landings hazardous at best, and rendered already precarious in-flight abort scenarios virtually impossible. With cost estimates ranging from $400 million to slightly over $1.6 billion per flight over its 135 flight history, the Shuttle cost per pound is an almost meaningless number for the reasons cited, but by any measure it was both much higher than originally predicted, and far too high to fulfill the promises of opening up a new era of space commerce. Not that it mattered. Once having returned Americans to space in 1981, it did not take long for the justification to keep flying the Shuttle to become the self-perpetuating rationale behind a flight program which launched hundreds, employed tens of thousands, and ultimately bored millions.

At one time, the Shuttle itself might have provided the means of launch for Elon Musk's Mars Oasis project. Prior to the Challenger accident, the Shuttle system benefitted from a national space policy which made it essentially the nation's only "commercial" launch provider, providing heavily subsidized launch "services" which came at the price of almost completely eliminating other domestic launch options, not to mention the incentive for developing new ones. [4] By 2002 however, as a result of the Challenger disaster, the Shuttle had long since been taken off the market for all but NASA launches. As Elon Musk debated entering the launch vehicle industry, the sheer size of the Shuttle program seemed to serve as a self-evident example that despite the best effort of some of the nation's smartest people, reaching orbit would always be a very expensive proposition.

Although there were other domestic alternatives to the Delta II he originally sought, including the Atlas 2AS, as well as the Atlas V and Delta IV both entering service that same year, each one was both a larger vehicle, and very much a product of the same industrial base and politically connected network which had produced the Shuttle, and it was reflected in their relative costs. Furthermore, it was a system both protective and perversely proud of its own inefficiencies, and highly motivated to vigorously resist the type of change Elon Musk was considering bringing to the industry.

Recognizing the inherent but artificial inefficiencies in the price basis of current U.S. launchers, the real question was how much was being left on the table, and was there enough of a margin to allow the recovery of research and development costs? It was clear for instance, that the European space consortium Arianespace, led by France, had gained a considerable market share at U.S. expense first with its Ariane IV launch vehicle, and then with its successor, the much larger and more powerful Ariane V. Furthermore, both Russia and China, benefitting from lower labor costs than could be found in either the U.S. or Europe, were easily able to undercut both. While Russia was on relatively new ground in marketing converted ICBM's for the class of small spacecraft launches which Musk originally sought, its launch history with larger vehicles such as Soyuz and Proton was extensive, longer in fact than that of the United States. Although any effort to enter the market would necessarily focus on smaller vehicles of the type Musk originally sought to purchase for the Mars Oasis project, the decision itself would have to be made within the context of the global launch vehicle industry, including the class of larger rockets which delivered communications satellites to geostationary orbit.

For anyone interested in the history and future of SpaceX as it relates to mankind's permanent expansion into space and the potential to become a multi-planet species, understanding the basic elements of the global launch industry which faced Elon Musk as he considered undertaking a rocket manufacturing effort in 2002 is an important component in considering the prospects for the future as they currently stand, as little else has changed in the intervening decade, other than the fact that with the exception of SpaceX, launch costs have climbed steadily higher. In the murky analysis of what it actually costs to launch a payload into space, there is more to be gleaned from the somewhat open competition in the commercial satellite launch industry than from the elevated, distorted and frequently hidden "costs" of national space programs. Nevertheless, with most "commercial" launchers originally developed by national governments, and generally operating out of government controlled military launch facilities, even those numbers are open to debate. In 2002, the average launch price for the soon to be retired Ariane 44LP was $105 million, while the much more capable Ariane V, with double the lift capacity and the ability to deliver two satellites at once was roughly $140 million. By contrast, the older Russian Proton K was nearly half that, at $72.5 million. Altogether however, the total market for the year was only 24 commercial launches delivering a cumulative total of 34 commercial satellites to Geostationary and Non-Geostationary orbits. [5]

A snap shot view of the industry as of 2002 reveals the one critical weakness in any proposal for new exploration based on existing launch systems. The goal of exciting the space community through the Mars Oasis would never work because even if the first mission succeeded flawlessly, just the launch costs alone of the follow-on missions Musk hoped to spur were simply too high to be sustained. This was the crux of the decision Elon Musk faced. To become a space faring civilization with permanent roots on two worlds, which was his real goal, would require much more than developing a single, small launch vehicle for an inspirational photo-op. It would require spearheading a massive reduction in launch costs on effectively a global scale, and developing a complete end to end transportation infrastructure which would make his, as well as countless other's dreams of becoming an interplanetary species a reality. The difference was that as a person with means, Elon Musk began planning to do precisely that. Not just building a better rocket, but doing it in order to go to Mars.

To many, espousing such ambition can easily reek of either child-like wistfulness or hubris on a scale usually reserved for Bond villains, but to a critical few, exasperated with 40 years of stagnation in deep space exploration, it was a long overdue recognition of the plainly obvious. What was also obvious was that the entire plan hinged on developing a critical foothold in a skeptical commercial launch market infinitely less concerned with grand schemes than with getting its satellites delivered to space, on time, on budget and with absolute reliability.

A Reality Check

One of the many good things about being a self-made millionaire is that your ideas, even the weird ones, frequently seem to carry a little more respect than that which accrues to the average daydreamer. Better yet, you also have the resources to hire other people to vet your concepts. If it works, you are still the genius, and if it doesn't, you already have someone to blame. Before Elon Musk seriously committed himself to the unlikely rocket project which was on his mind, he did two very important things. First, he outlined three critical questions which would help him make a decision as to whether to go forward with the project, and second, he brought in experts to help answer two of those questions.

The questions were simple, straightforward and could serve as guide for any prospective business venture.

Was there a market?

If so, was it achievable?

Was it interesting? (He already suspected he knew the answer to this one.)

Beginning in early 2002, and several months before the completion of the sale of PayPal which would deliver the resources to make a new project possible, Musk convened a sort of Saturday morning breakfast club to look at the problem. Present were several aerospace heavyweights. In his own words, "engineers which had been involved with all major U.S. launch developments over the last 30 years."

Was there a market?

As recently as four years ago the answer was a resounding yes, but in the intervening time much had changed, and at the moment the answer was not as certain. The question was how soon and to what extent it could be expected to rebound. Elon Musk had selected a rather inauspicious time to begin a launch development project. The previous five years had seen a virtual wipeout of what had only a few years ago seemed the most promising of times. The optimism was based on the projected rollout of a new system of satellite based mobile phones led by telecommunications giant Motorola, with the promise of others soon to follow. Satphone business plans, predicated on deploying large fleets of low Earth orbiting satellites at minimal cost, provided an incentive for a number of new launch vehicle development projects, and offered at least the potential of a market capable of attracting the investment capital required to bring them to reality. Rather than take off, the market imploded instead, leading to bankruptcy filings on the part of existing operators and a cancellation of plans for those which were just getting started.

The would-be satellite phone providers were on the losing side in an epic battle over the exploding market for mobile phones versus tower based systems. During the mid-1990's, as the mobile phone industry was beginning to pick up steam, it appeared that for higher end business applications, satellite based services which rapidly switched calls from one passing satellite to another, offered a realistic alternative to the limited coverage areas and poor international coverage of terrestrial cellular networks, which commanded the apparent necessity and expense of placing an endless series of cell towers across the landscape. The emerging Satphone business was led by Iridium Inc. a joint venture of Orbital Sciences Corporation and Motorola, originally planned to launch a constellation of 77 satellites into low Earth orbit, a number which provided the inspiration for the company's name, Iridium, the 77th element on the periodic table. As events unfolded, the original number was revised down to a still substantial fleet of 66 satellites, but perhaps wisely, no-one attempted to change the name to the 66th element, Dysprosium. While the comparatively small size and particular nature of the satellite constellations lent themselves to being grouped into clusters, and placed into orbit in a single launch by larger vehicles, which was generally the case, they also opened the door for a series of follow-on "maintenance launches" by much smaller vehicles to replace individual failed satellites. Taken together, it was an opportunity which held the potential of creating an entirely new market for launch services.

A number of factors conspired against satellite phones, not the least of which was that by the time the spacecraft were eventually deployed, the terrestrial cellular providers had substantially accomplished a "build out" of ground based towers. The combination of high initial costs, ($3000 for a handset), high connection fees and ugly phones with squat, bulky antennae ran counter to the ever smaller, cheaper and decidedly sexier cell phones constantly marketed on television and on the big screen.

As the companies went bankrupt and the existing assets, including satellites already in orbit were auctioned off, the need for maintenance launches to deliver on orbit spares or replace failed satellites subsided to nearly zero. A funny thing about bankruptcy in space is that it is damned difficult to seize the assets, and so even as the companies went under, the satellites in orbit sped merrily along their way. While the initial business plans failed, a fully functioning fleet of satellites available at bargain basement prices is still a business opportunity. It would not be long before new owners began to make new and perhaps sounder plans. Until such a recovery got fully underway however, the collapse of this particular industry meant that anyone hoping to introduce a new launch vehicle would need to go after every segment of the overall market, and still plan for a low flight rate for the foreseeable future.

The only other immediate opportunity was in government launches, and here the prospects were somewhat more encouraging. In the first place, dating back to 1996, the official space policy of the United States government called for the purchase of commercially available domestic launch services whenever possible. [6] This policy effectively blocked any widespread use of converted American ICBM's as competitors for governmental launches in the manner being undertaken in Russia. The same policy also essentially restricted NASA to working only on manned spacecraft or reusable launch vehicle development projects, while assigning the Air Force responsibility for expendable launch vehicles. Nevertheless, NASA was still a reliable customer for a variable number of expendable launch vehicles for space science missions, most of which took place on the Delta II, but also included a lesser number of flights on small launchers.

Yet another opportunity lay with the Department of Defense, which was beginning to become interested in the notion of small, inexpensive satellites which could be quickly deployed over developing "hotspots" around the globe to provide an entire range of militarily pertinent information and services. In the aftermath of the 9/11 attacks, then Secretary of Defense Donald Rumsfeld concluded that the nation needed to develop a faster, almost entrepreneurial mindset in dealing with asymmetrical threats. As a result, he established within the Office of the Secretary of Defense, the Office of Force Transformation. The OFT, working in conjunction with the Naval Research Laboratory established a program to test key technologies for rapidly deploying small satellites for less than $15 million, including launch costs. [7]

Finally, on the international front, the increasing capability of miniaturized spacecraft components and a growing awareness of the potential for resource mapping satellites was beginning to make it affordable for smaller, non-space faring countries to requisition and operate their own satellites. As frequently happens, once one country acquired the prestige of owning its own space based assets, its neighbors would often reach the same conclusion, leading to the slow but inexorable growth in a second tier of nations searching for affordable launch solutions.

Although any one source did not offer a promising market opportunity, the cumulative total of all sources suggested that the answer to the first question was yes, there was a market. Answering the second question; was it achievable, depended on an assessment of the competition.

While one doesn't ordinarily think of the FAA in the context of space launches, it, and not NASA, is in the licensing agency for non-governmental spaceflight in the United States, and for good reason. Until they exit or after they re-enter the atmosphere, launch vehicles and returning spacecraft ply the same airspace as both passenger jets and light airplanes and helicopters. The FAA office which oversees space activities is the Office of Commercial Space Transportation, and it has played the lead role in fostering the regulatory environment for launch vehicle development. As part of its mission, OCST gathers and presents data on the commercial launch market, providing definitions, and common terminology.

The FAA defines a small launch vehicle as any system which can place no more than 5000 lbs. into low Earth orbit. In early 2002, as Elon Musk contemplated entering the industry, this class offered the opportunity of metaphorically dipping his toe in the water, before committing to a larger development project which could almost certainly consume every penny he had. What it did not offer however, was a very large potential market, which was one of the reasons industry giant Lockheed Martin had withdrawn its Athena launch vehicle from the market the previous year, after seven launches (including two failures) dating back to 1995. In fact, 2002 saw only one launch in the small vehicle class, that of NASA's HESSI mission, which was conducted aboard Orbital Sciences Corporation's Pegasus booster. The only other active small booster in the United States, the OSC Taurus, did not fly at all.

Between the two entries, the most popular was the small, winged Pegasus rocket which offered a lift capacity of 977 lbs. to LEO. First launched in 1990, the Pegasus appeared to be something new under the sun, thanks in part to two very distinguishing characteristics. In the first place, its "launch pad" was a jet aircraft, initially a B-52, and then a Lockheed L1011, and just like the aircraft which served as its launch platform, and the mythical horse for which it was named, the Pegasus had wings.

Unlike the airliner, the horse and even the Space Shuttle, the wings had nothing to do with re-usability or returning to land. Instead, they were included merely to stabilize the rocket in the initial moment after release from its carrier aircraft, and then to provide first stage guidance through the thicker part of the atmosphere, contributing nothing in the way of actually recovering the fully expendable launch vehicle. Despite its unique appearance, the Pegasus did little to actually lower launch costs, and in fact turned out to be a surprisingly expensive booster. Its chief advantage lay in the simple fact that as the smallest available domestic booster, with a $13.5 million to LEO price in 2002 dollars, it offered the lowest total launch price available, but at $13,800 per pound, it was more than twice as expensive as any other booster by that measure. [8] With the United States government legally barred from purchasing far more affordable launch services on small Russian boosters, and NASA's own Scout rocket retired in 1994, the Pegasus had no realistic competition for the small, but credible piece of the launch market which it served and no incentive to lower its cost.

Orbital Sciences introduced the original Pegasus booster to the market in 1990 with a fixed cost to the Defense Advanced Research Projects Agency, DARPA, of just over $6 million, but following a series of changes which were instituted in response to early launch failures, quickly began to exceed that price with the addition of a more powerful version, an optional fourth stage and several other improvements. [9] By 2002, the price increased to $13.5 million per flight. Furthermore, following a path similar to that of the Shuttle, whereas the Pegasus was considered capable of a launch rate of once every two weeks, (the Shuttle was also predicted to fly every two weeks!) the actual high water mark came in 1998 with a total of 6 launches. Ten years later, at nearly $30 million per launch, Pegasus was a very expensive horse to ride. A depressingly familiar story, the cycle of increasing launch costs resulted in progressively fewer launches and still higher costs. Nevertheless, even with its small payload capacity of just under 1000 lbs., the combination of launch flexibility to any orbit by virtue of its carrier aircraft, and overall lower total price compared to larger boosters, the Pegasus was a much more popular solution than its stable mate, the land launched Taurus. Although the Taurus offered triple the capability of Pegasus, lofting just over 3000 lbs. to LEO for $19 million, resulting in a comparative cost per lb. of just over $6,000, the demand for the upper segment of the small launch class simply wasn't steady, and the Taurus, which debuted in 1994, had completed only six launches by the end of 2002, with the most recent being a failure on September 21, 2001.

In the case of the Pegasus and the Taurus, as well as the recently suspended Athena, the U.S. entries in the size class Musk would be targeting were based on a solid rocket manufacturing industry with deep roots in the defense establishment, and each was unable to shake off the built-in cost issues which came with that heritage. As Musk would observe time and again, "legacy systems bring legacy costs." They also bring with them the very real risk that in the constantly evolving ebb and flow of defense spending, cutbacks in one area can result in price increases in another, as defense contractors seek to transfer overhead to whichever account is deemed capable of bearing it. Based on the Pegasus numbers as a point of comparison, Musk's group estimated that they could field a higher performing two-stage liquid fueled rocket for less than half the launch price of the OSC product, while offering enough payload capacity to serve a broader range of the small launch market segment.

If there was a point of concern regarding cost competition, it came from the other side of the Atlantic, and in one case, from under it, in the interesting spectacle which was the continuing devolution of the Russian defense industry. In some ways it was similar to what was transpiring with American small launchers, but with several key differences. Whereas U.S. companies such as Lockheed Martin and Orbital Sciences sought to adapt defense components in order to build civilian launchers, the Russians took a much more direct approach, performing the minimal work necessary to convert whole intercontinental ballistic missiles into launch vehicles. At one time or another, no less than 4 different launchers were offered on the international market at prices substantially under what could be secured in the West. Interestingly, three of the four were based on liquid fueled ICBM's, reflecting Russia's long standing aversion to solid rockets. One, named Shtil was actually launched from a ballistic missile submarine, or "boomer" in the Barents Sea with the launch heavily subsidized by the Russian navy, on the grounds that it was a "training exercise."

"Was it Interesting?"

The conclusion of the series of meetings which Musk organized to consider the subject was that there was indeed a market for a new small launch vehicle, and also that there were sound reasons to believe that successfully entering the market was in fact, achievable. As for the final question, "was it interesting?" the future was a subject very much on Elon Musk's mind, and there was no doubt that if the opportunity really existed to change the structure of the launch vehicle industry, lower costs, and possibly set mankind on a path of renewed space exploration, it wasn't merely interesting, it was irresistibly compelling.

Even with the confidence that he stood a reasonable chance of success, Elon Musk still had to consider the fact that the enterprise he was contemplating undertaking was not just starting a company and building a small launch vehicle which could likely undercut two clearly overpriced competitors. It was something much larger. At stake was a crusade against the seemingly impregnable barrier of high launch costs which dominated all aspects of the launch industry from the smallest rockets to the very largest. If Musk kept to his closest held dreams, which were coalescing around a vision of Mars as a second branch of human civilization, and a "backup" hard drive for the human race, it could not stop with small launch vehicles. Although a 2002 survey by the Futron Corporation confirmed that on average, commercial launch costs had decreased during the previous decade, coming down from $18,000 to $12,000 per pound between 1990 and 2000, the prices were still far too high to encourage a market expansion with new entrants. While a nearly 30% reduction in launch costs was substantial, it provided very little real benefit outside of reducing the costs for limited existing market for commercial satellite operators and governmental efforts. [10]

The issue as seen from Musk's perspective, one which he shared with space enthusiasts worldwide, was that until launch costs dropped much further, the almost overwhelming sense of opportunity and wonder in exploring an entire solar system; for research, exploration, commerce and settlement, would be held forever at bay. After forty years of standing by, and listening to endless assurances, followed by one patronizing explanation after another about how difficult and expensive it was to launch anything into space, the frustration was nearly overwhelming. To a growing NewSpace movement, a collection of entrepreneurs, companies and organizations such as the Space Frontier Foundation, and the X-Prize foundation, the existing aerospace establishment was anything but the intrepid explorers on the cutting edge of technology which they so frequently presented themselves to be. Instead, they were the self-appointed gatekeepers for the very future of the human race, and it was well time to knock the damn gates down.

To the believers, the battering ram as such, was represented by the concept of a single, nearly mythical benchmark, a launch system capable of delivering both people and payloads to low Earth orbit for the comfortably round figure of $1000 per lb. One need not step on the scales wearing only a calculator to quickly realize that the long sought figure is still way too expensive for the average citizen to afford a trip into space even if they don't check baggage. In reality, the $1000 per lb. is actually more of a suggested tipping point than any hard and fast benchmark. As Robert Heinlein reportedly once observed, once you were to low Earth orbit, you were "halfway to anywhere." The assumption, yet to be tested, is that the $1000 figure will mark the moment at which the launch market begins in its own way to go viral, expanding with new entrants, new ideas and new ventures, all of which will spur new innovations to drive prices even lower, towards the next benchmark goal of $100 per pound, at which point a ticket to space would presumably be affordable to anyone capable of flying first class on a trans-oceanic flight.

Sometimes dismissed as naïve 'space cadets" hopelessly out of touch with the hard realities of actual spaceflight, promoting unrealistic and ill-considered space panaceas, (which historically speaking has often been the case) NewSpace advocates could point to a historical precedent for a transportation revolution sparked by a tipping point. The commercial airline industry as we know it today, took off in 1933 with the introduction of the Boeing's 247 twin engine aircraft, the first passenger carrier to attain a regular flight speed of three miles a minute. This figure, nearly 20% faster than the loud, uncomfortable and lumbering three engine Fokker and Ford Tri-Motors which were making their way across the country in long segments composed of short hops, convinced the general public almost overnight that air travel offered a viable alternative to passenger trains for long distance trips. The groundbreaking 247, which carried a maximum of 10 passengers, was in turn soon eclipsed as the Douglas DC-3 flew past it and into aviation history by achieving a modestly greater speed with a considerably higher passenger load. In the process, the legendary aircraft changed the economics of the airline industry for good, providing the tipping point which made commercial aviation affordable for the passenger and profitable for the carrier. [11] Achieving the same thing in space transportation meant finding a similar tipping point.

Elon Musk very explicitly took up the challenge of the $1000 per lb. figure without suggesting that radical new technology was required. As a goal for the existing launch vehicle industry the $1000 per lb. figure was certainly elusive, but perhaps not impossibly so. After all, Russia and China were each marketing rockets which averaged $2500 to $3000 per pound, almost half the going rate in the U.S. and Europe. While both nations benefitted from lower labor rates than Western competitors, the Russian technology base, although superior in some areas such as high pressure engines, was clearly lacking in others. Furthermore, inefficient design solutions dating back to the earliest years of the space age combined with the lack of resources to drive innovation, a stifling bureaucracy and pervasive element of corruption in a sprawling Russian space establishment which had yet to downsize from the heights of Soviet empire, strongly suggested that the comparatively lower prices being offered by Russia in no way represented a floor of what was possible within the bounds of existing technology.

In the United States, NASA was in the process of backing away from a launch vehicle development effort specifically targeting the $1000 per lb. benchmark, not because it was considered particularly unobtainable, but for the altogether more disturbing reason that the agency which had once taken up the cause of reducing launch costs as its battle cry, no longer seemed to care. In 1999, NASA had formed its Integrated Space Transportation Plan, intended as a "living document" to shape the development of existing and future launch systems. In addition to planned Shuttle upgrades, a key element of the ISTP was a new launch vehicle development program called the Space Launch Initiative, or Second Generation, which very practically aimed at producing a second generation of partially reusable launch vehicles, with the Shuttle representing the first generation. If anything, unlike the wildly optimistic forecasts of less than $200 per pound as NASA was seeking to secure the Shuttle program in the early 1970's, SLI and Second Generation took a more measured approach, selecting a $1000 per lb. target for LEO. Consensus seemed to suggest that the goal could be readily achieved by a launch system featuring a reusable first stage and with an expendable second stage. This was functionally the opposite approach to that taken with the Shuttle program, and it represented a far more affordable, and achievable, means of tackling a very hard engineering problem.

Second Generation hoped to select by 2006 at least two proposals for development and demonstration, with a fully operational system in place by 2012. [12]

Like so many other efforts over time, it went nowhere. Affordable access to space, once motivating rationale for NASA planners had lost its appeal, discredited by the economic failures of the Shuttle program, and disillusionment with more recent X-programs in the 1990's, all of which had led to naught. Before it really even got started, the Second Generation Reusable Launch Vehicle program was quickly abandoned in favor of other priorities such as a small orbital space plane designed to complement the Shuttle. Nowhere in the mix however, was a sense that the goals themselves were unattainable. If anything, a close reading of relevant documents suggests that they were so sufficiently conservative as to be considered boring.

The paradoxical truth facing Elon Musk was that although substantially lowering launch costs was universally recognized as both an important and necessary goal for the ultimate future of space exploration, science and commerce, neither NASA, DOD or anyone else was actually committed to trying to do anything about it. China was pursuing its own state plan in a methodical manner. Russia, the world's leader in the number of space launches conducted was financially incapable of doing anything more than buying time through selling its considerable services even as its infrastructure slowly deteriorated. NASA had other objectives, which for the moment still included flying the Shuttle. The legacy American aerospace contractors seemed content to live off the Department of Defense, even as commercial markets slipped away, and above it all, the European Ariane V program was settling into its long run as the world's undisputed leader in the commercial launch market. Worse, it appeared that as long as no party's own core interests were threatened, the status quo would remain just that, and another decade or two would quietly pass by. In short, it was a perfect time to begin a revolution.

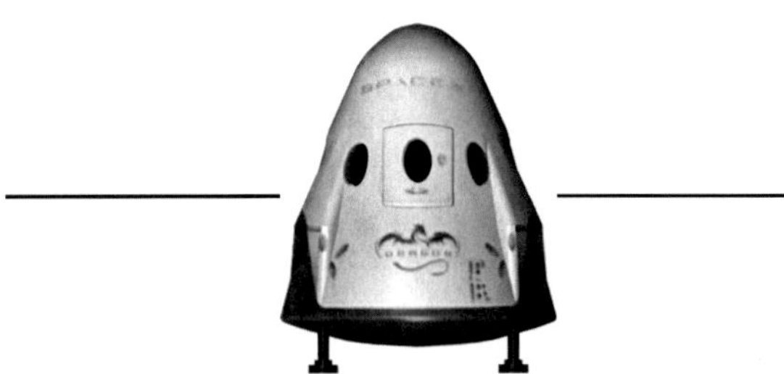

Chapter III: Starting SpaceX and Developing the Falcon 1

Convinced by the working group's results that a start-up launch vehicle project was economically and technically feasible, Elon Musk filed California articles of incorporation for Space Exploration Technologies Corporation on May 6, 2002. [13]

Although Musk provided both the financial clout and the driving force behind the project, the decision to form SpaceX, as well as its subsequent accomplishment would not have been possible without the unique contributions provided by one of the company's most influential co-founders, propulsion engineer Tom Mueller. A rocket is only as successful as the engine which propels it, and Mueller's expertise in the field was a key asset on which the newfound company's prospects would rise or fall. Elon Musk would later be frequently compared in the popular press to Iron Man's eccentric genius alter ego and creator Tony Stark, a comparison not without merit, and grounded in the fact that some scenes were actually filmed in the SpaceX factory, with Musk himself making a cameo appearance in the second Iron Man movie. Nevertheless, to the extent the analogy holds, Tony Stark is more a composite of two men, and the one in the lab, constantly tinkering with hardware is undoubtedly Tom Mueller. In fact, this is precisely what Mueller was doing when Elon first met him in January 2002.

At the time, with the decision of whether or not to go forward still hanging in the balance, and guided by his experience in vetting the Mars Oasis concept, Elon Musk was searching for an inner cadre of experienced personnel who had both the engineering chops to bring to the project, and equally important, direct hands-on experience with building and testing hardware. Finding someone possessing both skill sets was a challenge, and it was one which he pursued in an interesting way. Rather than hiring a recruiting firm to look for talent, he instead went to one of the few places likely to contain the type of individual he was seeking; the Reaction Research Society, an amateur propulsion group whose hobby was developing small, generally liquid fueled rocket engines. What set the RRS apart was the fact that a number of members were employed in the aerospace industry, and made a hobby of indulging a passion on the weekends; actual rocket development, which didn't necessarily come with the day job. In following this lead, Elon Musk in some ways reached back into the pre-history of the space age, when "private" research groups in Germany, the U.S. and the USSR laid much of the critical groundwork for orbital spaceflight and set the stage for publicly funded efforts which were soon to come. It was also setting the stage for one of the core requirements for SpaceX personnel, the ability to perform hands on work with flight hardware with the same degree of confidence and comfort that normally came with computer aided design.

Mueller, a propulsion engineer for TRW, was in the process of mounting a 13,000 lb. thrust pressure fed engine to a frame in a friend's rented garage on the day Elon Musk walked up. Getting straight to the point, the internet entrepreneur asked simply "Can you build something bigger?" [14] The answer was yes.

Four goals guided SpaceX from the beginning. First, the company was committed to developing a launch vehicle which was both simpler and more reliable than anything else on the market. Beyond that, Musk wanted a design which could be readily translated to larger vehicles, and most importantly of all, set the company on a path towards full reusability. Each of the critical decisions soon to be made in determining the key characteristics of the rocket meant to take on the 980 lb. capacity Pegasus would also be weighed against the ultimate goal of launching the first pioneers to Mars. It was only slightly less grandiose than if the Wright brothers had taken into consideration the runway requirements for the 747 when designing the Wright flyer. Within ten years however, SpaceX would be assembling a rocket which was the core component in a system required to field a minimalistic manned Mars mission, and even more incredibly it would be directly based on engine and systems which marked this initial effort.

SpaceX began both with the development of a rocket, as well as the development of a culture. If one was to be conventional, it was clearly the rocket. As for the culture, it definitely depended on your point of view, and if it happened to be that of the traditional aerospace establishment, neither the rocket, the company, nor the founder, particularly the founder, deserved to be taken seriously. If Elon Musk had been seeking outside financing for his project, this would have presented a serious problem that brought a familiar solution, staff the company with heavy hitters, preferably former NASA program chiefs, who were more adept at working inside the hallways of Washington D.C. than they were at working outside the comparatively tiny box which represented the SpaceX development budget. Musk, under no such constraints, took a very different path. The initial approach was to create a culture similar to the software development environment which he was used to, a very small, very young, vertically integrated team comprised of the very best in their field, but working to develop launch systems rather than launching the next killer app. The global epicenter of software development is without a doubt San Francisco and Silicon Valley, where Musk had spent the previous seven years, and for a time, he debated whether or not to base SpaceX in the region as well, just as he would with concurrent project Tesla Motors in Palo Alto, at the southern tip of San Francisco Bay. But if Musk wanted to begin "bending metal" in the aerospace industry, it meant shifting his life south to the center of American aerospace capability, Los Angeles, where he could recruit right out of the considerable local talent. Musk sought to staff his team with the very cream of talent, in his words the top 1-2% of engineers, brazenly stating in an early interview that if you were in that elite cadre "you shouldn't be working for Boeing or Raytheon," you should be working for us." The result was more like the fusion of an internet startup with a Formula 1 or NASCAR shop than the traditional aerospace companies dotting the Southern California landscape.

Although the arrival of SpaceX, with its little rocket and big plans garnered its share of press attention, the major story in 2002 was the debut launch of both the Boeing Delta IV and the Lockheed Martin Atlas V, competing entrants in the U.S. Air Force's Evolved Expendable Launch Vehicle Program (EELV). The successful introduction of both vehicles marked the end of a development effort which had begun in earnest in 1996, and the beginning of new era of publicly subsidized launch operations which would be a major counterpoint to the evolution of SpaceX. As for NASA, entering its third decade of flying the Shuttle, and with a constantly shifting notion as to what to do next, the agency was building little more than frustration. Furthermore the most recent company to make a high profile attempt to build a low cost launch vehicle, Beal Aerospace, had just thrown in the towel, citing the commercial satellite meltdown, the U.S. government's direct subsidy of the EELV program and NASA's multi-billion dollar program to develop its own reusable launch vehicle as its rationale for abandoning the effort.

As a result, when it came to recruiting aerospace talent, 2002 was the beginning of a very good era. Compared to the heady early days of the missile and rocket era when newly minted propulsion engineers had the good fortune to work on one system after another, the last thirty years had been a relative desert. Incredibly, the United States had only fully developed and launched one new liquid main engine, the Rocketdyne RS-68, since the Space Shuttle Main Engine (SSME) program began in 1970. In what had to be the penultimate insult to American aerospace prestige, after a long and successful career which ultimately recorded 576 launches and a span of 47 years, Lockheed Martin had chosen a Russian main engine, the RD-180 to power the two most recent editions of the nation's original manned orbital booster, the Atlas. [15] Entire careers had been started, worked and the gold watch handed out in the time frame since a previous generation of engineers had the opportunity to work a clean sheet rocket problem from the ground to orbit, and see it through. For American liquid propulsion engineers seeking to work on a completely new orbital system, SpaceX was the only game in town, and for the talented veterans who found a home there, it was as if they were released from 30 years of bondage.

When questioned about his plans for entering what was clearly a depressed market, Musk repeatedly observed that it is better to enter a down market and plan for the upswing than to enter an up market and instead encounter a downturn. While the comments were made in context of the overall satellite launch industry, it was equally true about the market availability of top talent. There was no better example than Mueller who brought with him a particular technological link from his company's past role in the Apollo program, one which would prove critical in allowing SpaceX to begin the comeback of the American launch vehicle.

"It was an opportunity I couldn't pass up," Mueller said. "It was a chance to develop some new rocket engines and a challenge to try to change history."

Mueller's comments, though made by an industry veteran, would be echoed time and again by the engineers, programmers, machinists and fabricators who would soon be coming to work for SpaceX.

In the light of later developments and the announcement of one increasingly bold plan after another, it is easy to lose sight of the fact that the biggest step of all may have been the leap of faith in starting the project to begin with. The idea that it was even remotely possible to design, build and introduce a new launch vehicle as a competitor to the Pegasus at half that vehicle's price was considered by many "experts" to be borderline delusional fantasy. Those who were not outright patronizing simply dismissed the project by confidently asserting of Elon Musk that "he doesn't know what he doesn't know." Ten years later during an interview with the CBS News Magazine "60 Minutes" Musk acerbically reversed the barb, asking the obvious question, how anyone could know what they don't know? Within months of the interview, it would become abundantly clear to the entire world that he obviously knew quite a bit more than most of his critics.

The antipathy, while disappointing considering some of the sources, who knew better, was also understandable. It seems to be a trait of human nature that people who for a time enjoy unfettered authority, whether parents, scientists or dictators, rarely behave well the first time they face a serious challenge to the basis for that authority. In starting SpaceX, although the initial goal was only one small launch vehicle, Musk was by extension challenging the entire infrastructure which supposedly represented the pinnacle of American talent and technical genius, and it was not as if he was the first to do so. Recent history was littered with the wreckage of various launch vehicle plans, all of which sought to bring down costs and open markets, only to end in failure and sometimes bankruptcy court. Each successive failure seemed to confirm the conventional wisdom that achieving orbit was such an overwhelming challenge that after forty years, it was only barely within our grasp, a task for nations, not entrepreneurs. It was all the more disconcerting then, when Elon Musk asserted with what seemed like such casual confidence that his company would change the launch industry forever.

It certainly begs the question, what made Musk and his team so sure it could be done? Perhaps more importantly, what made him sufficiently certain that in the event he found some measure of success with the diminutive Falcon 1, it could even be translated to larger rockets and bigger dreams? The answer was both simple and utterly damning. For anyone who cared to look, whether one worked for DOD, NASA or a major contractor, it was painfully obvious that the aerospace industry was mired in an outdated manner of doing business in which the inefficiencies were both pervasive and overwhelming. It was an industry which paid far more attention to self-perpetuation than any actual innovation, and even less to costs.

One of the most elemental problems was the inherent expense of the cost plus contracting method employed to develop much of the defense and space hardware which underlay the system. Cost plus is an effective method of developing unproven, technologically challenging or one of a kind systems for which there is no rational market mechanism for determining what the product should cost. There are a number of different variations of cost plus contracting, but in general the contractor is reimbursed whatever it costs to develop the system in question plus an additional percentage which serves as "profit." Frequently there may be incentives for reaching pre-agreed expense, timing or performance milestones aimed at containing overall costs. It is a legitimate and effective tool for incentivizing companies to produce goods or systems which no sane business person would undertake in a less secure format, but it is almost inescapably expensive. In the first place, for both the contractor and the agency managing the contract, the mere act of documenting every item in order to comply with contracting rules is itself labor intensive and very expensive. It also has a tendency to drive smaller and possibly more nimble and creative competitors from the arena, leaving only larger corporations equipped with teams of lawyers which become adept at gaming the system. Furthermore, with all "legitimate" costs being reimbursed, there is limited incentive to control labor costs, particularly when it comes to layers of management. From the contractor's point of view, the more personnel which can be effectively maintained in-house the greater flexibility to pursue other contracts when opportunities arise. Other times it is not the contractor, but instead the contracting agency itself which drives up costs by making changes to systems, both large and small after an initial contract is let. What does the contractor care? As long as it is reimbursable, they can make all the changes the buyer wants. On the other hand, protected by restrictive civil service regulations, it is not as if the agency is going to have to face the painful choice of firing personnel when the costs begin to mount. In most cases, work is delayed into the next budget cycle instead.

The problem also applied to components and services. Why for example, should a contractor produce a seal in house for $3.00 when they could procure it externally instead for $17.00 and charge the government 18% profit on their "costs." The answer was of course, they would purchase the component from an external vendor and book the higher profit while spreading the liability. Repeated throughout the supply chain, and in countless other aspects of the entire launch process, not just the vehicles themselves, and the result was an industry utterly separated from the economic reality of the private market which provides most of the goods and services we buy. Sometimes the problems are almost comical. For example, the U.S. Air Force pays United Launch Alliance to "broker" the fuel it buys at Cape Canaveral for its own launch vehicles. But the fuel itself is purchased from one government agency NASA, and then sold to another, the Air Force, by EELV operator United Launch Alliance, (ULA) which passes on a hefty broker fee for doing essentially nothing. [16]

One of the most insidious aspects of the cost plus system is that once the high cost is built in, it becomes a part of the price basis of the product and never comes back out. It was for this reason that a relative newcomer such as Orbital Sciences could introduce an innovative product like the Pegasus booster, but never effectively escape the high cost of the solid rocket motors which power it. Even in developing a liquid fueled booster, the only way to actually break the chain was to unwind the entire process and start with new components which did not include heritage costs. The first step towards changing the outcome was explicitly recognizing the problem posed by embedded costs from the outset. Coming to such a realization meant that rather than approaching the problem by offering radical new architectures, the key was to design a relatively conventional rocket, but one which was produced by a far more efficient launch vehicle manufacturing and operations company. Musk's advisors were well familiar with the status quo, and in a position to assure him that if the basic rocket was sound, the market itself was a target rich environment.

With reliability as the principal objective, one of the first steps the design team took was to arm themselves with a study conducted by the Aerospace Corporation which examined launch vehicle failure incidents between 1980 and 1999. Following a series of launch failures in the mid 1980's, which saw the human tragedy of the Challenger disaster occur in-between two high profile and very costly launch failures of unmanned Titan III rockets carrying DOD payloads, the Department of Defense began to become seriously concerned about the price and reliability of the heavy lift vehicles charged with lofting large reconnaissance satellites into orbit. After another high profile failure in 1993, this one of a Titan IV carrying a KH-11 spy satellite, the Department of Defense initiated a Broad Area Review to bore down into the problems. As part of that effort the Aerospace Corporation, a government chartered "private" corporation, technically labeled as a Federally Funded Research and Development Corporation, which provides independent analysis of major aerospace acquisition programs was brought in. The company produced an analysis of launch vehicle failures over the last several decades, grouping them by type. The conclusion was that 91% of all failures could be assigned to only three causes; propulsion (56%), separation (10%) and avionics (10%). A final category, unknown (20%) likely included additional propulsion failures coming from closed societies such as the Soviet

Union and China which regularly attempted to conceal potential embarrassments whenever possible. [17]

As one of the most recognized and respected authorities in the field, Musk's team paid particular attention to the Aerospace Corporation report, using its findings to guide an overall approach to designing their new vehicle architecture. The reasoning was that if they specifically accounted for the most common failure modes across all types of vehicles, and produced a design which minimized each of these factors, then the odds of success would be vastly improved, and it would be possible to build-in reliability, and thus market acceptance from the beginning. Even though Elon Musk was quite wealthy, he was contemplating investing a considerable percentage of his fortune in the effort, even as he was also in the process of forming Tesla Motors, it would not take many trial and error mistakes to burn through that fortune. For the man who wanted to lead humanity to Mars, it would be a sad consolation indeed if his chief accomplishment was to provide nothing more than a series of data points for the next Aerospace Corporation failure analysis study.

While the immediate goal was to achieve a substantial reduction in the cost of reaching orbit versus the competition, SpaceX also wanted to develop a pathway towards reusability beginning with the very first flight. Over the course of the previous decade, DOD and NASA as well as a number of start-up aerospace companies had all pursued Reusable Launch Vehicle (RLV) systems, but with nothing to show for the effort in terms of an operational system. While the concepts varied significantly, suggesting that there might be many paths to achieving the elusive goal, they were all so sufficiently removed from the existing launch systems that the design effort far outstripped available resources, which were often quite meager to begin with. SpaceX had resources, but they would need to be carefully marshaled. Respecting this fact, the design team chose instead to improve on an already conservative path mirroring that taken by contractors in the U.S. Air Force Evolved Expendable Launch Vehicle Program, which also relied heavily on input from the Aerospace Corporation. Development of these intermediate and heavy class launch vehicle families focused on modestly improving existing vehicle design, manufacturing and launch processing procedures as areas to address in building a more affordable expendable launch vehicle. The results, at least at first, and seen from the viewpoint of 2002, were credible. Both the Boeing Delta IV and the Lockheed Martin Atlas V, two very different rockets, incorporated substantial process improvements over legacy launch vehicles. Nevertheless, in both cases, it was a cautious effort which left a number of issues unaddressed, and this was the territory Elon Musk's effort would to look for bigger gains.

Rather than injecting radical untried technology, the goal would be approached through pursuing simplicity and reliability under the conviction that the result would be simplified manufacturing, from which low price would naturally follow, even if the vehicle itself offered only modest performance. Performance increases, and cost decreases, would come in their own time. Meanwhile, in a major departure from the norm, SpaceX would attempt to recover its first stage from its earliest flights through the seemingly simple expediency of parachute softened water landings. It was in some ways even a more radical approach than if the company had based its plans on game changing technology. Merely by applying sound business practices, it sought to produce a more affordable, and a more reliable launch vehicle than anyone in the world. It was a damning indictment of the existing industry if ever there was one, and sure to incite the ire of aerospace veterans. In following such a path though, SpaceX was actually heeding one of the most common maxims in the industry which nearly everyone said, but few ultimately followed, that 90% of the cost comes in the last 10% of performance.

As the proud owner of a McClaren F-1, one of the fastest production automobiles on the planet, Elon Musk chose to use the plight of a somewhat less fortunate buddy to illustrate the reasoning. A friend Musk observed, had purchased a new Ferrari only to have it break down on the way home from the dealership. Had he chosen to purchase a Honda or Chevy however, he might have reasonably expected to drive the admittedly less exciting new car for an entire year or more without ever needing a repair. The moral of the story, high performance is frequently the enemy of high reliability. Based on the price point of the launch vehicle he was targeting, the Pegasus XL, now coming in at over $13 million per flight, it was not necessary to achieve a cost break-through. He simply needed a reasonably priced rocket which was fundamentally reliable, something on the order of Honda Accord. Wisely, Musk did not address the topic of high maintenance companions which often seem to accompany both Ferraris and McLarens. [18]

Achieving high reliability would begin then, with a new beginning, a literal "clean sheet of paper," according to Tom Mueller. Doing so would mean avoiding the almost compulsory tendency to select old hardware over new, in spite of the industry tendency demonstrated time and again to do exactly that. To outsiders, one of the most surprising aspects of what would otherwise be considered a cutting edge business, the space launch industry, is the fact that it rarely elects to develop new technologies, preferring to perpetually recycle older, "proven" hardware, into new designs, even if doing so prevents new results. For example, after the massive buildup of infrastructure for the Apollo program; including the gargantuan crawler-transporters, and the Vertical Assembly Building, NASA, despite having studied far simpler integration architectures and techniques, felt compelled to utilize these same assets in the Shuttle program. When that program added large solid rocket boosters into the technology baseline, they too became part of the historical baggage, incorporated into NASA's next two projects as well.

Both could be considered examples of the "sunk cost" fallacy, an unfortunate tendency to pour good money after bad, and one which is still driving NASA policy today. Formally identified in 1968, when researchers visited a horse track to study how gamblers regarded the potential outcome of bets they just made, "sunk costs" have been a hallmark of human behavior since the dawn of history. In the landmark racing study, betters who had just placed a $2 bet on a horse consistently rated its chances of winning as higher that the next horse on which they were about to place an identical bet. Though the study focused on $2 bets, the results shed light on some of the most expensive blunders in human history. From Pickett's Charge to the Charge of the Light Brigade, military history is replete with examples of the sunk cost fallacy, and often the most successful leaders have been those who simply recognized when it was time to cut and run and live to fight another day. In aerospace history, adherence to the sunk cost fallacy led the governments of Britain and France to pursue the development of the Concorde supersonic jet even when the lack of a business model became abundantly clear before it ever took to the air. It was a mistake so clear cut that the whole sunk cost phenomenon is now frequently identified as the "Concorde Fallacy."

In a subtle way, elements of the sunk cost fallacy can be found in the predilection of aerospace companies for selecting "proven" components from within the existing supply base rather than embarking on a new development effort. It clearly influenced the pricing environment of SpaceX competitor Orbital Sciences Corporation. Even when a vehicle is comprised of "new" components, as was the case with the three Orion solid rocket motors employed by OSC to build its baseline Pegasus booster, the reality of heritage costs, passed along in this case by Shuttle Solid Rocket Booster vendor ATK, meant trading the short term gain of reduced outsourced development costs for the long term problem of being unable to control the supply chain.

Although any person or organization can fall victim to the sunk cost fallacy, and most of us do so daily in one way or another, it is critical to avoid it when your limited capital is being sunk, and conversely much easier to fall into when someone else is footing the bill, particularly the taxpayers. SpaceX was determined to avoid this trap by beginning with a clean sheet design.

A clean sheet it might be, but as the intended precursor to an entirely new family of launch vehicles with bold goals, including eventual human rating, the sheet contained some pretty tight margins. Even though the immediate objective was a low manufacturing cost, with the goal being reusability, from the outside it may have seemed that the company was pursuing mutually exclusive considerations. For one thing, if you achieved your first goal in obtaining low manufacturing costs, you were already beating the market, why even bother with re-usability? The answer, as Elon Musk would come to point out time and again, to investors, reporters and anyone who would listen, was that SpaceX wasn't created to make money, it was created to achieve a purpose. And, as anyone could tell you, getting into the launch vehicle business was a damn poor choice if your overriding goal was to make money. His company was created to lower the cost of getting into space. Making money was simply one of the requirements it would face along the way.

For Elon Musk and the close group of aerospace veterans establishing the basic parameters of what would become the Falcon 1, there was never any doubt as to a few of the key characteristics it would have, such as the type of main engine and what the fuel source would be. Still, it is important to understand why certain decisions were made, and what the alternative choices were, in order to fully grasp both the scope of change which the company was attempting to undertake, as well as alternative decisions not made, many of which led other launch vehicle efforts further and further away from the path to lower costs being charted by SpaceX.

The most elemental decision was what type of fuel would power the new rocket, and there were three basic categories from which to choose; liquid, solid or hybrid. Liquid fueled rockets are far and away the most common, and completely dominate the larger vehicle size classes. Compared to the alternatives, liquids offer more performance, but at the offsetting costs of expensive and troublesome pumps and associated valves and plumbing which are tasked with delivering tens of thousands of gallons to the engine in a matter of few brief, furious moments. If the choice was to be a liquid fueled rocket, the next issue would be the specific fuel and oxidizer combination, from which there is a surprising combination to choose, each with its own unique requirements.

An alternative selection, and the one most commonly found on smaller launch vehicles, was that of solid fueled rockets, which in the parlance of the industry are powered by motors, as opposed to engines. Ranging from the tiny Pegasus to the massive twin boosters used by the Shuttle program, solid rockets are a vital part of the defense industry, enabling the diverse array of offensive and defensive missiles fielded by surface, air and naval forces. In space applications, solids can be found almost everywhere, but one of their most effective usages is in the role of strap-on boosters, where the instant acceleration and rapid combustion characteristics are particularly helpful in getting heavy liquid fueled vehicles though the critical first few moments of flight. Solids are also especially appealing for smaller launch vehicles, where the comparative lower performance is offset by the benefits of lighter and simpler control mechanisms, namely the absence of fuel pumps, valves and piping found on liquid fueled vehicles. As rockets grow larger however, the overall percentage of the vehicle's mass which is taken up by control mechanisms decreases, and solids quickly lose much of their advantage. There are only two main suppliers of solid rocket motors in the U.S., Alliant Tech Systems (ATK) and Aerojet, with ATK being the largest. With a mix and match catalogue of SRM's offered in a variety of sizes, the Utah based company can provide a nearly customized launch solution for any new small or medium sized booster project, just as it had done for the Pegasus, Taurus and Athena launch vehicles, but it comes with the substantial drawback of relinquishing a great deal of control over the booster's cost.

The third option, hybrid fueled rockets, combine certain aspects of both liquids and solids. While the fuel itself is solid, the oxidizer is not contained in the mixture which forms the solid fuel core. Instead, it is provided in the form of liquid oxygen or nitrous oxide injected into the core of the otherwise inert solid fuel, making it possible to terminate thrust on command, thereby negating one of the biggest safety problems with conventional solids, namely the fact that once ignited, they cannot be turned off until they burn out. Hybrid rockets have found a home in the suborbital launch vehicle industry, powering notable craft such as X prize winner SpaceShipOne and Virgin Galactic's follow on SpaceShipTwo, but they generally offer the lowest overall performance of the three basic categories, and have yet to be used as the basis for an orbital launch vehicle.

For SpaceX, there was never any doubt that its family of boosters would be liquid propelled, and that the fuel selected would be a highly refined form of kerosene, called RP-1 (rocket propellant 1). It was the same source which had powered the first stage of the mighty Saturn V, as well as numerous other historically significant launch vehicles, including both the Atlas in all its various forms, as well as the Russian Soyuz. In selecting RP-1, SpaceX was turning its back on recent history, including the Space Shuttle and the Boeing Delta IV, as well as the world's premier commercial launch vehicle; the European Ariane V, all of which are powered by liquid hydrogen fueled first stages. Considering that recent trends seemed to indicate a shift towards clean burning and environmentally popular hydrogen technologies, the decision was an early sign of the practical approach SpaceX would take in addressing the many challenges in reducing the high costs of spaceflight.

For the most part, any liquid which will burn in the presence of oxygen or some other oxidizer is a potential rocket fuel and the range of choices runs from alcohol, which powered the German V-2, to exotic designer fuels which require specialized handling techniques and cannot be purchased from your local liquor store. What makes some better than others is the relative amount of energy they contain, which is expressed in terms of specific impulse, the theoretical maximum speed at which a given fuel combination can eject the products of combustion, and thus deliver forward motion. Offsetting the benefits of a given specific impulse is the amount of trouble involved in working with that particular fuel. Furthermore, it makes a difference whether the fuel under consideration is being used in a first stage which begins essentially at sea level and zero velocity, or in an upper stage where the working environment is little or no atmosphere and a vehicle which is already traveling at a substantial percentage of it maximum velocity.

It was for this reason that over the years some very notable vehicles including the Saturn and Atlas families were designed to take advantage of a duel arrangement utilizing RP-1 to fuel the first stage, and liquid hydrogen, which offers the highest specific impulse, for the upper stages, thereby gaining the relative benefits of each fuel at a different phase of the flight, but at the cost of increased complexity in conducting launch operations working with two fuels. While RP-1 is stored and used at basically room temperature, its most common oxidizer, liquid oxygen, must be maintained in a cryogenic state. Liquid oxygen, or Lox, is among the most widely used cryogenic fluids, is common in both industrial and medical settings, and can be found available in significant quantities anywhere in the world which has even a small hospital. Liquid hydrogen is a very different matter. It must be stored and used at a lower temperature than liquid oxygen, and its smaller atomic structure means keeping it contained offers a much more difficult challenge than presented by Lox. For example, liquid or gaseous hydrogen, which is highly flammable, (think Hindenburg) can pass right by a seal which is thoroughly adequate to contain liquid oxygen or other cryogenic fluids.

Among the many benefits of RP-1 as a fuel is that its relative density requires a lower overall volume of fluid for a given level of performance, and thus a smaller tank structure compared to liquid hydrogen. For example, one gallon of water weighs 8.34 pounds, and a gallon of kerosene weighs 6.82 pounds, whereas a gallon of liquid hydrogen weighs just .57 pounds. Consequently, even though liquid hydrogen offers considerable more performance in terms of specific impulse, the benefit is partially offset by the increased tank size. As the smallest molecule in nature, the hydrogen atom is both difficult to liquefy, and just as difficult to maintain in its liquid form, which requires being stored at -423 Fahrenheit. As heat invariably leaks into the storage vessel, a portion of the super cold hydrogen changes state from liquid to gas, gradually building pressure.

Maintaining hydrogen in its liquid form in a launch vehicle sitting on the pad requires a considerable amount of insulation which in the end can do little more than slow the rate of boil off. A separate issue is the formation of ice on the outside of the tank and any exposed surface which has been chilled down below 32 degrees. This is generally not a major problem, and even casual observers will recall most space launches accompanied by an initial shower of glittering white as accumulated ice is shaken off the suddenly dynamic rocket. The one tragic exception was the horizontally mounted Space Shuttle configuration, which placed the orbiter in the path of insulating foam breaking off the external tank and entering an airstream which was already moving at several hundred miles per hour.

For SpaceX, with an eye towards a very different two stage, partially re-usable architecture, the advantages of RP-1/Lox offered not only a more achievable engine development program, it also allowed for a straight forward approach to the vehicle design itself. Of critical importance from SpaceX's point of view was the fact that the comparatively oversized and necessarily more fragile tanks required to house liquid hydrogen would make eventual recovery and re-use an even more difficult proposition than it already was. Because the Falcon 1 was intended from the outset to have a recoverable and reusable first stage, it would need to be strong enough not only to withstand launch, but also a parachute deployment, water landing and recovery. With prospects for water recovery yet to be proven however, the trade off was to design a tank structure strong enough to enable recovery if it reached that point, but with a manufacturing cost low enough to still allow breakthrough pricing if it did not.

There is another class of storable liquid fuels which played a pivotal role in early space exploration, and is still a vital component of launch systems today, hypergolics. Although the world's first ballistic missile, the German V-2 was powered by alcohol and liquid oxygen, and the original American ICBM's, the Atlas I and Titan I, were powered by RP-1 and Lox, the difficulties of maintaining a state of military preparedness in working with cryogenics led designers on both sides of the Cold War to develop alternative, non-cryogenic fuel and oxidizer combinations which could be launched with relative ease at a few moment's notice. One of the key attributes of hypergolic fuels, and the characteristic which gives its name to the overall class, is that they ignite spontaneously upon contact, eliminating the need for complicated and finicky igniters or ignition fluid found in other launch vehicles, and making multiple in-space restarts a simple process. Being non-cryogenic, they can easily be stored for months or even years at a time. It was only natural then, that as both nations raced to develop orbital space launch capabilities, weapon systems such as the Titan II ICBM and Soviet Proton were adapted to new roles as space launch vehicles, bringing hypergolics alongside petroleum based fuels as common rocket propellants.

When the specific function of the missile you are launching is to travel most of the way around the world and deliver a warhead capable of extinguishing entire cities and wiping out millions of innocent civilians in a nuclear exchange which in all probability would result in the same thing happening to your own county, it is of little consequence that the fuel which is powering the warhead on the way to Armageddon is itself quite toxic and poses a threat to safe ground handling as well as to anyone on the vehicle's flight path if it should happen to crash without detonating the warhead. On the other hand, once the purpose becomes less nihilistic, then the use of highly toxic fuels is a different issue. Following the Titan series, which would serve the U.S. as a family of launch vehicles from the early 1960's all the way into the new millennia and the age of SpaceX, the United States abandoned hypergolics for first stage engines. They still remain in widespread use today as fuel sources for escape, orbital maneuvering and upper stage engines, where the characteristics of long term storage, multiple re-start capability and instant ignition are indispensable. Nevertheless, hypergolics are in a word, nasty, requiring complex ground handling procedures to prevent the launch area, (and any possible crash area) from resembling a World War I battlefield after a gas attack.

It was due to the inherent danger of hypergolics, and not a lingering fear of infection from Andromeda Strain style space germs which mandated that the first technicians to inspect a Space Shuttle upon landing dress in the very latest in contagion containment gear. It was a necessary precaution to avoid a potentially fatal inhalation of gases coming from either a possible leak or dispersion of hypergolic gases coming from recessed areas within the Shuttle's Orbital Maneuvering or Reaction Control Systems. Despite their obvious drawbacks, the relative simplicity of hypergolics compared to cryogenics in terms of successfully launching a rocket have led to it being a sort of starter fuel for new launch vehicle efforts in other parts of the world. Currently the Russian Proton and Chinese Long March, and Indian GSLV launch vehicles are all powered by hypergolic stages, and both India's Polar Satellite Launch Vehicle (PSLV) and the few remaining American Delta II's utilize a hypergolic second stage. Nevertheless, its overall use for main propulsion is on the wane, and even Russia plans to make the switch to RP-1 for its Angara series of launch vehicles slated to replace the Proton. Despite the fact that it would soon be competing against other low cost providers using hypergolics, there was never any question that SpaceX would seriously consider it as an option for the Falcon 1. It should be noted however, that several American companies, with NASA's encouragement are currently developing new "green hypergolics" which are neither toxic nor polluting, and offer enormous potential as an alternative, particularly in any manned spacecraft applications.

Two other alternatives deserve mention, the first of which is the use of hydrogen peroxide as an oxidizer in the place of liquid oxygen. Hydrogen peroxide is an interesting compound in that it can be used both as a monopropellant for low thrust applications such as station keeping, as well as an oxidizer when higher levels are called for. When pumped through a catalyst, such as a silver screen, it rapidly breaks down into steam and free oxygen. This was the combination chosen by Beale Aerospace in the mid 1990's for its abandoned commercial launch vehicle, the BA-2C. Although never launched, Beale successfully demonstrated its BA-810 second stage engine in the largest ground test of a single thrust chamber liquid fueled engine since the Saturn V's F-1. While offering the advantage of being a liquid at room temperature, highly refined hydrogen peroxide of the necessary purity is both expensive compared to Lox, and it is easily contaminated. Nevertheless it has adherents, and its exhaust, which is predominantly water, is cleaner than every possible fuel combination except hydrogen and oxygen. As such, hydrogen peroxide may still find its way into more widespread usage in the future.

One final possibility for liquid fuels, and perhaps the one which offers the most interesting long term potential is that of liquid methane. Although it has yet to be used on any orbital launch vehicle, liquid methane bridges much of the gap between RP-1 on one side and liquid hydrogen on the other. As a mild cryogenic, is offers superior performance compared to RP-1, but is much easier to store and work with than liquid hydrogen. It also is the lowest cost fuel available, a fact of negligible importance for expendable launch systems, but of potential enormous consequence for any future reusable launch system for which fuel costs would become an important variable, just as they are for commercial airliners. Of equal or even greater significance, methane has been detected in trace amounts in the Martian atmosphere, and can be readily derived from the CO2 which is that planet's most abundant gas. As popularized by Robert Zubrin in his landmark 1996 book, The Case for Mars, this one small detail is one of the most significant features of the inner solar system, one which offers the almost priceless potential of an on-site (in-situ) fuel source for an eventual round trip transportation system between Earth and the Red Planet.

In the final analysis, although some factors pointed to a hydrogen/oxygen architecture, due to its higher specific impulse and clean exhaust, the most important criteria were cost and practicality, leaving Lox/RP-1 as the clear choice to power the new launch vehicle. Furthermore, it was not lost on SpaceX that several of the more successful lower price commercial launchers, the Ukrainian Zenit and its Sea Launch variant, as well as the Russian Soyuz were all powered by RP-1 and Lox for both lower and upper stages. By contrast, there has never been a hydrogen fueled small orbital launch vehicle.

With the preliminary issue of the fuel source a foregone conclusion, the hard decision lay in designing an engine and stage structure which could achieve the desired overall performance.

From the outset, Elon Musk intended his new rocket to offer increased performance compared to its primary competition, Orbital Science Corporation's Pegasus XL, which was capable of placing approximately 980 lbs. into low Earth orbit. Leaving a comfortable margin, SpaceX set the goal for the new launch vehicle at a modest 1300 lbs. capacity to a 200 km orbit. The name, "Falcon 1" came from Elon Musk, in deference to science fiction's most famous commercial spaceship, the Millennium Falcon.

With the performance criteria in mind, and knowing the fuel combination to be used, it was a relatively straight forward matter to begin to outline the characteristics of the vehicle itself, the required first stage thrust, the number of stages, their dry and fully fueled "wet" weights and delivery capabilities to various orbits. All of these elements, and many more, can be plugged in to computer formulas based around the fundamental rocket equation, and using a database of performance hard won over 50 years of launch vehicle success and failure, it is possible to determine how the rocket you are designing will ultimately perform. Conversely by starting with the desired performance, and altering other variables, you can also determine how to configure the project to achieve that goal. Hypothetical calculations are the easy part, the difficulty lies in actually achieving the measures specified. Narrowly missing a margin can determine whether a rocket will reach orbit or fall ever so short of the necessary velocity, and come crashing back to Earth. In all cases the principle problem to avoid is the inevitable performance robbing growth in overall vehicle mass, or "dry weight" which is the mortal enemy of any launch vehicle planner as it moves from the computer to the launch pad.

For a two-stage, RP-1/Lox fueled rocket with conventional allowances for weight and performance margins, the basic rocket equation suggested the first stage would need to be capable of generating something on the order of 70,000 lbs. liftoff thrust. That number coincided nicely with a rule of thumb which suggested you can very roughly estimate a conventional expendable's launch vehicle's performance to LEO by calculating 2% of the total thrust at liftoff.

Not all rockets are conventional, or are comprised of two stages however, and this was an issue the design team would need to consider carefully. Historically, launch vehicles have been built in a variety of configurations, ranging from 1.5 stages such as the Shuttle and the original Atlas ICBM, all the way to four stages. Multiple stages make it possible to gain greater velocity by shedding unneeded mass as it accelerates on the way to orbit, and to an extent the inclusion of additional stages tends to make up for a lack of cutting edge performance on the part of any one specific stage. Each successive stage becomes a rocket launched not from sea level, which encounters the challenges of starting in an atmospheric soup of 14.7 lbs. per square inch pressure under a gravitational acceleration of 32 feet per second in the wrong direction, but from a starting point with (much) higher initial velocity, reduced or zero atmospheric pressure and decreasing gravitational attraction. Staging is the easiest way of "gaming" the gravitational system, and reaching orbit despite a less than ideal rocket.

For this reason, early in the space age, a number of boosters such as that employed for the United States' first attempted satellite launch, the Navy's unhappy Vanguard rocket, used three stages, and the first American rocket to achieve an orbital launch, Juno I, was comprised of four stages. Forty five years later, not much had changed. The Falcon's intended prey, the Pegasus is a three or four stage rocket, not counting the L-1011 aircraft which serves as the launch platform in which capacity it is called "stage 0." OSC's more capable Taurus launch vehicle, is a four stage rocket which is essentially a wingless Pegasus mounted on top of an ATK Castor 120 first stage derived from the Peacekeeper ICBM. [19]

While the act of jettisoning an expended stage significantly lightens the load and allows greater performance, it comes at the trade-off of the added structure, cost and complexity which comes with each of the additional stages. Compared to rockets with three or more stages, the two alternatives are either larger and more powerful main stages with ample performance in each, or the addition of parallel staged strap-on boosters which help lift a vehicle from the pad and propel it through the first few moments of flight. By the time such strap-ons, which can be either liquid or solid, are expended, the first stage will have consumed much of its fuel, and now suddenly much lighter, and having passed through the densest part of the atmosphere, rapidly accelerates towards orbit.

A two stage design with strap-on boosters would have been a highly unusual choice for a launch vehicle as small as the Falcon 1, but in a general sense it is a design solution which can readily compensate for any lack of intended liftoff thrust in the main engine, which is always a possibility when that engine has yet to be qualified. Given that almost every aspect of a launch vehicle grows heavier on the path from the design to operation, the potential inclusion of strap-on additional boosters offers a measure of insurance against such weight growth even when the engine performs as designed. Even when the first stage is everything you could want it to be, strap-on boosters still allow any launch vehicle to lift significantly greater payload than would otherwise be possible in its most basic configuration. Besides the Shuttle and the Ariane V, for which strap-on boosters are an integral part of the basic launch system, the Atlas V and Delta IV both utilize small solid boosters in a number of configurations, as does the Japan's primary launcher, the H-II as well as most of the Chinese family of Long March rockets. The Russian Soyuz launcher, which with over 1800 launches to its credit, has far and away earned the frequent flyer award over the history of the space age, consists of a central second core stage flanked by four first stage liquid strap-on boosters, and has flown in this configuration in a remarkable string of continuity since it was first launched as the R-7 missile in 1957.

With the latitude of a clean sheet approach, and confronting an array of choices, SpaceX settled on the simplest solution, a two stage vehicle. As the Aerospace Corporation study had indicated and SpaceX itself would soon discover twice before it achieved orbit, stage separation events are second only to engine malfunctions as the leading cause of launch vehicle losses over the last several decades, and for a good reason. Launching a rocket from a carefully controlled pad environment, complete with multiple sensors, redundant cameras, a hold down system, vibration suppression, a launch gantry of some sort, not to mention several thousand tons of reinforced concrete is difficult enough. Staging involves conducting a separate launch on top of a highly dynamic rocket which is screaming through the atmosphere in a maelstrom of heat, vibration and general hellishness. Worse, there is no way to "spin up" the second stage engine and verify that it is functioning normally until it is already cruising through this environment, and past the point of no return. After this, there is also the final challenge of successfully blowing apart the aerodynamic nose cone which was strong enough to survive the brunt of the experience in the first place. In the immortal words of Neal Sedaka, "Breaking up is hard to do." Best then to do it as few times as possible, unless you are a divorce lawyer.

It is one thing to design a paper rocket, but it is something else entirely to get an actual working launch vehicle out of the effort. NASA, its contractors and any number of would-be startups have in fact produced so many designs over the years that never made it to the launch pad, much less into space, that the effort has come to be derisively called "viewgraph engineering." If SpaceX wanted to avoid joining that list, it would need to establish and enact a manufacturing plan. The key question was just how much of the actual manufacturing SpaceX would undertake itself, and how much it would farm out. The initial plan was to start with a small engineering staff and farm out most of the actual hardware fabrication. As events progressed, it became increasingly clear that the preferred solution would not be the most efficient or affordable, setting the company on an ultimate path to undertaking the vast majority of its own manufacturing. This would turn out to be one of the single most significant developments in the evolution of SpaceX, and ultimately one which would change the nature of the launch industry. It began with the engine.

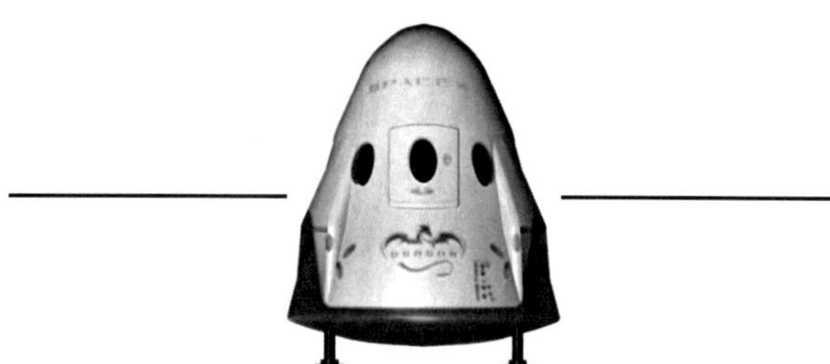

The Destination. Credit: NASA

Chapter IV: Developing the Merlin Engine

Every now and then, an industry is forever altered by the introduction of a product which is so revolutionary, that it changes not just the fortunes of one company, but the nature of the industry itself. Such is clearly the case with Apple's I-Phone, which almost single handedly altered the way the entire industrialized world relates to its primary communications device. Similar claims could be made of the Ford Model–T, the Douglas DC-3, and the Boeing 707. In much the same way cars and airplanes made the physically remote suddenly much closer, the widespread adoption of personal computers enabled by the Windows operating system, opened the way for the internet and "connectivity" to become a seemingly indispensable part of everyday life.

Frequently the agent of change is not the first, the biggest, the fastest or even the most capable member of its class, but the one, which for reasons which are hard to predict, seemingly catch lighting in a bottle, and subsequently go on to achieve iconic status. In recent years, as the world continues to revel in the digital revolution, the consumer products which capture the imagination are generally electronic in nature. It was not so long ago however, that for a least a certain demographic, the engines which power our cars and trucks were held in similar esteem. Except for a few holdouts, it is a rare thing in the 21st century for anyone to be able to readily identify the model of engine in their automobile, much less actually work on it. To the extent we think of engines at all, they are something we either use to search the internet, or regard somewhat apprehensively as we are about to board a plane.

Nevertheless, engines are still the basic building blocks of the world as we know it. The roots of the industrial age, and the digital age which followed, can be traced back specifically to the development of the first steam driven pumps to remove water from mines in England and Wales. Before that, before the ability to apply motive power wherever it was required, useful work could only be performed by harnessing the forces of nature. From the most humble of origins to enable the most dangerous and miserable of jobs, wresting coal from the deposits ever deeper underground, came the power to lift jet liners into the air and to blast launch vehicles into space. Along the way, particularly in recent years, we have become so good at building engines, that we rarely have cause to think about them. A car built in the 1950's needed maintenance on a regular basis and people worked on them because they had to. Spark plugs required regular replacement and distributors (remember those?) required timing adjustment. Whereas marine, automobile and aircraft engines evolved into the mass-produced and extremely reliable power plants which move people and goods around the planet today, rocket engines have yet to reach a similar state. While at least generally reliable, they are extremely expensive and essentially built by hand, and with the exception of the Space Shuttle Main Engine, are good for one-use only.

As they set out on an unlikely quest to develop the first new American liquid fueled hydrocarbon rocket engine in decades, Elon Musk, Tom Mueller and the rest of the SpaceX team intended to bring the rocket engine into a new era by challenging each of these three characteristics. At the same time however, they were also focused on ensuring success by taking an evolutionary path rather than by attempting to win superlatives with a breakthrough design. Such an approach may seem obvious, but it certainly isn't common. Historically, liquid propulsion projects have taken on a life of their own, and whether by design or out of necessity depending on the launch vehicle it is to power, invariably push the state of the art.

With no appropriately sized liquid fueled engine available on the commercial market, there was never any question that SpaceX would have to develop its own main engine, and in keeping with the automotive analogy put forward by Musk, SpaceX needed to design an engine more suitable for powering a Chevy than a Formula 1 race car. For the company to succeed and meet its price goals for the Falcon 1 launch vehicle, it would need to develop the aerospace equivalent to an engine familiar to every weekend mechanic over 40 and every dirt track dotting the landscape in locations somewhat less glamorous than Monaco or Cape Canaveral – the Chevrolet small block.

In the automotive world, the Chevy small block is pure Americana. From old reliable work trucks to blown out Saturday night racers, the small block has been installed, pulled, rebuilt, and modified by more people than perhaps any other engine in history. Built from 1955 until 2004 in a production run of an astounding 90 million units, and offered throughout GM's lineup as well as in myriad marine and industrial applications, where it is still being built today, the small block came to define nearly a half century of automotive power. It remains a favorite of auto enthusiasts and racers of all types due to four primary attributes: simplicity, affordability, reliability, and adaptability.

In most of the commercial world, it is normal to develop your product without having to develop the engine to power it. The ready availability of a common set of industrial engines in different power ranges is one of the fortunate facts which allow both individuals as well as companies of all sizes to produce a bewildering array of machines nearly as quickly as the mind can imagine them. Even in applications where the final product is a highly specialized device built in very limited production numbers, the power plant itself is generally common to the point of mundane. This, however, is obviously not the case that applies to space launch systems, and one might postulate that the long-sought era of space transportation for all will be marked by the moment certain rocket engines become so commonplace, we begin taking them for granted as well.

Conceptually at least, a rocket engine is an amazingly simple device. Fuel is introduced into the combustion chamber, ignited, expands, and performs all of its work in the simple act of expansion. Furthermore, the rest of the "drive train" generally consists of nothing more than a set of arms to "gimbal" or steer the engine to direct the rocket's path, much as one turns a small outboard motor to steer a boat. Compared to say, a car's engine which likewise also starts with the combustion of fuel, but then takes that product and channels it though a complicated set of individual cylinders into a crankshaft, turning it into rotational energy, transferring through a clutch or torque converter to an even more complicated transmission; then taking that variable output, splitting it at least two ways, sometimes turning it 90 degrees and channeling to a set of tires which must respond to an ever unpredictable road surface. The difference of course is the volume of fuel consumed, the terrible heat generated in the effort, and the blazing speed in which events occur.

In the course of nine or so minutes of acceleration towards orbit, a rocket engine the size of a single Merlin consumes more propellant than your car will use in its lifetime. Consequently the challenge lies in not just managing the heat and pressure of such a conflagration, but also delivering that much fuel from the tanks to the combustion chamber in such a brief period of time. There are two potential solutions. The first is to use high pressure gas to force both the fuel and the oxidizer from their respective tanks into the combustion chamber. The second, and more complicated, is the use of turbo pumps to deliver the fuel. The former method has the benefit of extreme simplicity, but offers the considerable drawback of both very heavy tanks with which to contain the pressurizing gas, as well as the requirement for more robust fuel tanks than otherwise necessary to handle the pressurization effort without rupturing.

Because of these drawbacks, it is an approach generally reserved for either smaller suborbital rockets, such as the Aerobee series of sounding rockets, or the upper stages of smaller orbital rockets. Nevertheless, it was a possible avenue for SpaceX based on the persistence of one of the most enduring alternate and untried paths to orbit of the space age, that of the "Big Dumb Rocket(BDR)." Designing rockets is an exercise in tradeoffs, shaped by the imperative to create a final product with the highest performance and lowest dry weight possible. It is an effort which perpetually seeks thousands of incremental gains in vehicle structural efficiency existing in a delicate balancing act with the corresponding tendency towards increased cost and complexity. The Big Dumb Rocket idea stands all of that logic on its head, and proposes instead the development of much larger and more robust vehicles which sacrifice efficiency to gain the benefits of a brute force approach. Over the years, BDR proposals have generally featured pressurized fuel delivery systems in combination with stronger tank and construction techniques which sometimes more closely resemble that of a ship than a rocket. In fact one of the most famous such proposals, put forward by aerospace legend Robert Truax, and called Sea Dragon, envisioned a two stage rocket half again as large as the Saturn V, which would be built by a shipyard and towed out to sea. Once on station, its pressure fed upper stage was to be fueled by hydrogen generated on site through electrolysis provided by the reactor of a nuclear aircraft carrier. The specific ship referenced in the original proposal was appropriately enough, the USS Enterprise. [20]

The world's first semi private attempt to develop a liquid fueled commercial booster was a variation on the BDR concept. Begun in 1975 by German aerospace engineer Lutz Kayser, at the request of the West German government, OTRAG, a duel language acronym for Orbital Transport and Rockets, Inc. , consisted of a number of simple steel tubes, each housing an individual pressure fed unit, welded together to constitute a stage assembly. The effort was serious enough that it attracted the participation of both Wernher von Braun, as well as former Kennedy Space Center Director Kurt Debus, but it came to grief as much over political considerations as it did technical problems, when Kayser sought to establish a launch facility in the African nation of Zaire. Thirty years removed from World War II, neither France nor the Soviet Union was particularly enthused at the prospect of a German booster which could also function as a ballistic missile, and both shared concerns that the technology might be seized by Zaire. West Germany bowed to political pressure and terminated OTRAG's domestic operations, but undeterred, Kayser moved his development efforts to friendlier confines. Regrettably, the new destination was Libya, which did nothing to improve the company's prospects. The OTRAG concept is still very much around, and has been the focus of several credible U.S. based small development projects, on which Kayser himself has consulted.

The BDR concept came halfway off the drawing board and closer to an actual flight with Beal Aerospace, whose abandoned BA-2C design planned to incorporate helium pressurization to fuel its hydrogen peroxide/kerosene engines. Had Beal persisted, and successfully demonstrated a large, orbital class pressure-fed engine, aerospace history might well have taken a very different direction, and it is still not out of the question that the BDR could see its day in the sun, particularly if, ironically enough, as a result of the work performed by SpaceX, the business case for truly large boosters ultimately emerges.

Despite the appeal of pressure fed simplicity, SpaceX elected to proceed with the more conventional approach for its fuel delivery system, the high powered turbo pump. The word "turbo" derives from the Greek word to spin or mix, and finds its' most common application in a "turbocharger" for automotive or heavy diesel engine in which the engine's own exhaust is used to spin a pump which increases air flow to the engine, and thereby boosting available horsepower. When applied to rocketry, a turbo pump similarly uses a stream of expanding gas to "spin up" the pumps which are responsible for feeding the fuel to the rocket's engine at astonishingly high rates.

Modern rocket engines driven by turbo pumps fall into one of two general categories, depending on how the engine treats the expanding hot gas after it has passed through the pump's assembly. In the simplest application, an open cycle, the rocket's own fuel is ignited in a small combustion chamber, and after routing through the turbo machinery, is emptied into the engine's exhaust stream. This waste gas is sometimes used to provide vehicle roll control though a steerable nozzle. The chief drawback to the open cycle design is that it essentially wastes the potential thrust which is contained within the fuel, resulting in less than optimal engine performance.

The alternative is a closed cycle, or staged combustion engine in which the stream driving the turbo pump is directed not into the exhaust, but into the main combustion chamber, where it can do the most work. This approach results in a more powerful engine with a higher thrust, but comes at the cost of both increased complexity in terms of the plumbing and the routing of hot gases, and more importantly the addition of carefully designed structures to manage the increased operating pressures, as well as increased mass. Because the gas is being introduced into a main combustion chamber which is operating at high pressure, it must enter the chamber at even higher pressure or risk a possible catastrophic buildup of back pressure. Both designs have their advantages, and both can be regarded as particularly efficient depending on which measure is used to gauge effectiveness. Whereas specific impulse measures a fuel's potential, a different metric, thrust to weight, provides a guide as to how effective an engine actually may be at getting the job done. A third type of engine, an expander cycle, uses the heat absorbed by pumping the fuel through the chamber walls to drive the turbo pump, but it is limited in size due to the square-cube rule, and as a result, finds its best application in upper stage engines.

Among historical engines, the Space Shuttle Main Engine is a highly complex cryogenic staged combustion engine, whereas the massive Saturn F-1 was an open cycle engine which made up for somewhat lower performance through its huge size. Among other space faring nations, Russia was, and still is, the undisputed leader in hydrocarbon staged combustion engine technology. Beginning in the early 1960's as part of its N-1 lunar rocket program, and then extending nearly up to the moment of its collapse, the Soviet Union developed a family of staged combustion engines to power the strap-on boosters for its Energia heavy lift vehicle, designed to launch the Russian counterpart to the Space Shuttle, the Buran. Unlike the Shuttle, the Buran orbiter did not carry its main engines, which were instead attached to the expendable hydrogen/oxygen main stage. The Energia's four liquid fueled boosters each featured a single large turbo pump driving four separate combustion chambers. During the course of its development, Soviet engineers made significant advances in perfecting the complex metallurgical requirements demanded by its oxygen rich fuel cycle. Even though the Energia booster only flew twice, once with Buran and once with a test payload, the Russian space industry emerged from the effort with what was a remarkable family of main stage engines.

After the Energia/Buran program was cancelled, the strap-on boosters found new life as the first stage of the stand-alone Zenit launcher, a formidable rocket in its own right, which today serves as the basis for the commercial Sea Launch vehicle. The Zenit's four-chambered RD-171 engine is also manufactured in a de-rated 2 chamber version, the RD-180 which Lockheed Martin adopted to power first the Atlas III and then the Atlas V booster (there was no Atlas IV). The fact that one of the premier U.S. launch vehicles utilizes a Russian built main engine underscores one of the most bizarre aspects of the market SpaceX sought to enter. Despite the overwhelming technical success of the Apollo program, and the undeniable awe which comes from visiting one of the three remaining Saturn V launch vehicles in their respective museums, the United States abandoned the development of high power hydrocarbon rocket engines when it embarked on the Shuttle program, and in the process became a second rate player in this vital element of propulsion technology.

In addition to the RD family of staged combustion engines, there was another Soviet era engine family in existence and potentially available in the U.S. This was the NK-33, a smaller staged combustion engine originally designed for the cancelled Soviet lunar program and the giant N-1 booster. Commissioned after the initial NK-15 engines proved problematic, a large quantity of NK-33 engines were built, tested, and then ordered destroyed after the N-1 program was cancelled. Instead of following orders, an unidentified technician instead arranged to have the engines placed into storage in hopes of a better day. Although that day did eventually arrive, it might not have been the one he envisioned. During the course of a trip to Russia in the 1990's an American team of engineers from Aerojet was shocked to learn of the existence of the engines, still in storage. Recognizing the value, Aerojet arranged to both buy a number of the engines, reportedly for just $1 million apiece, as well as securing a joint licensing agreement to enable U. S production of new engines when the inventory ran out. [21] The NK-33 originally found a U.S. application as the power plant for the Kistler Aerospace K-1 Reusable Launch Vehicle project in the late 1990's, and was later refurbished and re-branded by Aerojet as the AJ-26, where it has since gone on to join the RD-180 as the second Russian built staged combustion engine powering a U.S. launch vehicle. While smaller than the RD-180, the AJ-26 was still far too large for use in the Falcon 1, even if SpaceX had wanted to acquire an engine externally. Furthermore, the combination of a finite supply and control by an external authority made it a dead end.

SpaceX elected to keep it simple, choosing an open cycle design for its main engine project. Besides simplicity, there was another reason which led the company to go with the lower performance engine design, and it was the same reason that would drive many of the decisions involving the Falcon 1; re-usability. While closed cycle engines offer a higher level of performance, they are also prone to a phenomenon known as "coking" in which carbon rich deposits from the turbo pump exhaust stream build up within the engine during operation. In fact, in the early days of designing the F-1 engine which would eventually power the Saturn V, the difficulties encountered with coking ultimately led to the decision to proceed with a less troublesome open cycle design. While no longer a sufficient problem to affect a single flight's performance, it is a serious issue for a multiple use engine, and mitigation would require an expensive and time consuming teardown and rebuild. The Space Shuttle Main Engines were not affected by the coking problem due to the simple fact that they ran on liquid hydrogen and oxygen, for which the only by-product is water. Nothing in space comes without a tradeoff however, and the primary solution to coking, an oxygen rich fuel cycle, results in increased wear due to the corrosive effect of pure oxygen on metal, and it was in mitigating this effect that the Russians had gained such an expertise.

Besides the limited alternatives available, there was also a significant business factor behind SpaceX's decision to develop its own engine. Following any other course would have both put the company at the mercy of whomever it chose as a supplier, and perhaps even more importantly, prevented or forestalled its internal capability to manage its own destiny through gaining an immersive understanding of its propulsion systems. And of course, in financial terms, just as the engine is the heart of a rocket, controlling its expense offers the best opportunity to control the expense of the rocket.

Surprisingly, SpaceX needed to look no further than an engine designed, built and tested by NASA's Marshall Space Flight Center for confirmation that a new engine close to the performance requirements for the Falcon-1 was readily achievable at a reasonable cost.

The engine was called Fastrac, and in some ways it was an effort which did everything you might want a government development program aimed at promoting new commercial investments to do; absorbing expensive R&D costs to develop a new and efficient technology, while offering industry an attractive option for pioneering new markets.

Originally conceived as part of the Low Cost Technologies effort for Marshall Space Flight Center's Advanced Space Transportation Program in 1996, and billed as "one of the world's simplest turbo pump rocket engines" the Fastrac was a remarkable achievement, a high performance engine consisting of a minimal number of parts with low production costs, but promising the potential for high production volume. With NASA engaged in a number of R&D efforts aimed at advancing the development of reusable launch systems during the mid and late 1990's, the agency needed a small, low cost engine which could be counted on to provide a reliable propulsion source for several ongoing programs, beginning with the X-34 suborbital space plane. With an estimated production cost of less than $1 million per unit, and open licensing rights because it was developed by NASA rather than a contractor, the Fastrac appeared to be the perfect engine to power a small, low cost launch vehicle. [22] As it turned out, the traditional aerospace industry wasn't really all that interested.

Specifically Fastrac was a 60,000 pound thrust engine with a pintle injector, an ablative combustion chamber, a simplified rugged turbo pump and an overall minimal parts count. At the time, NASA estimated the original per unit production costs at $1.2 million with the possibility of falling as low as $350,000. Interestingly, besides its intended purpose for the atmospheric X-34 flight test vehicle, NASA also estimated the engine as being capable of launching a 500 lb. payload into low Earth orbit. In a surprising choice for a NASA field center which is often criticized for being overly attached to its past, the Fastrac team chose Barber-Nichols of Colorado to develop the turbo-pump. The choice was notable because Barber Nichols was not a traditional aerospace vendor, and was best known as a producer of automotive and industrial pumps. The engine came together in a remarkably short time for a NASA project, and was test fired in less than three years. Unfortunately, after a series of successful demonstrations and three production units, the cancellation of the X-34 left the Fastrac an engine without a rocket.

Although SpaceX would go on to design its own main engine which did share some commonality with the NASA project, the mere existence and demonstrated firing history of the Fastrac provided a clear indication that a basic power plant could be developed and produced at a price point necessary to enable the planned Falcon 1. Fastrac was in its own way, existence proof which suggested that if a reliable first stage engine could be produced for the right price, then the rest of the launch vehicle should be achievable within budget as well. While not guaranteeing success, it argued that the ultimate outcome of the project was within their means to control.

The Fastrac episode provides a bit of interesting insight into the NASA/contractor culture.

If a small team of facility personal could design and test fire a low cost, space capable engine with substantial contributions from vendors outside the normal supply chain, it might have been a signal that something was very amiss within the dominant aerospace prime contractor culture. Perhaps to some it was, but for the time being at least, the agency chose to look the other way and proceed with business as usual, consigning the remarkable little engine to apparent obscurity.

Some people, including those participating in the SpaceX engine program paid attention however, and readily applied the lessons learned from the now abandoned program.

Was the Merlin developed directly from the Fastrac? SpaceX maintains it was not, and at times has been rather emphatic on that point. After all the Merlin sea level thrust is substantially greater than offered by Fastrac, and subsequent Merlin engines would yield performance results nearly three times that of the little engine from MSFC. There is however, one critical link between at least the first generation Merlin and the NASA engine; both are powered by turbo pumps built by Barber-Nichols. This became a regular point of contention for some of SpaceX's early gallery of critics, who offered the Fastrac as proof that Elon Musk's effort was hardly innovative, and was in fact simply living off government largess. The more relevant fact was not that SpaceX benefitted from the NASA program, which it clearly did, but instead that its key personnel were willing and motivated to make the most out of readily available resources in pursuing a more affordable small launch vehicle, several years after the established aerospace industry essentially turned its nose and looked the other way.

With the development program for what was the SpaceX main engine getting underway in 2002 and guided by Tom Mueller, the first unit came together quickly. Dubbed Merlin 1A, the baseline version was configured to generate 78,000 lbs. of thrust, powered by the Barber Nichols turbo pump and featuring a simple ablative thrust chamber which by design, charred away during the course of firing and afterward could be replaced, cartridge style, to enable a new firing. The most innovative feature however, was the pintle style injector. One of the most difficult tasks in designing a rocket engine is the makeup of the injector assembly, the critical component which introduces both fuel streams into the combustion chamber. The challenge is to introduce the fuel in such a way that the streams of oxidizer and propellant successfully intermix without inadvertently creating hot or cold spots in the combustion chamber, the former of which can lead to chamber failure and explosive destruction. A cold spot leads to poor combustion efficiency and a loss of thrust. Yet another problem, one which has plagued liquid fueled rocket engine development from the very beginning is the creation of a series of micro explosions in the combustion chamber which rapidly build in intensity, once again leading to catastrophic results. Precise placement of the varying orifices and the creation of impinging fuel streams are the primary determinants of success or failure, and even today it is still very much a trial and error process because even the best flow dynamics software is not able to precisely model the release of energy in a rocket's combustion chamber. The general result in many development programs is a lengthy and very expensive testing program in which first one injector assembly after another is tested in relation to a given combustion chamber in order to find a goldilocks solution which is "just right." Unfortunately, the problem seems to grow substantially as engines increase in size and power, a challenge which led both the original German V-2 engineers, as well as later generations of Soviet designers to substitute a series of smaller combustion chambers driven by a single piece of turbo-machinery instead of a larger mono chamber engine driven by the same pump.

Early in the Apollo program however, even as engineers working on the Saturn V's massive F-1 engines struggled with combustion chamber difficulties, the TRW team working on the engine for the Lunar lander took an entirely different approach, and it was one TRW exec Tom Mueller would bring to SpaceX. In a system comprised of millions of parts, the two components which worried planners the most were the engines which would allow the Apollo astronauts to first descend to the lunar surface, and then make the ascent back to lunar orbit to join the awaiting command module for the ride home. A failure in either situation could lead to a national nightmare, with doomed astronauts crashing towards the Moon, or even worse, marooned, running out of air and waiting for the end as the entire drama played out on network television. TRW was assigned the task of designing the more problematical of the two, the descent engine, and making sure this did not come to pass.

Both engines were fueled by hypergolics, eliminating the need for an igniter, and both were pressure fed, with pressurization enabled by inert helium gas. To that point, the design was nearly fool-proof. The one remaining characteristic to be accounted for was smooth combustion, and to achieve this, TRW engineers developed an ingeniously simple solution. Rather than struggle with a complicated injector plate for a descent engine which would have to be deeply throttled, they elected to follow an approach familiar to anyone with an older style garden hose which adjusts the volume of spray by sliding a tapered tube inside a slightly larger tube. The inner tube, called a pintle, allows for a somewhat free form pattern which prevents the problems associated with conventional shower head style injector plates. In addition to its inherent reliability, the pintle injector is far less expensive to both build and test, offering a natural and obvious alterative for low cost engine development programs. Nevertheless, after TRW and Apollo, with the exception of Fastrac, it was not employed again until selected by SpaceX for the Merlin.

With an appropriately sized turbo pump available and a readily fabricated pintle injector assembly designed in house, SpaceX was ready to begin its initial test firing only 9 months after development began. While SpaceX's El Segundo headquarters was a great place to acquire aerospace talent and design rockets, it was certainly no place to test them. Given the fact that engine tests are very noisy even when everything goes right, and dangerous when they do not, SpaceX required a testing facility which was both remote, and preferably private. For many in Southern California's aerospace industry, this meant the relatively accessible grounds of the Mojave Desert which has played host to many aerospace startups including X Prize winner SpaceShipOne, built at Burt Rutan's Scaled Composites. SpaceX however, had something more remote in mind.

First factory in El Segundo. Elon Musk's McClaren F-1 has plenty of room. Credit: SpaceX

In 1942, the Army Ordnance Core established the Bluebonnet Ordnance plant on 5,700 acres outside the small town of McGregor, Texas, 16 miles from Waco and midway between Dallas and Austin. Following the Second World War, the plant closed in 1946, and a significant swath of the property was deeded to Texas A&M University for a research station. It was not long before the demands of a new conflict prompted the Air Force to take control of the remaining portions of the Ordnance plant in 1952 to set up manufacture of Jet Assisted Takeoff (JATO) solid motors to get heavy aircraft off the ground. The Navy took ownership in 1966 establishing the site as Naval Weapons Industrial Reserve (NWIRP) McGregor, a government owned, contractor operated facility which produced a wide range of air launched solid fuel rocket motors for stand- off weapons such as the Phoenix, Sparrow and Sidewinder missiles.

With the end of the Cold War, the Navy began making plans to divest itself of the McGregor facility, a process complicated by the fact that environmental testing found that excessive amounts of perchlorate used over decades of solid rocket manufacturing had leached into ground water. Following a remediation plan that featured the first ever widespread use of buried biowalls to intercept and treat the contaminated water, the Navy began transferring sections of the facility to the City of McGregor for development as an industrial park. One of the earliest tenants was none other than Beal Aerospace which selected McGregor as the site for testing its planned series of hydrogen peroxide powered rocket engines. Over the course of its brief tenure, Beal built a horizontal test stand capable of supporting up to 3 million pounds thrust; a 50,000 lb. thrust vertical test stand, and began construction of a massive, tripod shaped 135 foot tall concrete test stand, visible for miles across the flat Texas landscape. After Andrew Beal abandoned his effort in frustration in 2000, it was a rather unique facility sorely in need of a new tenant, presenting a one of a kind opportunity for SpaceX. Signing a lease with the City of McGregor in 2003, and beginning with just 3 employees, SpaceX began the process of converting the hodgepodge of infrastructure, including a partially buried block house and firing room built in 1960, which was once home to multiple tape drive computers, each the size of SUV's and connected by miles of cable to a single test stand. [23]

What gradually emerged at McGregor was not simply a modified test stand for a fairly small, single rocket engine tucked safely away in rural Texas, but instead a sprawling rocket development facility which would be in many ways the very heart of the SpaceX effort, and the key to its success. Space launches are innately dramatic image driven affairs which can for a few moments capture the attention of the press, and make it seem as if places like the Kennedy Space Center are the epicenter of the effort. In reality, launches tend to represent the final, brief conclusion to an act which has played out in relative obscurity over a span of months and years in carefully guarded labs or the comparative isolation of remote testing facilities.

For solid rocket motors such as those built at McGregor over the years, it is possible to test a sample of the batch of rubber-like compound which is formed into segments and is poured into the shell of the motor, but it is not possible to test any given motor which will later be used in flight since it is consumed in the process of testing, a limitation which bears a funny resemblance is cake batter before the cake is placed in the oven, you have a pretty good idea what the final product will be like, but you can never say for sure. Hybrid rocket engines are somewhat different, given that it is possible to test the pump, tanks and plumbing assembly which is used to inject the oxidizer into the solid fuel motor, but it is still a limited capability compared to the full range of testing which is achievable with liquid fueled rocket engines, one of their most appealing characteristics. As the old saying goes, "fly like you test, test like you fly." The more testing you can perform, the higher the likelihood of success on launch day.

Dual Merlin test stands at McGregor. Credit: SpaceX

On March 19, 2003, nine months after engine development began, SpaceX conducted the first test firing of the prototype Merlin engine, which briefly roared to life producing 60,000 lbs. thrust with an initial combustion efficiency of 93%, only ½ percent less than the fully developed 1.5 million lb. thrust Saturn F-1. [24] Although the numbers were impressive, leading the company to believe that it could be in a position to conduct a first launch by the end of the year, it did not take long to determine that the engine was in fact quite a ways removed from being able to withstand a full length test burn, let alone an actual launch. Still, the propulsion team at least appeared to be on the right track with a first stage engine which could take it most of the way to space. What it still needed was a smaller second stage engine to complete the journey.

If the Falcon 1's Merlin engine was comparatively simple, by most conventions the second stage engine, named Kestrel, was even simpler, bordering on primitive. Maintaining the RP-1/Lox fuel combination, Mueller designed the second stage Kestrel engine to produce 7,500 lbs. of thrust by foregoing the complexities of a turbo pump and relying instead on the simpler option of a pressure fed system complete with pintle injector. It was an ideal choice for a small, second stage which would need to be capable of both deep throttling and multiple starts for precise orbital injection, two tasks which present difficult challenges for turbo pump driven engines. For most launches it is common procedure to initially establish a basic "parking" orbit and then gradually achieve the intended final orbit by one or more placement burns. Second stage engines, while far smaller than the boost engines which drive a rocket off the ground, are typically required to burn for a much longer period of time, and must be designed with rugged durability in mind.

Like the Merlin, the Kestrel engine featured an ablatively cooled combustion chamber, but one noticeable difference was the inclusion of a much larger bell assembly or "expansion skirt" which would shape the exhaust stream and allow the engine to function efficiently in the near vacuum of the upper atmosphere as well as in outer space. The expansion skirt was fabricated from the extremely light weight and heat resistant metal niobium. Although the first stage of the Falcon 1 was intended to eventually be recovered for reuse, the second stage and the Kestrel engine were designed to be expendable only.

At 7,500 lbs. thrust, the Kestrel engine actually fell within the performance range of several amateur liquid fueled rocket engines, including the one Tom Mueller was working on in his garage on the day Elon Musk introduced himself. Although it would be governed by a considerably more sophisticated flight control system than found on amateur rockets, the role it played in allowing the Falcon to eventually reach orbit, are a testimony to both the SpaceX commitment to simplicity as well as the surprising accomplishments made by amateur rocket builders which are often overlooked in an environment dominated by behemoth government contractors.

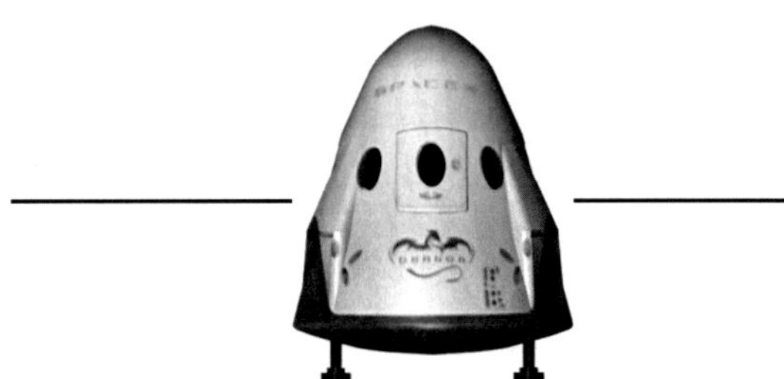

Chapter V: Designing the Stage Structure

In the broadest sense, SpaceX was required to make three major and interrelated decisions regarding the development path for the Falcon-1, and from each of these; the type of fuel to be used, the nature of the engine, and finally the arrangement of the stage structure, thousands of smaller decisions would follow. For a vehicle intended to both shatter launch costs and to cut a direct path towards partial re-usability, the design team made what could be considered surprisingly conservative decisions for the first two categories. Kerosene/oxygen offered the most straight forward fuel combination, and an open cycle, ablatively cooled first stage engine fed by a pintle injector appeared to offer a conservative path with the lowest development cost. If the Falcon 1 project was to prove a game changer, then it would be necessary to establish a departure from the norm somewhere in the fundamentals of the design. Almost by default, that distinction fell to the stage structure, an element heavily influenced by the other two choices already made.

Viewed in terms of their overall bulk, rockets are little more than flying fuel tanks. All fuel tanks are not created equal however, and intimately tied to the initial decision of selecting an engine; the choice of the particular fuel combination has a major impact on the size and nature of the fuel tanks, on how those tanks are built, and how they are combined to form one integrated vehicle.

One critical industry term for the overall efficiency of a launch vehicle's stage structure is called the *propellant mass fraction,* which is expressed as the ratio of the weight of fuel contained in a given stage, versus the total mass of the stage structure plus the fuel. The higher percentage of a vehicle's mass which can be assigned to propellants, as opposed to the structures which contain and channel them, the greater the potential for overall performance. The amount of useful goods actually delivered to orbit as payload is called the payload mass fraction, and it is closely tied to both the propellant mass fraction, and the relative performance of the fuel combination selected. Keralox systems, with their lower specific impulse, require a propellant mass fraction of roughly 94%, whereas higher performing cryogenic hydrogen and oxygen can theoretically enable a lower propellant percentage approaching 83%, and thus seemingly more payload. Unfortunately, it comes at the cost of increased complexity in the vehicle structure. In general large launch vehicles are more efficient, but even then wider and larger tanks bring their own problems in increased aerodynamic resistance. Regardless of the fuel system, one thing remains the same, one still has to leave room for the payload fraction, which is the percentage of initial mass actually delivered to orbit as payload. The numbers are depressingly small. Two percent is considered average, five is outstanding and ten percent would be a miracle. Getting closer to the latter number is one of the most enduring challenges in launch vehicle design, and many an engineer has laid awake at night trying to come up with a new way to shave just a little more weight off the overall vehicle structure.

One of the most fundamental solutions, and one which is a major point of departure from even the fastest supersonic aircraft, is the fact that unlike aircraft, an orbital rocket's fuel tanks are not contained within the overall structure, they are the structure, which means that in addition to containing thousands of gallons of fuel, they must also be able to endure the full thrust of the engines and transmit that force safely up through the entire structure in order to hurl the payload into space. That was the crux of the challenge facing SpaceX, designing a tank structure robust enough to handle the job, but light enough to avoid eating into the available payload margin, which is a very easy thing to do. In approaching the problem, SpaceX, combined elements from two historical programs to arrive at a unique solution.

For most of the space age the material choices for stage structures were limited to either high grade aluminum, or to stainless steel. Beginning in the 1990's however, impressive advancements in the development of Kevlar and graphite epoxy materials led to the development of composite fuel tanks, offering another potential design path. Unlike metal tanks, composites can easily be formed into irregular or unusual shapes, and even when the shape is conventional, they offer the potential of lighter and stronger structures, and thus better mass fractions. It was the first of these characteristics though, which attracted the attention of one of NASA's boldest attempts at lowering the costs of space transportation after the initial disappointment of the Shuttle system.

The X-33 project was an effort to develop a 1/2 scale single-stage-to-orbit demonstrator intended as the precursor to a fully functional SSTO spacecraft, the Lockheed Martin VentureStar. Given that the end goal was the same as that which Elon Musk was pursuing through the formation of SpaceX, the introduction of affordable rapidly reusable transportation to low Earth orbit, in many ways the project encapsulated all of the challenges Elon Musk would eventually face with SpaceX, and the differences between the two attempts are notable in almost every way.

The X-33 was to be powered by a radically different form of hydrogen/oxygen main engine, and intended to be capable of reaching Mach 13, or roughly half the speed required to reach orbit. The new engines were aerospikes, which differed from conventional rocket engines in that they did not have a traditional bell nozzle at all, but instead featured a circular acceleration "ramp" which ends in a tapered spike in the center, with a series of injection assemblies placed around the ring. For the X-33, the spike was "unwrapped" and formed instead as a straight line with the injectors placed on either side, resulting in a "linear aerospike." In either configuration, the theoretical advantages are two-fold, lighter weight and a performance curve which self-tailors to every atmospheric environment from sea level to space. Equipped with a pair of linear aerospikes, the X-33 gained the added advantage of differential steering by varying thrust between the two power plants, eliminating the need for gimballing hardware. The aerospike concept has been around for a long time, and like Big Dumb Boosters, represents one of the many untried or unfinished development paths of the space age. In the case of the X-33, though, it was not the engines, but the tank structure which proved the project's undoing.

The X-33, and the orbital VentureStar which would have followed it, were conceived as wedge shaped lifting bodies which pushing the very limits of mass fraction design, contained all the fuel, engine, frame and payload in a single stage. In order to provide adequate room for fuel within the structure, designers incorporated a specially formed, multi-lobed composite tank for the cryogenic hydrogen fuel. As frequently happens with projects which seek to combine multiple technological breakthroughs, difficulty in developing one, in this case the composite tanks, led to the entire effort unraveling. Specifically, NASA and Lockheed Martin were unable to prevent the composite liquid hydrogen tanks from undergoing destructive delamination during pressure testing. What was both curious and disturbing about the cancellation was the fact that as a technology demonstrator, as opposed to an operational system, there was no reason NASA could not have proceeded with an alternate metallic tank for the liquid hydrogen, while still making significant strides in retiring other areas of risk, such as the engines, airframe and metallic tile heat management systems. [25]

Although the SSTO project collapsed under the weight of its extreme engineering challenges, the appeal of composite tank construction was also a major draw for Beal Aerospace in the design of its BA-2 rocket, even though NASA's support of Lockheed Martin in the concurrent X-33/Venturestar program was one of the reasons Andrew Beal cited as his decision to close shop. [26] The two stage BA-2 rocket was initially designed to be a fully expendable system, but the long term goal was to recover the first stage by parachute for re-use. The Beal rocket featured composite fuel tanks built by the world's largest filament winding machine, a design choice made due to the perceived simplicity and comparatively lower cost of construction rather than the need for an unusual tank geometry. Because the effort was aimed at producing a launch vehicle powered by kerosene and hydrogen peroxide, it neatly avoided one of the principal drawbacks of composite tanks, which is the relative difficulty in containing super cold cryogenic fluids.

Although Beal Aerospace abandoned its efforts well before it had a flight ready vehicle, composite structures continue to enter the mainstream in both aviation and aerospace applications, as evidenced by the Boeing 787 Dreamliner and the Airbus A-480 commercial jetliners, as well as the all composite SpaceShipOne, SpaceShipTwo and Sierra Nevada Dream Chaser spaceplanes. With both NASA and private companies continuing to research cryogenically capable composite tanks, it seems likely that they will eventually find a starring role in launch vehicles as well. Early in the process, the Falcon 1 design team considered composite tanks, as well as several more exotic oxygen resistant "super alloys" before determining that the pursuit of simplicity and reliability argued convincingly for a more conventional aluminum alloy tank structure.

In the early years of rocket development, Convair took this concept of lightweight metallic tanks to its extreme limit in designing the Atlas ICBM, which employed a unique stage and a half structure propelled by three engines at lift off, two of which were designed to separate from the ascending rocket only a few moments into flight. The real key though, was the Atlas tank structure itself, which Convair designed as an ultra-thin balloon skinned tank out of stainless steel. The resulting structure is called monocoque, coming from the Greek word for one, mono, and the French word for skin, coque. Unlike contemporary rockets, the Atlas did not have a skin and stringer construction to provide strength, but instead gained strength when pressurized for flight. During ground operations, which had to be handled very carefully, the tank structure was kept pressurized by nitrogen gas in order to prevent the rocket from collapsing under its weight.

From a technical standpoint the pressurized balloon approach succeeded brilliantly, and was so light in fact, that as a part of 1958's Project Score, a complete Atlas 1st stage launched itself into orbit, becoming in the process the world's first fully, and regrettably disposable, single stage to orbit rocket (minus the two discarded engines.) Due to its outstanding efficiency, the Atlas went on to become the basis for America's first orbital manned space launch vehicle, launching John Glenn's historic mission, as well as the four remaining Mercury flights which followed, not to mention one particularly unhappy chimpanzee Enos, which preceded them. The original Atlas, with a variety of upper stages would go on to a stellar career as a government and commercial expendable launch vehicle. As NASA began to adapt the Atlas ICBM for civilian purposes, Convair extended the balloon skin architecture used on the missile to create the high performance, all-cryogenic Centaur upper stage, where it is still in use on the present day Atlas V.

Despite its efficiency, balloon tank construction was eventually surpassed by a weight saving technique called Isogrid, a technique which depends on the unique properties of the triangle as one of nature's strongest shapes. Isogrid construction involves carefully milling thicker stock metal components to a minimal acceptable thickness, using the inherent structural strength of triangles and other geometric shapes (in which case it is called "orthogrid") to provide the needed stiffness at reduced weights. Used on the Boeing Delta IV, the Lockheed Martin Atlas V and the European Ariane V main tank assemblies, in recent years it has become "state of the art" in the launch industry. While the technique is very effective in achieving the goal of fabricating strong, lightweight structures, it is also both time consuming and very expensive, requiring highly specialized machining equipment.

Although SpaceX would eventually adopt isogrid construction for its Dragon capsule, it was not the solution selected for the Falcon 1. Instead, the design team selected a partially pressurized tank configuration which was conceptually similar to the original Atlas, but manufactured to considerably stronger tolerances and able to withstand ground handling without resorting to pressurization. For flight, the stage would be pressurized with heated helium, providing the extra strength needed to survive the stress of launch and attempted recovery. The result was termed a graduated monocoque design, and it was an approach which allowed for a rapid and inexpensive fabrication technique using conventional aerospace aluminum sheets rolled into barrel sections with internal stiffening and welded seams.

In addition to the mass savings enabled by the partially pressurized architecture, SpaceX was also able to incorporate another significant weight saving feature in both stages, a common bulkhead between the fuel and liquid oxidizer tanks. This would have proven much more difficult to obtain with a liquid hydrogen fuel combination due to the temperature difference between the liquid hydrogen at -423, and liquid oxygen being stored at a comparatively balmy -297 degrees. This temperature differential between two super cold fluids is sufficient to make a material performance nightmare for any common structure, and it was one which forced both the Saturn V and Space Shuttle designers to employ separate fuel and oxidizer tanks, which had the added effect of lengthening the overall tank structure and requiring what amounted to an extra dome.

With an eye towards the next generation launch vehicle which it was already contemplating, SpaceX did ultimately borrow one critical feature from the Space

Shuttle external tank for the Falcon 1 upper stage tank structure for later models of the booster, a special alloy of aluminum and lithium which offered both higher strength and lower weight than traditional space grade aluminum. The need was driven in part by the use of the pressure fed Kestrel engine, which would place added internal pressure on the tank structure compared to a turbo pump arrangement. In defining its approach to developing the Falcon 1, SpaceX was also evaluating possible production techniques for the larger rockets which would follow, and as part of this effort, elected to only utilize the lithium alloy for the second stage as it sought to gain experience with some of its more challenging characteristics.

Three Dragons under varying stages of construction. The unit in the foreground is the parachute test article for Commercial Crew. Credit: SpaceX

Having defined the fuel, engine and basic tank design for the Falcon 1, the major remaining structural components required to make a complete launch vehicle were the thrust structure connecting the engine assembly to the first stage tank, the interstage adapter between the first and second stages, and finally the payload mounting platform and jettisonable fairing.

The primary function of the interstage adapter is to connect the first and second stage, (and additional stages when required) of a launch vehicle. They are generally fabricated out of aluminum into a truss assembly and covered with a thin aerodynamic skirt, although in the case of some Russian launch vehicles, the truss is left completely open. Besides serving as the critical connection juncture between what are basically two separate rockets, interstage adapters function as shock absorbers to prevent the vibration and acoustics from the highly dynamic first stage from damaging the second. Following first stage burnout, the adapter manages a vital role in one of the most trouble prone phases of any flight, stage separation. One of the most important decisions in considering a rocket's design centers on the moment when you no longer want to keep the stages together, and instead need them to separate as quickly and cleanly as possible, and how that can be accomplished.

In designing the Falcon-1 interstage adapter, SpaceX originally began with a conventional aluminum structure, but as the engine testing program in McGregor began to experience a mounting series of delays in achieving full length test fires, the structural engineering team took the opportunity to rethink its solution, and ultimately departed from its tried and true approach to incorporate instead a graphite epoxy structure similar to that found in high performance athletic equipment. The stage separation mechanism was also a departure from the norm, incorporating an entirely new pneumatic pusher design which initiates stage separation with conventional explosive bolts, but then incorporates gas driven cylinders to gently shove the second stage away from the first. Attached to the now expended first stage, the interstage adapter has one more vital role to perform, serving as the mounting point for the stage recovery parachutes.

The brains of any launch system, the flight control computer, is typically mounted on a vehicle's upper stage, where it can continue to oversee the climb to orbit after the first stage has separated and is falling back to Earth. For the Falcon 1, engineers took advantage of recent advances in computer technology to pioneer an innovative solution to vehicle control. Following convention, the flight computer and its related avionics were mounted on a special plate immediately above the second stage, where they employed for the first time an Ethernet bus rather than a thick bundle of copper wires to control the first and second stage engines as well as the rocket's other functions. Eliminating miles of wiring not only saved both weight and expense, it also removed a potential failure point in terms of the potential for human error in producing and connecting the complex wiring bundles in the first place. The use of an Ethernet system also greatly simplified the failure testing analysis which would play a critical role in certifying the vehicle for flight. Perhaps more importantly, it also allowed for the inclusion of a system on the Falcon 1 which could be readily adapted and upgraded to larger launch vehicles already on the drawing boards.

Following the initial prototype engine firing in March of 2003, Elon Musk speculated that his rocket might be ready for launch by the end of the year, and for a while hoped to conduct the first Falcon launch on 12/17/2003, the 100th anniversary of the Wright Brothers historic flight. Unable to keep this schedule, SpaceX instead loaded up a preliminary version of the Falcon 1 and its mobile launch system on a tractor trailer and drove it to Washington, D. C. for an official unveiling ceremony on Independence Avenue, across the street from the National Air and Space Museum. In setting out across country on the back of a simple flat-bed truck, SpaceX sought to draw a distinction between the simplicity and durability of the new Falcon 1 with its partially pressurized design, and the older balloon skinned Lockheed Martin Atlas IIAS, which as Elon Musk pointed out, could not have endured the trip. [27]

Back in Texas, following the initial success of early test firings, continued testing of the Merlin engine revealed several significant issues which would have to be addressed before it could be declared flight ready. What SpaceX encountered was a phenomenon all too common in rocket engine development and an indication of just how narrow the margins really are. Testing usually begins along a very gradual curve, with short firings at reduced thrust levels. As often as not, they are successful. The real problems tend to occur as the tests begin to approach the pressure, thrust and duration of an actual launch which of necessity occur at the very limits of tolerance. Because of "the tyranny of the rocket equation" and the extreme challenge of wresting enough performance from a given fuel to narrowly escape the pull of Earth's gravity, everything occurs at the margins, and it takes a great deal of testing to determine where those limits fall.

Reflecting on the experience, Musk called it a time of "many delays, and a few setbacks," including a few "RUD" rapid unscheduled dis-assembly's, or in the common vernacular, explosions. During an initial October 2003 60 second test run, roughly 1/3 of the duration necessary for reaching orbit, the near 6000 degree exhaust began to melt the metal in the engine's throat, its most vulnerable area. At the same time, heat buildup began to degrade feed line seals. As each issue was resolved, longer burn times led to more unwelcome discoveries, such as the appearance of stress cracks in the Merlin's cast aluminum fuel manifold which mandated a switch to a much more robust manifold made of Inconel, an alloy of iron, chromium and nickel. Exhaustive testing of the engine's ablative thrust chamber also revealed a vulnerability to burn-through; a problem addressed by increasing film wall coolant flow rate, a technique which uses the fuel itself as a barrier between the heat of combustion and thrust chamber wall, it comes at the cost of reduced performance. Next, a related problem arose when the first stage pintle injector failed to produce the desired level of combustion efficiency, an issue addressed by increasing propellant flow, and thereby raising total thrust from an initial target of 72,000 lbs. to a final flight level of 78,000 lbs.

The net result of the string of difficulties in producing a fully qualified engine was a major delay in schedule which took the company all the way through 2004 and into early 2005 before a flight capable Merlin could be mounted to a first stage for full duration, full qualification testing. Elon Musk appeared to be learning, as so many predicted, that designing launch vehicles was very, very hard work. Rather than fixate on the one obvious issue standing between the Falcon 1 and launch day, SpaceX instead took the opportunity of an additional year to evaluate and test the various subsystems, and wherever possible, find ways to shave off excess weight and improve performance. As a result, the rocket which gradually began to emerge as the flight ready Falcon 1 was substantially more capable than its initial version.

One of the most significant changes was in the design for the thrust structure, the frame which mounts an engine or cluster of engines to the base of a launch vehicle. A thrust structure must be able to withstand the full force of launch, serving as a strongback and mounting point for the engine guidance control mechanisms, evenly distributing the forces imparted to the base of the fuel tanks and stage assembly. As planned and originally built, the Falcon 1 thrust assembly was made of steel, but during the process of overall upgrades, it was subsequently altered to titanium, a metal both lighter and stronger than steel, but considerably more expensive. In what could be taken for an excess of optimism, Musk noted that the new thrust structure was so strong and durable it could be used for potentially hundreds of thousands of flights, "provided it is recovered." It would be, but unfortunately not in the way he intended. Musk could perhaps be forgiven for the exuberance, as the new thrust structure was in fact a remarkably strong assembly, weighing slightly under 75 lbs., but able to endure loads of 150,000 lbs. force, twice the maximum flight load, with zero deflection. Besides the thrust structure itself, other upgrades included changing both the engine gimbal joint as well as the helium pressure plumbing from steel to titanium.

One of the largest single improvements came from the effort to redesign the interstage adapter from aluminum to carbon fiber. The switch in materials, along with an optimization in design, cut the overall weight of the interstage structure in half. Pleased with the results of the home grown effort in composite engineering, engineers began making plans to switch from aluminum to carbon fiber for the two part payload fairing as well, with the intention of introducing a new, lighter weight fairing after the initial qualification flights for the Falcon 1 launched with the original design.

Considering the starting point, which was essentially at zero, the setbacks SpaceX encountered early in the development process for the Falcon 1 were hardly unusual, and had the company been a little more circumspect in publicizing its plans, perhaps both the development effort and its outspoken leader might have come in for less criticism than what was routinely sent its way.

At the same time however, SpaceX's comparative openness, highlighted by regular internet updates from Elon Musk himself, began to build an enthusiastic following for the young company, which at times only served to further aggravate its critics even further. What some may have failed to notice however, was that a number of issues arose not because engineers had chosen a poor design, but because the company doggedly carried out an exhaustive testing regime with an eye towards future upgrades. By the early fall of 2004, the Merlin 1A test engine was regularly putting out more thrust than the baseline Falcon 1 could reasonably handle at 81,000 lbs. which actually resulted in a need to detune the engine for its first flights. It would not be until January of 2005 though, that the engine was able to regularly sustain a full mission length firing without encountering a burn through.

Even as Merlin engine testing was acting as the long tail on the program, the rest of the Falcon 1's structures passed through their own qualification tests with little difficulty in a rigorous series of tests taking place on a newly constructed structure test stand in Texas. Inaugurating a process which would become routine in the coming years, completed stage sections were shipped from California to McGregor, hoisted in the air and then strapped into a tower resembling something the Spanish Inquisition might have devised if it ever needed to test the devout faith of 100' foot tall giants. First and second stages were tested by applying pressure from a series of hydraulic cylinders to simulate the various stresses which would be encountered in a nominal flight. Passing those criteria, the stages were then stressed at levels which significantly exceeded the worst case situations that one could ever expect to encounter on a day when everything went bad at once. On the other hand, anticipating a day when everything went as well as could be imagined, engineers put the first stage through more than 150 pressure cycles simulating nominal launch conditions to verify that it was capable of being re-flown multiple times.

Structural qualification testing drew to a close on March 31, 2005, meaning the next remaining obstacle was acceptance testing, to determine if the specific vehicle being prepared for the company's maiden flight was ready to launch. As part of acceptance testing, fully assembled stage sections underwent full flight duration firings on their respective stands. Upon successful completion, the Falcon 1 having demonstrated that it was mechanically fit and capable of performing all the work necessary to achieve an orbital launch, would be formally "accepted" for that assignment. The next step was another trip across time zones, this time back to California and on to the launch pad at Vandenberg Air Force Base and the Western Test Range, where the company intended to conduct its initial launch, for a final series of tests which would be familiar to anyone who has been in a school play, the dress rehearsal. In this case however, it would be a "wet" dress rehearsal, which would precisely simulate all the steps in launching the rocket from rollout, to fueling (the wet part) and finally a countdown up to and including ignition and a brief burn of the main engine. If everything went as planned, there was little else to do except to wait for the formal launch approval, repeat the practiced steps of the dress rehearsal and then hopefully, go make history.

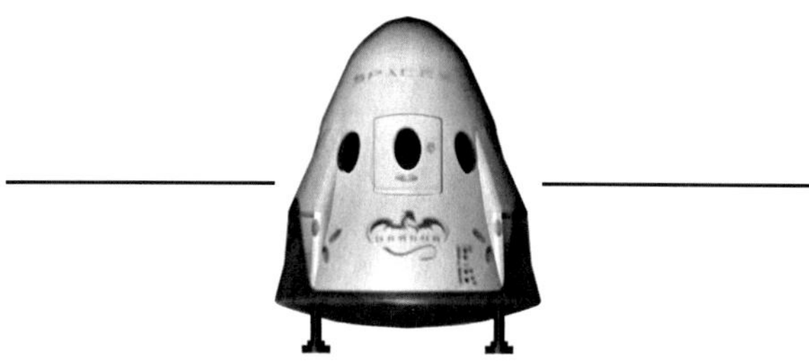

Initial Falcon I delivered to Vandenberg for testing. Credit SpaceX

Chapter VI: The First Falcon Flight

As the Falcon 1 testing program finally neared its conclusion in May of 2005, SpaceX originally planned to conduct its maiden flight from Vandenberg Air Force Base on the California Pacific Coast. Jutting out into the sea between Los Angeles and San Francisco, Vandenberg, the West Coast counterpart to Cape Canaveral, is a cooler, misty place, very different from the more famous Florida launch facility, and the nature of its launches are different as well. Vandenberg is used for missile testing over the vast Pacific Ocean, and as a launch site for spacecraft heading south over Antarctica and into polar orbit, where they offer the unique advantage of covering almost the entire Earth's surface in their sweeps. Throughout the Cold War, surveillance satellites launched out of Vandenberg enabled the U.S. military to spy on Soviet military assets, many of which were located at much higher latitudes than could be easily overflown from launches coming out of Cape Canaveral. In turn, the U.S.S.R. launched most of its spy satellites into polar orbit over the North Pole from the Plesetsk Cosmodrome located close to the Arctic Circle.

Although the end of the Cold War certainly did not put an end to the mutual spying, new conflicts brought new challenges, and it was in response to the needs of the rapidly changing fronts of the post 9/11 era that the military commissioned the first planned SpaceX payload, TacSat-1, a small satellite designed and built by the U.S. Naval Research Laboratory for the Office of Force Transformation, located within the Office of the Secretary of Defense. Carrying infrared and visible light cameras, as well as machine to machine communications gear, TacSat-1 was intended to be placed into a low polar orbit, where it would hopefully pave the way for a new series of rapidly deployed, customizable satellites which could serve military assets in real time literally anywhere in the world. [28]

While Vandenberg is an excellent choice for polar orbits, and has even been used for retrograde orbits heading due west, it is essentially useless for the more common eastern launching flights like those out of Cape Canaveral or Wallops, which take advantage of the Earth's rotation to provide an extra measure of launch velocity. For SpaceX though, the Vandenberg launch site offered the decided advantage of being only a few hours' drive from its El Segundo headquarters and the location of the factory where the Falcon 1 was built, and was the perfect location for an entirely new rocket's first launch attempt. Depending on traffic, a SpaceX truck transporting a launch ready Falcon 1 could pull out of the factory and onto the launch pad at Vandenberg in less than four hours, a proximity which offered enormous potential benefits in time and money which might otherwise be lost on the far more circuitous commutes common to most launch vehicles.

In addition to the unique design and manufacturing approach for the Falcon 1, SpaceX was also intent on making a change in how rockets were processed for launch, and as with much of the rest of the system, it was both geared to lowering recurring costs, and serving as a test case for the larger launch vehicles which were to come. Historically, at least in the U.S., launch vehicle assembly has been performed on or near the pad in a slow and expensive process in which one stage was "stacked" upon the next, gradually building up the complete launch vehicle and frequently tying up individual pads for months at a time. Adapting a phrase from aircraft history, the worst offenders became known as "pad queens" and just south of the launch site the Air Force had originally assigned to SpaceX, Vandenberg's SLC-3 West, reigned one of the all-time winners in that dubious category, a Lockheed Martin Titan IV.

In the halcyon budget years of the Cold War and the space race, the practical response to the slow pace of rocket assembly had simply been to build as many individual launch pads as required, with each tailored to the needs of its specific rocket. As often as not, once a rocket reached the end of its operational life, the pad would lie neglected and unused, sometimes for many years, until the infrastructure was either adapted, or demolished and replaced, generally at considerable expense, to support the next rocket to come along. In the case of the pad SpaceX would be using, Space Launch Complex 3 West, the most recent use had been the last flight of the Atlas E/F on March 24th, 1995. [29] Although years had passed since the last launch, and the pad was in a state of disrepair with not much left of the original structure, the site condition was of minimal importance because SpaceX would employ a launch procedure which had more in common with America's Cold War Soviet rivals than it did with the vehicle which had last called this pad home.

Cape Canaveral steams in the Florida heat and humidity, and the air itself can offer menaces large and small, from hurricanes and lightning to ever present mosquitoes. Vandenberg is frequently wet and soaked with a heavy fog which rolls in from the Pacific on summer nights, but absolutely nothing compares to the miserable weather which is all too frequently a staple of Russian space launches, whether carried out on the wind-swept steppes of Baikonur in Kazakhstan, or from the deep freeze of Plesetsk. Adapting to these conditions, Soviet era engineers quickly developed the art of assembling and prepping their launch vehicles horizontally and indoors, and then rolling them out for launch with as little time exposed as possible, even though the vehicles themselves were unusually hardy, designed with the prevailing weather conditions in mind. Throughout its history, Russia has routinely carried out launch procedures in harsh conditions of high winds and driving snowstorms which would be almost unimaginable for NASA. In fact, for its one and only flight, the Soviet Space Shuttle, Buran, translated appropriately enough "snowstorm," took off and then landed 200 minutes later in snow flurries driven by 55 MPH winds. [30] Despite these conditions, the completely automated Buran rolled to a stop on the runway less than six feet from where it was predicted. It would be nearly 25 years before the United States achieved a similar feat with the far more modest unmanned X-37 space plane.

The Falcon 1 was no slouch either, designed to take off in winds up to 46 MPH when fully pressurized, a feature which prompted Elon Musk to observe that unless impeded by a weather system formidable enough to have its own name, his rocket would be able to fly, "if you could stand up, you could launch." The procedure, dubbed "ship and shoot" called for Falcon 1 rockets to roll out the factory door in what amounted to a ready to launch scenario. As SpaceX would stress repeatedly, personnel costs were a major driver in launch operations costs, and small armies of technicians camped out at the launch site for weeks at a time prepping a rocket, waiting on better weather or a range delay were an expensive proposition. So too was maintaining and staffing a launch control center, and compared to other launch operators, SpaceX adopted an eminently practical solution which looked more like a NASCAR race trailer than anything else, converting and outfitting a standard 53' semi-trailer into something very different from what can be found at the nearest truck stop. While perhaps inglorious, it was also ingenious, and as befitting a man who had made his fortune in advancing e-commerce, the trailer functioned as a state of the art relay center staffed by 12 people, allowing the much greater bulk of launch personnel to remain completely in the loop back at the home office in Los Angeles. All told, the entire launch team consisted of less than 25 people.

Common 53' trailer outfitted as Falcon I Mission Control Center. Credit SpaceX

Launch operations were equally simple. In the first place the Falcon 1 did not require a fixed launch tower. Instead, the same steel and aluminum box truss transport cradle which supported the rocket as it left the factory on the tractor trailer served double duty as the launch tower.

Once arriving on sight, the trailer would be parked on the launch pad and protected from the elements by a lightweight movable hanger sitting on a short railway. After pre-launch preparations were completed, the hanger would be retracted along the rails to a safe distance and the launch cradle transitioned from horizontal to vertical by a single hydraulic cylinder, in a manner similar to the raised bed of a dump truck. After the rocket was elevated to vertical, technicians secured the launch structure to the pad via four tie downs, with the Falcon-1 remaining securely in place by a hydraulically operated c-clamp near the top of the tower. Altogether it was a remarkably simple process, which ironically enough according to Elon Musk, was partially inspired by an old Thor launch vehicle his team had observed preserved as a museum piece at Vandenberg.

Arriving from the Texas test facility in early May of 2005 after passing acceptance testing and conducting a full length mission firing, the Falcon 1 was on the pad and ready for its first full hold down test by the end of the month. This brief firing, lasting only a few seconds, would verify that the Merlin engine was operating at full thrust, and sensors were working and accurately sending vehicle health data. If successful, it would mark a milestone in the program, and mean that there would only be a few more tasks at hand before it was time to metaphorically hit the shiny red button.

The initial firing attempt quickly revealed one of the most common issues which cause launch delays, a problem with the vehicle's ability to monitor itself. It is much the same as driving thousands of miles with a check engine light on your car. Most times the problem is not the engine, as demonstrated by the fact that you are still driving, but instead with a bad sensor. In the case of the very first hold down test, the problem lay with the engine ignition sensor. As the countdown proceeded, cameras revealed a small puff of smoke indicating the engine torch igniter had fired, but because the sensor reporting that fact failed, the engine never got the message, and the test was automatically terminated. As it turned out, the root cause was actually a location design flaw, which allowed the igniter to literally "torch" the sensor.

The next effort also ended prematurely, when a pad technician, one of only 8 on site, accidentally left a helium ground supply valve in the closed position, preventing the on-board tank from developing adequate pressure to spin up the gas generator. Again, the result was no ignition. As Elon Musk observed prior to the next attempt, a total of two problems, one with the vehicle and one with the pad, was pretty good performance for an entirely new system. On May 27th, at 9 am local time, with both problems corrected and the Falcon 1 held firmly in place, the Merlin engine roared to life for a five second firing, enough time to verify that it was operating at full power. For the people who had been working on the launch vehicle project, the firing was a brief but thrilling moment which led to an unavoidable conclusion that the Falcon 1 was ready for its moment of truth.

All systems appeared go for launch, but the range was not. Several miles south of the SpaceX location, the final Lockheed Martin Titan IV to be launched from Vandenberg sat on the pad waiting through a seemingly endless series of delays for what would be its final West Coast launch. Although the Titan IV was the most powerful expendable booster in the U.S. fleet prior to the debut of Delta IV Heavy, it had become outrageously expensive and a poster child for everything that was inefficient with expendable launch vehicles of the previous era. Worse than its $420 million dollar price tag was a checkered launch history marked by several high profile failures. After a series of incidents in the late 1990's, as the program was winding down, the Department of Defense commissioned a Broad Area Review to look into the problem as part of a larger assessment of the nation's space infrastructure. The conclusion regarding the Titan failures was that they stemmed from completely avoidable human error relating to "going out of business syndrome" due to the rocket's impending cancellation and replacement by the much newer Boeing Delta IV Heavy. As a consequence of the findings, the Air Force initiated an exhaustive program of checks and rechecks on every conceivable system to ensure that the remainder of the program "flew out successfully." For this final flight, the process of trying to end the Titan's West Coast career on a positive note would drag on for nearly half a year. [31]

Unfortunately for SpaceX, because the unproven Falcon 1's planned flight trajectory would carry it over the Titan IV launch pad, and its classified billion dollar payload (a KH-11 spy satellite), the Air Force sensibly insisted that the new booster wait until the launch occurred, a caution which would prove remarkably prescient. Unfortunately, the Titan IV was making a long goodbye, and as the delays built up, time dragged on. Initially, the delay appeared manageable because the Air Force would still need to go over the test data supplied by SpaceX before it could issue a launch permit. Furthermore, even though SpaceX had completed a test firing and successfully validated ground system interface and countdown procedures at the Vandenberg site on May 27th, the Falcon 1 was missing its second stage engine, which was still back in Texas undergoing qualification tests, leading SpaceX to estimate a late summer launch date.

Atlas V. Note the complicated vertical assembly structure. Credit: NASA

Consequently, by the time the sun set on the evening of the 27th, the Falcon 1 was safely secured on its trailer and was already most of the way back to Los Angeles. As for the Titan IV just down the road, it was going nowhere fast, and would remain on its pad until the fall. By the time it finally lifted off on October 19th, SpaceX had gone to its backup plan.

Beautiful but sparse, the landing dock on Omelek Island in the Kwajalein Atoll. Credit SpaceX

In early June it became clear that the Titan IV delay would not be resolved quickly, prompting the company to make what amounted to a change in the batting order. Rather than possibly waiting out the remainder of the year to conduct its first launch out of Vandenberg, SpaceX decided to switch its manifest and proceed with what had originally been scheduled to be its second flight, one of two sponsored by the Defense Advanced Research Projects Agency, DARPA. The payload was a small satellite called FalconSat-2, built by cadets at the U.S. Air Force Academy and originally slotted to be launched from the Space Shuttle Atlantis in March 2003, but left waiting on a ride to space following the loss of Columbia and subsequent stand down of the Shuttle fleet. Instead it would become the primary payload on a new launch vehicle lifting off not from Cape Canaveral, but from the Kwajalein Atoll in the middle of the Pacific Ocean.

While the Titan IV delay was the critical factor leading to a sudden change in launch sites, another more troubling development came to light when Lockheed

Martin fired the opening shot at SpaceX in what would become an epic Goliath versus David battle. It seemed the aerospace giant had suddenly discovered that it didn't want the start-up company anywhere near the adjacent SL3-East launch site which it had recently begun converting to support the Atlas V booster. The development was somewhat ironic considering the fact that only a few years before, having won a Department of Defense contract for two launches out of Vandenberg, Lockheed Martin actually declined the award because it did not want to shoulder the expense of preparing a launch pad.

Falcon 1 on the pad at Omelek. Credit: SpaceX

By 2005 however, everything had changed. Now, Lockheed Martin wanted the site, and the Air Force, rarely one to say no to its favorite contractor, concurred. After a $7 million dollar investment for which it was never compensated, and more than a year spent negotiating its way through the maze of local, state and federal agencies, all of which wanted to have a say in the process, SpaceX was being tossed out of Space Launch Complex 3 West to make room for a possible second Atlas pad which was never built. Although the Air Force was willing to offer an alternate site, one which SpaceX would ultimately need if it wanted to fully compete for DOD launches, the company was already well behind schedule, and with other customers waiting, was feeling the pressure to launch. Fortunately, the U.S. Army had already proved somewhat more accommodating, providing the Falcon 1 with a new, albeit sparsely furnished home on tiny Omelek Island, located halfway across the Pacific Ocean, in the Kwajalein atoll in the Marshall Islands in the western Pacific Ocean, 9 degrees north of the Equator and west of the International Date Line.

Although remote, the Marshall Islands have long been a sort of inflection point between worlds.

Ferdinand Magellan's original expedition to the Moluccas passed to the south within several hundred miles of the island chain in 1521. Only four years later, as part of a follow up expedition which had departed Spain with a fleet of seven ships under the command of Garcia Jofre de Loaisa, the Santa Maria De La Victoria, which was now the only remaining vessel, first sighted an atoll in what would eventually become the Marshalls on August 21, 1526. Named by Captain John Marshall of the Royal Navy during the first planned expedition to the islands in 1797, they were claimed as a part of the Spanish East Indies for 350 years, but were only recognized as such in 1874. They were then sold to Germany in 1899 following the breakup of the Spanish Empire as a consequence of the Spanish-American War. Now a German possession, parts of the island chain were overrun by Japanese troops ostensibly fighting for the Allies during the First World War. Following its defeat in that conflict, Germany was forced to dispose of its overseas possessions, and in 1920, the League of Nations awarded former German holdings in island territories north of the Equator to Japan. It was an ominous development for the indigenous population, when as the Second World War erupted; the Marshalls were established as the outermost defensive perimeter of the Japanese Empire. The results were predictably horrific, as the islands became the scene of intense combat in 1944. Kwajalein was successfully captured by an American amphibious assault preceded by some of the most concentrated artillery shelling of the entire Second World War, during which the island was essentially flattened.

Following the war, the U.S. retained possession as a "ward" of the island chain which it had wrested from Japan. The benefits were debatable however, as the United States proceeded to utilize a number of the remote atolls for atomic weapons testing during the course of which it conducted no less than 63 detonations, including the largest above ground nuclear test ever conducted. As a result, parts of the chain have been described as the most contaminated place on Earth.

Finally granted self-control in 1979 and full sovereignty in 1986, the Republic of the Marshall Islands leases the Kwajalein atoll to the United States Army, where it is home to the Ronald Reagan Pacific Testing Range. The Reagan Test Site, as it is generally called, serves as a launch site for testing anti-missile systems, and as a massive radar and space tracking installation.

Lying nearly 5000 miles west of the continental United States, and more than 600 miles from the nearest island chains, the atoll is a perfect place for testing systems out and away from prying eyes. Perfect that is, unless you are a startup aerospace company encountering the added expense and complexity of conducting your first launch thousands of miles from where you originally planned. Given its recent history as a testing ground for global destruction, it was perhaps appropriate however, that due to events far away in California, Kwajalein would become something else, the starting point for a family of launch vehicles founded not as instruments of Armageddon, but instead with the intention of safely transporting and securing human civilization on a new world altogether.

The move was not a wholesale surprise. SpaceX had been planning to use the Pacific launch site, affectionately dubbed "Kwaj" for three of its first four planned launches in any case, but not for its all-important first launch. As events unfolded,

Omelek would soon turn into SpaceX's very own Survivor's Island, a crucible from which Elon Musk and SpaceX would only emerge after surviving an onslaught of trial and error mistakes combined with a seemingly endless stream of bad luck which very nearly derailed the entire enterprise. Perhaps with a sense of foreboding, Musk observed of the switch that the Pacific site offered one of the more corrosive salt environments imaginable and that "If we have three consecutive failures. . . it's not clear to me that we know what we're doing and maybe we should go out of business." [32] An optimist might point out that despite the added expense, the isolation of the Pacific site offered a welcome respite from the added distractions which would ensue from operations conducted within the vicinity of two major US cities, and if by chance things did go awry, it might offer just the sort of team building exercise upon which lean organizations tend to thrive.

The process of relocating from Vandenberg to Omelek was complicated by the fact that even though SpaceX had been granted use of the Kwajalein facility and Omelek island over a year before, it had been on the back burner with very little to show for the effort in terms of physical infrastructure other than a few partially poured concrete footings. Besides a small dock, a landing ramp and a few items of broken down equipment, the tiny island offered little more than spectacular views and world class diving in the surrounding waters. Even for an organization as minimalist as SpaceX, this would be a challenge. Following the June decision to relocate, a small team was dispatched to begin the herculean task of setting up a remote launch facility. Within a few short months the team cleared vegetation, poured a concrete pad, installed generators, lightning towers, a liquid oxygen plant, fuel tanks, an erector structure, a water deluge system and a vehicle hangar along with a clean room for integrating payloads. They also installed nitrogen pressurization and helium purge systems as well as remote cameras and fiber optics communications relays back to the control room which being prepared on Kwajalein island in a huge concrete facility once used as a target for space based laser testing as part of President Ronald Reagan's "Star Wars" anti-missile defense program. Amazingly, only two short but grueling months later, SpaceX was able to erect the Falcon 1 at its new home for the first time, beginning the process of conducting the same round of fit and cold flow tests which had already been performed at the Vandenberg site. The initial launch date was planned for September 30th.

That's when the trouble began. Back in Texas, as a final part of the qualification program for the Merlin in which technicians ran the engine with different fuel ratios and thrust settings for off-nominal conditions, production unit number 13 experienced a chamber burn through. Although the test parameters created conditions which should not occur during a normal flight, which the previous 12 engines had all passed, SpaceX elected to run a new series of qualification tests with elevated film wall cooling to counter what appeared to be a manufacturing variance. The resulting stand down left the Falcon 1 sweltering in tropical conditions for an initial delay to October 31st which was then extended to November 25th. "Ship and shoot" had become something closer to ship and sit.

Launch procedures finally got underway on November 23rd, as technicians installed the Merlin 1A's ablative thrust chamber, personally signed by the team which built it. At 4:00 AM the next morning, Elon Musk received a phone call advising that the Army had suddenly scheduled "work" near Omelek, and that the launch would have to be pushed back yet again, this time by a day. Infuriated by the seemingly pointless procedure, SpaceX re-set the launch for 1:00 PM California on Saturday, November

26th, 9:00 AM local, opening a six hour window granted by the military. Launch day began early, with the pad crew arriving on Omelek by boat shortly after 3:00 AM to scour the beach for sea turtles, and begin a final check before departing to the safety of a 3 mile exclusion zone as LOX loading got underway. With the launch control team monitoring by video link from Kwajalein itself located 22 miles away, the countdown initially got off to a smooth start, with the one area of concern being possible high winds, which threatened to warm the booster's on board liquid oxygen beyond acceptable limits.

The first sign that everything would not go smoothly came when the ground supply liquid helium pump began to slow under the uphill effort to transfer helium from its storage tank at 1,400 PSI to the Falcon 1's onboard tank at 4,900 PSI, resulting in a 1 hour delay as the overworked pump slowly completed its task. Once completed, the countdown proceeded until a pre-programmed hold at T-10 minutes, with all indicators looking good.

As the countdown picked back up, pressure readings indicated that a valve on one of four liquid oxygen ground tanks failed to close properly. Before the pad team could return to the launch site to correct the problem however, SpaceX had to produce a written procedure proposal for the military addressing the problem, and then wait for it to be approved, while all along mounting winds sped up the loss of oxygen supply, threatening the launch effort. Moments after submitting the plan, the military granted approval to return to the pad, setting off a desperate race against the clock to reach the island, address the problem and depart once again in order to enable an attempt which was now facing the twin deadlines of the closing launch window, and mounting liquid oxygen losses.

The pad crew quickly discovered that a manual vent valve on one of the auxiliary liquid oxygen fill tanks had been incorrectly set in the open position, allowing oxygen to escape. Re-setting the valve was a simple matter, but the next challenge was in pressurizing the backup tanks so that they could begin supplying LOX to Falcon 1 to replace the quantity lost to boil-off. In an effort to reduce the rate of LOX boil-off, the crew took the added step of also de-tanking 15% of the Falcon 1's onboard kerosene fuel supply to slow the warming effect it had on the adjacent LOX tank due to the nearly 400 degree temperature differential, a move which reduced the rate of loss from 8 to 4 gallons per minute, saving 750 gallons of the now dwindling oxidizer. With the ground supply systems successfully reset, and the pad crew once again safely off the site, SpaceX reset the liftoff at 12:50 pm local, leaving just over 2 hours remaining in the launch window.

The countdown resumed yet again, with tensions climbing even higher in the control room, as well as the SpaceX facilities in California and Texas monitoring the events. This time as the count approached the final 10 minute hold, everyone understood that if another problem arose, the day was over. Entering the final hold at T-10 minutes, the sole remaining task required to get the rocket absolutely ready to launch was once again, the completion of liquid helium loading.

It proved to be the showstopper. By the time the countdown reached this point, enough LOX had escaped that it was no longer possible to keep the on-board liquid helium, which was stored in small tanks submerged within the main liquid oxygen tank, in a liquid condition. Instead, the helium having reverted to a gas, vented back to the ground system. No worries, the next step would have been to top off the LOX from ground storage provided by the onsite oxygen plant. Unfortunately, over the long wait as SpaceX went through engine testing in Texas, the always cranky island plant, laboring in the tropical conditions, died. Anticipating just such a problem, Elon Musk had taken the practical step of ordering more LOX, procured in Hawaii and delivered to Kwajalein by barge. As luck, which was mostly bad, would have it, the barge arrived with only 20% of the liquid oxygen which had been ordered, the remainder steadily bleeding off into the atmosphere on the slow boat ride from Hawaii. Even all of this might have been overcome, except for the fact that as the Falcon 1 sat on the pad, the winds blowing at 85 degrees across a cryogenic tank, accelerated the boil-off of liquid oxygen just enough to rob an amount which could not be readily replaced due to one final problem, the helium could not be loaded faster than the oxygen boiled off.

The loss of liquid oxygen was actually one of two impediments to SpaceX conducting its first launch that day, the second being a reboot in the engine flight computer which was also driven by the specific circumstances of the island location. Due to a change in the launch configuration from what was in use at the Vandenberg site, SpaceX found it necessary to almost triple the length of the umbilical which supplied the rocket with electrical power as it sat on the pad. During the countdown process, prior to switching to onboard power, the electrical current flowing along the umbilical encountered increased resistance due to a combination of the longer cord length and the hotter temperatures. The result was a slight drop in voltage to the engine flight controller, causing it to reboot. Subsequent tests on the rocket's batteries revealed the problem to be with the ground system only, suggesting that if absolutely necessary, liftoff could have taken place anyway, an option rendered moot by the oxygen deprivation.

Although the launch scrub was disappointing, it was hardly surprising given the novelty of both systems, a point Musk stressed to followers on the internet in regular postings from the isolated island. Following the initial launch attempt, SpaceX sought to insure that the next time, now scheduled for December, would find more than enough LOX on hand for any foreseeable circumstance. To make sure of it, Musk chartered a C-17 transport plane to fly his two good LOX tanks to Hawaii for refilling, and together with higher quality tanks sourced on Hawaii, placed them on the next available barge bound for Kwajalein. In addition, technicians repaired the existing SpaceX plant on Omelek, as well as made plans to wrap the first stage oxygen tank in a protective blanket which would be pulled away by trip wires as the Falcon ascended. As for the electrical flight control problem, the answer was to simply up the voltage to account for increased resistance.

As part of its contract with the Air Force and DARPA, SpaceX was charged with developing and demonstrating techniques for reducing the time required for preparing the vehicle for launch between rollout and liftoff. Consequently, the launch team took the opportunity afforded by the wait for the fuel barge to review its countdown procedure. As a result, the next attempt would be accelerated by the use of better ground equipment, and both stages would be fueled at the same time. Furthermore, SpaceX streamlined the process by substituting computer controlled commands for manual procedures in a number of areas. The net effect was a reduction in total time required for countdown from 5 hours for the initial test firing in Vandenberg, to 4 hours for the first launch attempt on Kwajalein, with a goal of achieving a three hour countdown on the upcoming attempt.

After the initial scrub, some members of the launch team remained on Kwajalein, some flew back to California, and still others only went as far as Hawaii for a brief respite before it was time to re-assemble on the island for a new attempt. Following a successful prelaunch checkout on December 18th, launch was set 7:00 AM local for the following day, and this time there would be an abundant supply of liquid oxygen on hand. Naturally then, a problem occurred instead with the other component of the vehicle's propellant supply, the kerosene fuel system. During the course of fueling the Falcon 1, wind speed suddenly picked up from a placid 7 knots to an undesirable 28 knots, four knots above the temperature driven acceptable limit of 24. With the countdown holding at T-15 minutes, and the internet web-cam shut off due to excessive traffic, SpaceX evaluated its options, hoping the winds, which were once again rapidly accelerating LOX loss, would die down. During the hold, the launch team attempted to partially detank the fuel, which had cooled excessively due to its proximity to the now insulated liquid oxygen. As the kerosene was being pumped from the rocket back to the storage tank where it could warm up, one of three redundant pressure sensors on the tank shorted out, causing a vent valve to remain in the closed position, creating a vacuum within the tank. The resulting suction collapsed the thin-walled first stage fuel tank, resulting in structural damage which could not be repaired on the island. Launch would not occur in 2005.

Elon Musk, who must have been getting tired of chartering airplanes and barges across the Pacific when a simple 4 hour truck ride out of Vandenberg would have otherwise sufficed, dutifully chartered another plane to fly an entire new first stage from California to Hawaii, where it was loaded on yet another barge bound for Kwajalein just before the New Year's Eve holiday.

With the replacement first stage arriving in early January, SpaceX scheduled a new attempt for February 10, pending the successful outcome of a wet dress rehearsal and static fire to be conducted the day before. During pre-launch tests on the evening of February 4th, engineers noticed that power distribution boards on the Falcon 1 were malfunctioning, likely requiring the installation of new capacitors which would have to be tested prior to launch. Early Sunday morning, technicians lowered the rocket to the horizontal position, separated the two stages and removed the distribution boards. The boards were quickly taken by a SpaceX engineer in time to board the regular transport plane which would take him back to California. At the same time, the home office in El Segundo put an intern aboard another plane, this one bound for Minnesota and the only capacitor supplier which was open on Sunday. Both arrived back in California by Monday afternoon, where the engineers integrated the boards with the new parts, put them through an 8 hour series of testing, and then, carried by the now exhausted engineer, back on the next plane out, bound for a 6:00 AM arrival in Kwajalein. Eleven hours later, the Falcon 1 was re-integrated and lifting once again to the launch position.

The hold down test fire was scheduled for 8:00 AM local, with fuel loading once again beginning during the early hours of the morning. This time, the countdown reached T-1 minute, but ended in a hold, when a mistake in auto sequencing reset a flight controller during the changeover from ground power to the rocket's on board batteries. After diagnosing the problem, the count restarted, progressing all the way to T-0 and the split second of engine ignition, but instead of a brief roar, there was only silence. Going back into a hold, engineers ultimately traced the problem to a minor (millisecond) difference in igniter timing, which had stopped the count. At the same time, an errant pressure reading in a second stage tank demanded attention as well. Unfortunately, despite all the preparations, the need to perform another hold down firing on the 10th meant that there would not be enough liquid oxygen on hand to support a launch attempt on the 11th, which was now a moot issue because this Kwajalein launch window only extended through the 10th. Even though the launch attempt was off for now, and could not be rescheduled until March, SpaceX went ahead with the planned test fire on the 10th, and for the first time in the launch campaign, it went off without a hitch. Unfortunately, although the static firing appeared to go quite well, subsequent analysis revealed a small leak in the second stage. Although the leak did not require a major repair, it was also not one which could be readily performed in the field. Now, the second stage would have to be swapped out as well. If there was any good news, it was that the much smaller second stage would fit inside the standard DC-8 aircraft which made the regular run from Hickam field in Hawaii to Kwajalein, at least alleviating the need to charter yet another military transport.

Upon the second stage's arrival at Kwajalein, the launch campaign which seemed like it would never end got underway once again, with a new static fire and a new launch date set for what was essentially an entirely different rocket than the one which had originally shipped to Kwajalein the previous August. The first attempt came on Saturday, March 18th, and proceeded all the way to engine startup, but ended prematurely when a helium ground supply quick disconnect fitting disconnected a little too quickly, and prevented the engine from achieving full power. The following day, a second attempt at conducting the test fire was called off when one of the video cameras charged with monitoring the rocket malfunctioned.

The next attempt came on Tuesday, March 21st, and for once everything went smoothly, with Merlin 1A conducting a clean three second burn with all systems reporting "Go." Following a launch readiness review, SpaceX added another day to the schedule for additional analysis, and convinced that they were as ready as they would ever be, set the launch attempt for two days later on March 24, 2006 at 1:00 pm Pacific. After four long years, and far more testing and expense than ever anticipated, Elon Musk and SpaceX were at long last ready to aim for the heavens. Unfortunately, the rocket was aimed for a shallow reef just yards off shore.

The initial countdown proceeded smoothly, with both SpaceX and Range Control establishing firm locks on the Falcon 1 booster as it slowly came to life, charging its batteries, activating its avionics, switching to on-board telemetry via C band transponders aligning its navigation system. With the booster appearing ready for flight, the next step was for the pad crew to open vent valves on the ground supply LOX tanks and make their way to the workboat to depart the area before LOX loading began.

At 12:01 pm PT, the launch director issued a hold at T-64 minutes, an event triggered when range safety detected that a commercial vessel on hand to recover the first stage, was found to be within the limits of the booster's exclusion zone. The launch would now be on stand-by until the vessel could make its way clear of the line of demarcation. Just before the original launch time of 1 pm, the recovery craft finally exited the safety zone, allowing the countdown to restart at T-75 minutes. After another, brief delay for a Collision Avoidance (COLA) to allow the International Space Station to exit the Falcon 1's flight path, SpaceX began the final countdown for a liftoff now scheduled to take place at 2:30 pm PT, and this time, with calm winds and the first stage protected by a blanket of insulation designed to pull away as the rocket took off, the countdown proceeded smoothly.

With the launch team polled at T-10 minutes, and all hands reporting an enthusiastic "Go!" for the historic attempt, the remaining minutes steadily ticked away, with the strongback tilting out of the way at T-5 minutes. At T minus 90 seconds, the Falcon 1 switched to internal power, and 30 seconds later, the auto sequence engaged to control the final 60 seconds of the countdown. At the 30 second mark, the deluge system began to spray the pad with water to cancel acoustic vibrations.

After all the previous drama, engine ignition and liftoff was picture perfect, and as the Falcon 1 slowly ascended through its first 24 seconds of flight, telemetry showed the vehicle to be performing exactly as expected. Suddenly and without warning, engine thrust terminated and the Falcon 1, now without direction, began a sickening plunge back to Earth, or to sea actually, landing on a dead reef 250 feet offshore. The FalconSat-2 satellite, with grim comic flair, bounced high into the air upon impact, and somewhat improbably, returned to Earth through the roof of the island's small machine shop, coming to rest next to its own shipping container.

Although it would be several months before an official review board could be convened and release its results, it only took one close look at high resolution photos of the launch to reveal the basic problem. Twenty five seconds into the launch, a fire broke out at the top of the Merlin engine, its bright orange glow an unwelcome addition to the perfectly normal thrust cone from the tail of the rocket. It only took a few seconds for the fire, the source of which was not yet clear, to burn through a helium pressurization supply line. As soon as the escaping helium dropped the system pressure below a critical threshold, spring loaded fuel valves feeding the engine automatically slammed shut, terminating thrust and sending the Falcon to an inglorious demise. [33]

A disappointed Musk sought to put a positive spin on the day by pointing out quite correctly that this was a test flight, and that prior to the fire, the flight met a number of its objectives, including liftoff, initial guidance and verification of the safety shutdown mechanism, but there was no hiding the disappointment. Although there was no doubt that a new launch attempt would be forthcoming, it would take much longer than anyone expected, due not to the Falcon 1, but because of a rapidly unfolding series of events which would forever change the nature of the U.S. space program, presenting Elon Musk and his young company with the opportunity of a lifetime, but one which came with a set of challenges far greater than those on Kwajalein. For the moment, the initial launch failure, although hardly either surprising or unusual would have to take a backseat to bigger changes afoot.

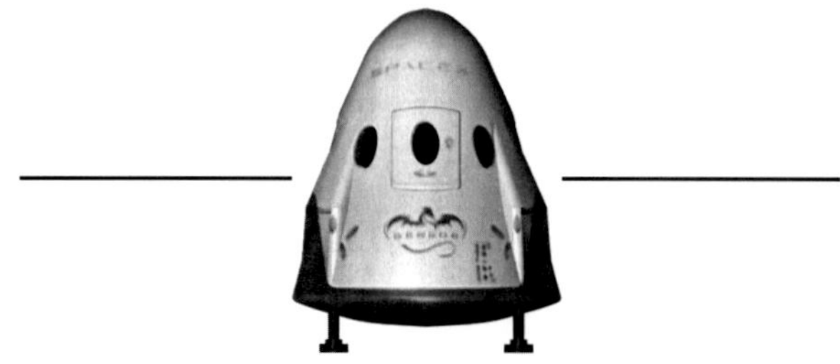

Chapter VII: Setting the Stage; Columbia and the Vision for Space Exploration

The failure of the Falcon 1 on its first launch attempt was a clear setback for SpaceX, one which possibly could have had a chilling effect on the company's long term prospects had it not been for a major change in the landscape of the American space program which was gradually taking place during the four years between early 2002 when Elon Musk founded the company, and March of 2006, when Falcon 1, Flight 1 came crashing back down into the Pacific Ocean.

Most improbably, the company Elon Musk started because he had lost faith in NASA's ability, or even willingness, to address the challenges of actual space exploration and to recognize the necessity of reducing launch costs, was already well on the way to becoming an integral component in the American space agency's plans to do both. Consequently, as SpaceX struggled through the development process for its smallish rocket and nearly toy-sized engine, there was much more at stake than the personal fortune of yet another somewhat eccentric internet millionaire. How that came to pass was the result of a bizarre chain of events as unusual as anything that has ever occurred in the space age.

After more than three decades of almost myopic focus on the system which dominated nearly every aspect of American spaceflight, the Space Shuttle, NASA was forced to re-evaluate its priorities following the destruction of the Space Shuttle Columbia over the skies of Texas on the fateful morning of February 1, 2003. Much like the Challenger disaster, the loss of Columbia was principally a human failure, both entirely predictable, and entirely preventable. As the Columbia Accident Investigation Board would come to conclude, NASA missed multiple opportunities, eight to be exact, to uncover the true extent of the foam strike which mortally wounded America's first operational Space Shuttle in the moments of its final launch. Incredibly, despite the legacy of the Challenger disaster, seven lost lives and billions of dollars in return to flight costs, the agency had once again become beguiled by the Shuttle, increasingly blind to its shortcomings which the ensuing accident investigation board would assess as an "aging system which was still developmental in character."

Unlike the Challenger tragedy which had come seventeen years before, and relatively early in the Shuttle program's flight history, the Columbia accident occurred against the backdrop of a space program which although irrevocably committed to building the International Space Station, was struggling to find a path, and a purpose beyond ISS. In addition to the human tragedy and the destruction of a priceless piece of American history was another more practical problem, NASA no longer had access to the International Space Station it was only halfway through the process of assembling.

That the loss of a second orbiter and its crew would prove a turning point in American space policy was hardly a surprise, but the fact that in a few short years it would lead to a startup company such as SpaceX assuming a vital role in the immediate post Shuttle era was something else entirely, an unintended consequence of NASA maintaining a nearly unshakable affinity for the Shuttle system; its hardware and its culture, even when the flights remaining to the Orbiter itself became numbered. Quite simply, because NASA had inextricably tied itself to the Shuttle and Space Station project as the focus of its manned spaceflight programs, when the Shuttle system was once again grounded due to a catastrophic failure, the space agency was forced to consider alternatives which it would, and had, dismissed out of hand in earlier days.

In the weeks and months following the Columbia disaster, at least one thing was clear, because the Space Shuttle was an irreplaceable tool in assembling the ISS, unless the nation was prepared to abandon the Space Station project altogether, a prospect some would have welcomed, the Shuttle would have to return to flight as soon as safely possible. The criteria for determining that point fell to the Columbia Accident Investigation Board. Surprisingly, the CAIB did not conclude that the Shuttle program need necessarily be cancelled. Instead, it recommended that after completing assembly of ISS, if NASA wanted to continue flying the Shuttle, it would need to have the entire system re-certified, an expensive and time consuming process which depending on the results, could potentially lead to an endless series of mandatory upgrades in an environment with a renewed, and presumably permanent, focus on safety. [34]

From the CAIB Recommendations:

R9.2-1 Prior to operating the Shuttle beyond 2010, develop and conduct a vehicle recertification at the material, component, subsystem, and system levels. Recertification requirements should be included in the Service Life Extension Program

Despite the recent tragedy, many felt that pending a resolution of the foam strike problem, recertification was precisely the course to take. After all, the airframes of the three remaining orbiters, Atlantis, Discovery and Endeavor were still in pristine condition and nowhere near the end of their design life spans. Furthermore the Space Shuttle offered capabilities which simply weren't attainable with any other launch system on Earth, nor any in development.

On the other hand, the CAIB had also issued a strongly worded admonition that in the future, NASA should seek another source for missions which did not absolutely require the Shuttle's unique capabilities.

"Because of the risks inherent in the original design of the Space Shuttle, because that design was based in many aspects on now-obsolete technologies, and because the Shuttle is now an aging system but still developmental in character, it is in the nation's interest to replace the Shuttle as soon as possible as the primary means for transporting humans to and from Earth orbit."

While it declined to suggest a specific architecture for any successor to the Shuttle, it issued one of the most strongly worded rebukes of the entire 743 page document in addressing the fact that there was no plan already underway to do so, calling the failure to develop a successor as "a failure of national leadership."

The CAIB also went on to suggest a common denominator in why other recent launch vehicle development efforts, such as the National Aerospace Plane (NASP), the X-33 and the recently terminated Space Launch Initiative had failed to produce anything useful, their reliance on developing breakthrough technologies. (Arguably, SLI was never even given the chance.) Consequently, if NASA was to follow the board's advice and develop a replacement crew and cargo vehicle for the International Space Station as soon as possible, it would need to turn to a simpler system concept of operations, one which was not developmental in nature, or based on 40 year old technology.

The most immediate prospect was a program which was already underway at the time of the accident, the Orbital Space Plane. Even prior to the Columbia tragedy, NASA managers understood that once the Station was essentially complete, using the Shuttle as the primary means of crew exchange was an expensive proposition. Consequently the agency was already looking into the possible configurations of a crew transfer vehicle which could work alongside and complement the Shuttle well into the mid 2020's. What that system might actually look like was the subject of a hot debate with a long and tortured history dating back to the earliest days of the ISS's predecessor, Space Station Freedom, and one of the major vulnerabilities regarding the Shuttle system and its role in supporting a fully functional ISS staffed with a crew of seven. In that regard, the one function the Shuttle could not perform was that of an emergency return vehicle. It was a limitation due to the fact that the Shuttle, restricted by the performance of its fuel cells, could not remain on orbit for extended periods of time, and despite early predictions to the contrary, could neither fly frequently enough, nor be launched on demand to reliably serve as a rescue vehicle.

Dependent on Russia to provide emergency return capability in the form of the cramped three person Soyuz crew vessel, NASA wanted a different system, one which could linger on orbit, attached to the station for months at a time, and be called upon if needed for an emergency evacuation of the entire crew or to return a stricken astronaut needing immediate medical attention on Earth. Prior to the Columbia accident, as long as the Shuttle was the primary means of reaching the station, there was no particular requirement that a rescue vehicle had to serve as crew transport as well. Instead, it could simply be carried unmanned to ISS by the Shuttle, or launched by another vehicle for that matter, functioning only as an "escape pod."

After pursuing and then dropping several different proposals over the years, generally due to budgetary concerns, NASA appeared to settle on a small lifting body, the X-38, an updated version of the Martin Marietta X-24A lifting body which first flew in 1969, as a pre-cursor to the Shuttle program. The X-38 would have been carried to the station in the Shuttle payload bay, left in orbit attached to the Station, and if called upon, return the full complement of seven astronauts to Earth autonomously, gliding in to a soft landing virtually anywhere, safely cradled under a massive parafoil which deployed in 5 stages. A total of three test versions of the X-38 were built by Mojave California based Scaled Composites, successfully completing a series of a drop tests from a B-52 at the Dryden Flight Research Center. After what appeared to be a promising start, the program was cancelled in 2002 however, as NASA changed its plans and began to focus instead on a two way transportation system originally called the Orbital Space Plane.

Compared to the X-38, the OSP was considerably more capable, initially being conceived as a mini-shuttle, and intended to be launched on top of either an Atlas or Delta EELV. Like many other projects, the OSP soon began to increase in cost and complexity, and as a result, the requirements shifted as well. What began as a defined program to develop a true winged spaceplane, was subsequently revised to look at a number of options, including a simpler crew capsule based on the venerable Apollo design. In addition to serving as the Moon program's Command Module, the Apollo capsule had also served admirably as the crew transfer vessel to low Earth orbit for all three missions of the Skylab program, as well as the American side of the Apollo-Soyuz test project, and in the latter role, had been fitted with a specially designed docking collar which would later become the basis for American docking systems on the Shuttle, Mir and ISS programs. Interest in the capsule's potential for enabling a new era of simpler space transportation become so intense that at one point, teams of engineers could be found in the nation's space museums, carefully examining the remaining artifacts of the historic system they were considering reprising.

As NASA entered 2003, it appeared to many analysts that this would be the most likely outcome of the OSP project, and the alternative to the Shuttle for basic crew transport to ISS. Despite its retro look, one of the advantages of an Apollo derived design was the fact that even though the nation had no immediate plans to resume voyages beyond low Earth orbit, if that were to change, the basic capsule configuration with its broad heat shield was readily adaptable to deep space missions and their resulting high return velocities in a way winged solutions were not. Nevertheless, the OSP, regardless of its actual design, was far from an assured project, and faced significant opposition from the powerful Shuttle only constituency which perceived any alternative as a potential threat to the program which was in some ways finding its best destiny as the indispensable construction platform for ISS. Consequently, as NASA accelerated the construction of ISS in the first few years of the new millennium, both by design and by default, there was no ready alternative for either crew or cargo transport if anything were to change the Shuttle's status as it entered its third decade of flight.

Of course, everything did change, and while one might have thought that the loss of a second orbiter and seven more lives would have necessarily spelled the end of the Shuttle program, the U.S. was now so deeply entrenched in the ISS program, maintained by a web of international obligations for which it was primarily responsible, that it was basically beyond the point of no return. Pending the results of the Return to Flight Program and the efforts to eliminate future foam strikes, the Shuttle would continue to fly, at least as long as necessary to complete its role in constructing ISS. Beyond that point however, NASA faced a real dilemma. Because it had no further plans, there was no set context within which to fix the design requirements for the OSP.

The dilemma was resolved on January 14, 2004, when President George Bush introduced a new program, and a new direction for America's manned space program, the Vision for Space Exploration. If the CAIB was correct in assessing the lack of successor planning for the Shuttle as a "failure of national leadership," here it was, at least for the moment. The new policy called for the U.S. to safely return the Shuttle to flight in order to complete construction of the ISS by 2010, whereupon the Orbiter fleet would be permanently retired. The centerpiece of the policy was a renewed program of deep space exploration, beginning with a return to the Moon to take place no later than 2020 and possibly as early as 2015. This time, the purpose was not just a series of brief excursions, but as the beginning of a new phase of exploration in which the U.S. would establish long duration stays, and ultimately a permanent presence, conducting research, and applying many of the skills which would be needed to move on to the plan's long term goal, the first manned trips to Mars. Along the way the nation would invest in a number of critical technologies, including new high performance main engines, and in-space nuclear propulsion, the combination of which would set the stage for extending human presence throughout the solar system. [35]

Compared to President Kennedy's far more famous (and effective) "go to the Moon speech," the agenda laid out in President Bush's Vision for Space Exploration represented a seemingly stronger commitment to a permanent program of human exploration and expansion into space in order to advance the nation's "scientific, national security and economic interests" very different from the specific and limited goal of landing a man on the moon, and returning him safely to earth "before the decade is out." Compared to any previous American space policy, it established an open ended goal which was only a step short of going "where no man has gone before," even if it started out by going to the one destination already marked by American flags and footprints.

Following a well-established tradition, Bush quickly signed an executive order establishing a Presidential commission to determine the best way to enact the overall policy goals contained within the VSE. The commission, formally titled the "President's Commission on Implementation of United States Space Exploration Policy" was chaired by former Secretary of the Air Force, Edward C. "Pete" Aldridge, and included members such as Carly Fiorina, Paul Spudis and Neil deGrasse Tyson. Notably absent were any astronauts with flight experience. Aldridge had once trained as a mission specialist for the planned first Shuttle mission out of Vandenberg, but lost the opportunity when his flight, STS-62-A, was cancelled along with the rest of the military shuttle program in the wake of the Challenger accident.

Although it would receive mixed reviews, particularly from the Democratic side of the aisle in Congress, several of the specific points called for in the President's request and addressed by the commission in its final report delivered in June 2004 titled "A Journey to Inspire, Innovate and Discover" represented a remarkably cogent view of the problems facing the agency and the need for sweeping change if it were to successfully undertake this new program. One of the critical areas of concern was in the way NASA related to the commercial sector, with the findings suggesting that the agency should seek to involve private industry in a much more direct manner in acquiring space services, and in fact should retain for itself only those functions which private industry demonstrably could not perform. First among the assignments going to commercial providers was unmanned launch and operations to LEO, including cargo transportation to ISS. [36]

As for the primary effort to develop an exploration architecture for returning to the Moon, the commission called for a "go as you pay" approach enabled by a modest increase in annual budgets during the development phase, followed by a steady budget adjusted for inflation after that point. In some ways, it was a conservative approach to a bold goal, but it absolutely depended on the prospect that the stakeholders involved would place a higher premium on the going part, than on the paying part, and insist on more efficient allocation of resources than had happened with either the Shuttle or Station programs. Absent a major push by the Administration to form a consensus with Congress and the American people to change the basis of the nation's space infrastructure, and forge a unified vision which could last through successive administrations, and break the legacy contractor chokehold, none of which happened, the plan might as well have called for warp drive and transporters as well.

Beyond an additional increase in NASA's budget, funding would come from retiring the Space Shuttle in 2010, and then, ending the U.S. commitment to the Space Station program around 2016, with a presumption that ISS would follow the path of the Skylab and Mir stations, right into the Pacific Ocean. Freed from the twin obligations of the Shuttle and Station programs, the gradually opening funding wedge would allow a spiral "system of systems approach" to hardware development in which successive elements could build on each other, while allowing rapid entry of emerging technologies as they were developed. Development of major components would be handled by a lead systems integrator whose primary responsibility was to manage subcontractors rather than perform its own hardware development. This was a technology development approach which appeared to be paying large dividends in other highly complicated programs such as missile defense and the Joint Strike Fighter, and it was that program's director, retired Navy Rear Admiral Craig E. Steidle, which the Bush Administration tapped to become NASA Associate Administrator and head of the Office of Exploration Systems, where the new program, Project Constellation, would be housed.

As originally laid out, Project Constellation leaned heavily on a concept originating from the computer software industry, called "spiral development" in which a design process could incorporate, and build on, new technological developments even as the overall system evolved towards a mature design. Constellation consisted of three spirals, the first, Spiral 1, addressed the requirements to regain domestic crew access and re-supply for the International Space Station after the Shuttle's retirement. Spiral's 2 and 3 addressed the Moon and Mars respectively, and permeating the whole approach was the premise that developments in one spiral could positively benefit all three.

After delivery of the commission's final report on June 16th, NASA Administrator Sean O'Keefe announced that the agency would issue solicitations for a competitive proposal for a next generation manned spacecraft, now called the Crew Exploration Vehicle. Following another of the Aldridge Commission's guidelines, and Steidle's own experience, NASA elected to pursue a two stage competitive process for securing the Crew Exploration Vehicle, which began with a draft Statement of Work in December 2004, followed by a formal Request for Proposals issued on March 1st 2005, with the deadline for submittal on May 2nd. The path selected by O'Keefe, a former DOD budget analyst and acting Secretary of the Navy who had won accolades for getting the Space Station budget on track after first taking the position of NASA Administrator, would have resulted in something new to the agency, but common in military aircraft procurement, a fly-off to maximize risk reduction and to determine who won the final bid for an operational system expected to differ considerably from the initial winner. The two finalists were a team led by Boeing and Northrop Grumman, and another headed up by Lockheed Martin and consisting of members from Orbital Sciences, EADS Aerospace and Hamilton Sundstrand amongst others.

To some, the competitive development approach placed the VSE in the precarious position of getting off to a slow start which would not see a first crewed flight until 2014, four years after the planned retirement of the Shuttle. Equally absent was any means to transport supplies to the International Space Station during that period as well. As for crew transport, with the Station program scheduled to phase out in mid-decade, the emergent Crew Exploration Vehicle (CEV) would make its first flights to the orbiting laboratory, but with a design optimized for missions to deeper space. Viewed another way, the CEV would be both overqualified and overly expensive for Station crew transfer and re-supply, two conditions which helped to set the stage for a re-evaluation of the agency's plans.

In addition to beginning work on the CEV, NASA would also invest in a development effort to improve the state of the art for a new heavy lift launch vehicle by funding research into both hydrocarbon and cryogenic liquid fueled main engines, one of many aspects of the American aerospace technology base which had seen little advancement in the last 30 years, a major point of emphasis for the Aldridge Commission.

Following the release of the Commission's results, NASA issued a Request for Information, and received 11 proposals for carrying out the first lunar return mission, along the "go as you pay" format in October of 2004, many of which made use of existing launch vehicles to achieve the VSE's objective of a Lunar return before 2020. [37]

It was not to be.

Events came to another turning point when the cautious Sean O'Keefe, who had guided his agency with a solemn purpose through the difficult days following the Columbia disaster, and then played a key role in re-organizing NASA to enact the Vision for Exploration, resigned from his post on December 13, 2004. The previous January, O'Keefe raised the ire of many in the space community and the nation at large with a controversial decision to cancel a planned final Hubble telescope servicing mission on the grounds that the mission profile did not offer an option of any safe haven for the crew in the event another foam strike or other debilitating problem crippled an Orbiter. In the difficult months following the Columbia tragedy, O'Keefe had pledged to take the CAIB's recommendations to heart, and as he interpreted it, the Hubble servicing mission simply could not be justified. Others saw it differently, and reportedly, the broad based and unusually harsh reaction to dooming the exceedingly popular astronomical platform played a factor in O'Keefe's decision to leave NASA and take a position as the Chancellor of Louisiana State University. Regardless of the reason, O'Keefe's departure set the stage for a major shakeup in the way NASA went about implementing the President's plan.

Both the pace and the scope of VSE changed dramatically with the swearing in of a new NASA Administrator, Michael D. Griffin, on April 14, 2005. Griffin came in determined to fast track the Vision for Space Exploration by eliminating the competitive development strategy and proceeding directly to sole source contracts in order to shave two years and an estimated billion dollars off the CEV program. Griffin's approach would have two major effects, and very mixed results. Oddly enough, SpaceX would benefit from both.

Cancelling the dual procurement strategy for the CEV, Griffin assigned the contract for construction of Crew Expedition Vehicle to Lockheed Martin. Furthermore, under his direction the agency issued a "Call for Improvements" to radically de-scope the original Lockheed Martin proposal, which had transferred much of its prior work on the Orbital Space Plane into an initial concept for VSE which offered a capable Crew Vehicle comprised of three separate segments; a wedge shaped re-entry vehicle which would make a parachute landing, a small hab module which offered increased room and extended duration mission capability, and finally a propulsion service module. Under the CFI, the design would be transformed into little more than a larger Apollo Command Module, presumably offering the benefits of enhanced safety, simplicity and reduced cost.

Another, bigger issue lay in selecting a rocket to launch the newly named Orion capsule. Under the original Lockheed Martin proposal, each of three segments would have been launched individually aboard an EELV, and joined together in LEO for flights into Cis-lunar space. Though achievable, and having the distinct benefit of using existing launch vehicles, the architecture was not necessarily optimal. A potentially better solution, a true heavy lift vehicle, would not be available for a number of years, depending on both the development of a new, more efficient main engine, as well as the funding wedge necessary to pay for it. One complicating factor, in following the CAIB recommendations, future launches would take place very differently from the all up method employed during the Apollo program, Instead, deep space launches (anything beyond low Earth orbit) would need to take place within a dual launch architecture in which an unmanned heavy lift vehicle would be lofted to LEO, where it would be joined by a crew launched aboard a separate vehicle optimized for a level of safety an order of magnitude greater than that achievable with the Shuttle.

With a clearly inevitable delay in designing and building a heavy lift booster, driven in part by the ballooning costs in the Shuttle Return to Flight Program, [38] Griffin focused his initial attention on the smaller crew launch vehicle and its potential for reducing the "gap" in U.S. launch capability to the International Space Station after the Shuttle was retired. The next step was to form an internal study group, originally designated to work for 60 days, on what was termed the Exploration Systems Architecture Study, to review potential mission architectures and recommend the best way to proceed. One might have thought that the obvious answer lay in the Evolved Expendable Launch Vehicle program which had begun flight operations in 2002 and represented latest improvements in launch technology. This had been the basis for the original version of Spiral 1 and 2 under previous administrator Sean O'Keefe, with flights by the existing Atlas V, Delta IV, and Delta IV Heavy, as well as the yet (and never) to be developed Atlas V Heavy.

Griffin however, had another solution in mind, based on preliminary work he had performed as part of a study group for the Planetary Society entitled "Aim for Mars," and it was this plan which would define his tenure as NASA Administrator. Rather than avoid the sunk cost fallacy, Griffin's plan embraced it wholesale to produce two new launch vehicles. The first, a crew launch vehicle, combined elements of the Shuttle stack into a new two stage expendable launcher based on a Space Shuttle Solid Rocket Booster as the first stage, with a second stage powered by a Space Shuttle Main Engine. Sitting safely on top the stack was a CEV equipped with a retro style launch escape tower. Based on its unusually tall and spindly architecture, the booster, named Ares-1, in a nod to VSE's ultimate goal of reaching Mars, quickly became known as "the big stick."

Although it made a certain amount of sense, and reflected potential SRB based designs dating back for many years, it also drew a tremendous amount of criticism from across the aerospace community, including the Russians, who were simply stunned that the Americans would even consider placing astronauts on top of a solid rocket. Brushing aside those concerns, the ESAS conclusions validated Griffin's preference for a solid rocket booster and even presented analysis which predicted that despite its role in dooming the Challenger, it would be twice as safe as a booster based on an EELV. Anecdotally at least, this seemed a curious conclusion considering the fact that there had never been a manned flight failure resulting in loss of mission or crew for liquid fueled boosters in either the Mercury, Gemini, or Apollo programs. The ESAS was able to reach this surprising conclusion by imposing conditions which were questionable at best, and to some, bordered on naked fraud. To establish the safety record of the SRB's, the study only counted launches after the SRB redesign following the Challenger failure. Furthermore, each Shuttle launch was counted as two successful SRB launches, even though the strap-on boosters were hardly independent rockets, lacking both guidance and roll control as well as an upper stage. For the EELV flight histories however, the study included historical data derived from booster families dating back many years to launchers which bore almost no relationship to the current and future EELV variations under consideration. It was the equivalent of cumulatively judging the average number of falls for Olympic ice skaters by rating one skater from the time of their Olympic trials, and another since the day they first stepped hesitantly onto the ice as a 6 year old child. Other considerations also, such as the presumed necessity to build expensive new launch facilities for EELV based boosters also served to influence the final decision in favor of the preferred architecture. Although the solid rocket booster based Ares-1 might have turned out to be a perfectly safe and reliable booster, it was simply stunning that a space agency which had lost two orbiters and 14 astronauts through known and preventable failure modes would seek to begin its next big project by deliberately skewing the terms of a decision based on safety. Adding fuel to the fire, NASA declined to release several appendices to the ESAS study detailing the comparison criteria, only to face new questions when they subsequently showed up on Wikileaks. The space agency then took the somewhat unusual step of releasing a white paper in 2008 defending its decision process and re-iterating the arguments for the ever controversial Ares-1. [39]

The second booster offered for development was the Cargo Launch Vehicle (CaLV) a massive rocket intended to be capable of launching approximately 188 tons to LEO and just over 63 tons to the Moon. Designated the Ares-V, the new super rocket would consist of a cryogenic first stage which was essentially a lengthened and enlarged (33' versus 27.5' diameter) version of the Space Shuttle External Tank, and powered by 6 Delta IV RS-68 engines, mounted on a thrust assembly at the base and flanked by two 5.5 segment Solid Rocket Boosters. The second stage, intended to serve as the Earth Departure Stage for Lunar or Mars missions, would have likely been powered by a single Saturn V era J-2 cryogenic engine, now designated J-2X. Under a typical lunar launch scenario, an Ares-1 would loft a CEV to Earth orbit, where it would rendezvous with an Ares-V launched Earth Departure Stage (EDS) containing the two stage Altair lunar lander and ascent vehicle, and then depart for the Moon. From this point on, it would mostly follow an Apollo mission profile, with astronauts returning from the Moon via Lunar Orbit Rendezvous with the waiting CEV, followed by journey back to Earth and parachute recovery preferably on land rather than water. It was almost literally "Apollo on steroids," and compared to some of the more modest modular concepts utilizing orbital assembly and existing launch vehicles offered under the original mission proposals the previous year, it represented a major change for the agency in the wholesale adoption of the mega-booster strategy as the only acceptable way to return to deep space.

A Delta IV lifts off carrying the GOES-P satellite. Credit: NASA

Another major limitation on the architecture options set forward in the ESAS study was the fact that any new "clean sheet" booster design was precluded from the outset, based on two reasons, both of which would be proven to be demonstrably false by SpaceX. First, the ESAS study presumed that any new launch vehicle development effort must necessarily be a time consuming affair certain to stretch out the delay in U.S. access to space well past the Shuttle's retirement. The second reason cited was the belief that there simply would not be funding sufficient to support a new development effort. To avoid these two scenarios, the only available options were constrained to derivatives of systems already developed, namely the Shuttle's SRB's and the EELV program's Atlas V and Delta IV with a major advantage accruing to the Shuttle derived components being that they were already "man-rated," whereas the EELV's were not. Nevertheless, both EELV entrants were already operational, and beginning a long string of successful launches which would stretch on for more than a decade. Furthermore, as subsequent events and common sense also bore out, converting the Atlas V booster to a "man-rated" version was a far less expensive and time consuming option than that which NASA ultimately chose.

We may never know the full extent of what really went into the decision to select the Ares-1 as opposed to an EELV based crew launch vehicle, but as a major turning point in recent NASA history, and one which soon became an enormous embarrassment for the agency, it is one which warrants closer scrutiny. One thing is for certain however, both SpaceX, as well as Orbital Sciences Corporation were exceedingly lucky that NASA chose to go the way it did. Had the agency instead tapped either or both of the EELV boosters for the lead role in a post Shuttle environment, it would have almost certainly resulted in a very different trajectory for America's space program. As it turned out however, subsequent developments, and the Ares-1 program's eventual cancellation cast a very dim light on both the decision making process and the pervasive influence of the Shuttle program's far flung constituency in effectively hijacking what was to have been a fresh start for NASA with the Vision for Space Exploration, not to mention the CAIB's caution against employing 40 year old hardware. With the Aldridge Commission's recommendations dismissed just as easily, America's space program echoed the Who's famous line, "Meet the new boss, same as the old boss."

Whatever one makes of the actual technical decision to prefer Ares over EELV, and the even more significant decision to eschew any form of modular architecture utilizing orbital assembly in favor of the Ares-V mega booster, the net result would turn out to be a programmatic disaster, with one glaring exception, Michael Griffin's simultaneous decision to initiate what amounted to a backup plan and long-shot gamble to develop an alternate line of supply to the station via private industry, an item called for by the Aldridge Commission. What Griffin could hardly have been expected to know at the time, was that grave miscalculations in designing and implementing the Ares-1 would soon leave the nation with little choice but to place its short term hopes for the American space program entirely on private industry and an unlikely effort called Commercial Orbital Transportations Services, or COTS.

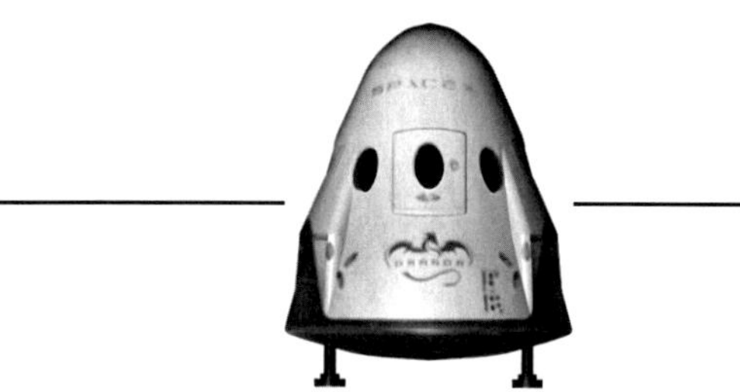

Ares-I tower. NASA spent nearly as much on the tower as it initially allotted for the entire COTS program. Credit: Stewart Money

Chapter VIII: The Commercial Orbital Transportation Services Program

The origin of COTS lay in a program which preceded both the Columbia accident as well as the formation of SpaceX. Originally called Alternative Access to Station, the program examined whether or not private industry was technically capable of providing limited, supplemental cargo supply to ISS, almost on the order of a space capable UPS or Federal Express, including the ability to conduct a launch within a week's notice. The policy origin for Alternative Access was the Commercial Space Act of 1998, which called for "the use of free market principles in operating, servicing and adding capabilities to the Space Station." Tellingly, the program did not originate within NASA, but came instead from the Office of Management and Budget and was placed within the broader context of the Space Launch Initiative, managed out of the Marshall Space Flight Center as part of the Second Generation Reusable Launch Vehicle Program. [40] It was an altogether awkward fit.

At the time the program was begun in the year 2000, with the Shuttle expected to continue operations into the 2020's, the services foreseen by AAS were supplemental delivery of limited supply quantities which were able to be called on and launched on a much quicker schedule than could be achieved with Shuttle flights. Establishing a more rapid supply capability would not only allow for timely delivery of items such as replacements for failed components, which was a regular problem on ISS, but also more frequent delivery of smaller scientific research experiments which were eternally at the mercy of the Shuttle's notoriously unreliable flight schedule. While the man hours available for actually conducting research were quite limited while the station was being built, once ISS was completed, the focus would shift from construction to operations, creating a need to significantly increase the volume and velocity of experiments coming and going from the facility. In addition to providing an alternative to the Shuttle and helping to accelerate the pace of microgravity research, Alternate Access would also offer the opportunity to improve the quality of life for astronauts aboard the station with the delivery of fresh fruits and vegetables as well as personal items. Beyond the immediate benefits to the Station, the AAS program, advocated by groups such as the Space Frontier Foundation, and championed in Congress by California Congressman Dana Rohrabacher, was ultimately intended to incentivize private industry to expand the scope of space commerce in LEO by providing an anchor tenant customer in the U.S. government.

Although sometimes easy to forget when considering the vast scope of protected interests represented in the large contractor dominated Shuttle and Station systems, NASA is far from a monolithic entity, and some elements within the agency understood that a different approach, one driven by expanding space commerce, was needed if the United States was to ever break out of the cycle of high costs which kept it trapped in low Earth orbit. It was a fact plainly apparent even before the Shuttle first launched, but overcoming politics and inertia took nearly 20 years before the agency officially addressed the issue in terms of operational policy.

Under an innovative new approach, Alternate Access to Station would not seek to fund the development of new systems, but instead pay builders for providing operating data on the systems they developed with guidance from, but not under the direction of NASA. After an initial round of proposals from 17 bidders, in August of 2000, NASA Marshall awarded four study contacts under small business guidelines to perform 90 day feasibility studies budgeted at a total of $902,000. [41] The winning bidders were Andrews Space and Technology, Microcosm, HMX and Kistler Aerospace.

Tellingly, most of the bidders produced a design featuring at least a partially re-usable system. One, HMX, proposed a recoverable and reusable capsule to be launched out of the remaining inventory of retired Titan II ICBMs. Having once served NASA flawlessly in the Gemini program, and still in service as the booster core of the heavy lift Titan IV rockets, and as a stand-alone launcher until it was finally taken out of service in 2003, the Titan offered a readily available, but finite inventory of boosters. Two other competitors, Andrews Space and Microcosm, took the opposite approach, and offered proposals based on new launch vehicles they hoped to develop. The most intriguing proposal however, came from Kistler Aerospace, and featured a two stage, fully reusable launch vehicle, the K-1, which had been partially constructed in the late 1990's as part of the RLV boom. In 1999, Kistler approached NASA with an unsolicited offer to provide Space Station resupply with a fleet of 5 fully reusable launch vehicles that the company anticipated building. Although NASA did not pursue the offer at the time, Kistler, led by legendary Apollo program manager George Mueller, played a significant role in laying the groundwork for what would first become Alternate Access and then COTS.

Following completion of the 90 day contingency cargo study, the proposals essentially went nowhere, evoking a blistering response from HMX CEO Gary Hudson who charged that NASA had never intended to pursue the program which had been foisted on them by the GAO and Congress, and that Johnson Space Center was adamantly opposed to any program which could take emphasis away from the Shuttle. Furthermore, although the AAS was placed within the Space Launch Initiative out of Marshall, Hudson went on to suggest that the SLI program director also wanted nothing to do with it. [42] Although Hudson's rather dramatic assessment could be written off as a particularly bad case of simply being a sore loser, what happened next certainly raised eyebrows.

After soliciting the contingency cargo delivery studies in August, NASA began a much larger solicitation in October as part of the "bottom up" technology maturing effort in the Second Generation program. As part of that solicitation, in 2001 NASA made a provisional award, worth up to $135 million, to Kistler for the purchase of data, both pre and post flight, from the K-1 vehicle. In order to earn anything from the award however, Kistler would need to raise the money to finish the vehicle and actually get it into the air. Unfortunately for Kistler, the timing could not have been worse. The satellite market's collapse between 1998 and 2000 was soon followed by the "dot. com" market crash, and sparking a recession, took with it any remaining venture capital appetite for the sort of high risk financing Kistler required. Without the means to complete the K-1 vehicle, the provisional award was essentially useless.

By July 2002, NASA had swung back to a more traditional approach for AAS, and issued four year-long study contracts totaling $10.2 million to Lockheed Martin,

Boeing, Andrews Space and Technology and Constellation Services International. What NASA, and particularly the Marshall Space Flight Center really wanted to build though, was the Orbital Space Plane, a program whose top level requirements were established in February 2003. Morphing into a major development program, with a final design and development decision not expected until late 2004, and a first flight no earlier than 2010, AAS and the goals of the Commercial Space Act of 1998 were essentially ignored, lending credence to Gary Hudson's complaint.

With the destruction of Columbia in early 2003, and facing an undetermined stand down in the Shuttle program, it might have seemed that an obvious response would be increased support for commercial re-supply of the Station, but as the study contracts neared their delivery date, NASA appeared willing to let them expire without substantive action. It was only at the insistence of Congress that the agency added six more months and an additional $4 million to continue work on the proposals, with the final product due in early January 2004. At that point it was hoped, NASA would presumably choose a single winner for fast track development.

With the introduction of the Vision for Space Exploration on January 14, 2004, events took yet another turn. The four study contracts for what was now called "Assured Access" were set aside, as the agency now focused on its new set of deep space ambitions. Interestingly, several of the bidders on the original round of lunar return proposals for VSE took the expedient route of incorporating previous work performed under AAS into the new project. With NASA for the moment looking beyond LEO and essentially leaving ISS resupply to its international partners, Kistler suddenly re-entered the picture. Following a high level visit to Kistler's Kirkland, Washington headquarters, NASA announced that it would make a firm commitment to exercise the options put forward in the original 2001 contract to purchase data regarding the development and operation of the K-1 launch vehicle. The contract modification, awarded in February 2004, offered to pay the company as much as $227 million for data gathered over a series of 12 launches. From Kistler's standpoint, this would be the critical link that would help the company secure funding to complete the vehicle.

For Kistler, it was too late. The advent of SpaceX was still several years away when the AAS program began, but by 2004 the company was fully engaged in developing the Falcon line of boosters, and in no mood to see its opportunities limited by questionable contracts. In the first of what would become a repeated refrain in the SpaceX story, Musk's company filed a protest against NASA's sole source award to Kistler with the General Accounting Office.[43] Elon Musk subsequently took the opportunity of an invitation to testify before Congress on May 5, 2004 to argue his case against the Kistler deal. Musk's objection was that NASA had not issued the contract on a competitive basis, while NASA contended that the Kistler launch vehicle was the only system offering the capabilities the agency was seeking. Advised by a GAO attorney that the ruling would likely go in SpaceX's favor, NASA rescinded the Kistler deal in early July, dealing a serious blow to the company which was desperately trying to emerge from Chapter 11 bankruptcy filed the previous year.

Although the ruling had a chilling effect on Kistler, and one particular avenue for commercial resupply, at the same time it also ultimately opened the door for the COTS program and a full scale adoption of the idea that not only might it be possible for a private company to supply the station, it was also conceivable that two or more companies could compete for the business, with NASA and the American taxpayer as the beneficiaries. For the moment however, the immediate effect was to terminate what remained of the spirit of AAS, even as the loss of Columbia and the lack of any immediate option for space station resupply argued heavily that the conditions for pursuing a program like AAS had never been better. NASA, for better or worse was now focused on developing options for the VSE, and with the station headed for an untimely swan dive into the Pacific in 2016, Assured Access was relegated to a low level study of docking mechanisms, and was essentially defunct by the time Mike Griffin swept into office.

Shortly after assuming the Administrator's position, Griffin delivered a speech unveiling his plan to address the twin challenges of continuing to support NASA's operations in low Earth orbit while maximizing the resources available for a program of exploration which would place demands not seen since the Apollo program. Addressing the Space Transportation Association on June 21, 2005, after only two months in the job, the new NASA Administrator observed that with the development of the International Space Station, for the very first time there would exist a "strong, identifiable market" for routine transportation service to and from LEO. Griffin went on to say that rather than dictating the specific terms and hardware that NASA was seeking from private industry, (a business as usual approach which would soon wreck its BEO counterpart, Constellation) the agency would instead state its requirements "to the maximum extent possible in terms of performance objectives, not process." [44]

Griffin's plan was to dramatically re-invigorate what had been the AAS program in a new effort which would later be formally introduced as Commercial Orbital Transportation Services. In unveiling the concept, Griffin drew on his experience in the commercial satellite industry and fixed price purchasing. He went on to point out that in the context of the broader economy, the way NASA normally did business was the exception, rather than the rule where goods and services were procured. For example, no rational person would consent to purchase a new car at an undefined price, but merely on an agreement to be handed a bill by the manufacturer for the total cost (including a pre-defined profit margin) upon delivery, yet that was effectively the way space hardware and services were acquired in the prevailing cost plus contracting culture.

At the same time though, Griffin cautioned that private industry had been approaching the agency for years offering various solutions to lowering the costs of access to space, if only NASA would just fund their particular proposal. Such propositions were generally met with withering criticisms from the established aerospace industry, which defending its perennial high prices would offer "you get what you pay for" as a catch all justification for the status quo. The offers and the typical response highlighted a Catch-22 at the heart of the launch vehicle dilemma. When a private company proposing a new venture sought financing, inevitably from financiers who knew little about the launch vehicle business, they commonly turned to NASA or sources within the established aerospace industry which would in turn suggest that building rockets was hard business, and the companies under consideration were in way over their head. In reality of course, most probably were.

Griffin's proposal deftly navigated this problem. As part of the new initiative, NASA would provide a limited amount of money for achieving certain accomplishments which demonstrated the capabilities necessary to safely re-supply the station, but would require the provider to raise and invest a significant amount of their own capital as well. In this way, the provider would not effectively be in the black until the goal was accomplished, but would not be starved for cash to the point that it failed along the way. To increase the odds of success and avoid dependence on a single provider, NASA would seek to develop at least two independent vendors, and possibly more, with the goal that they should compete with each other both for NASA's business as well as any subsequent commercial opportunities in low Earth orbit. The new program would supplement ongoing NASA efforts to incorporate Space Station re-supply in its Ares-1 development plans, but hopefully decrease the gap between the last Shuttle mission and the first Ares-1 flights, as well as create the potential for significant cost savings if private industry could live up to at least some of the many promises it had made through the years of being able to find a better way. In any event, the agency would continue to pursue the ability to support the ISS with the Ares-1 program, but with the added preference for using commercial providers whenever possible. Ares-1, while capable of supporting ISS or even other LEO operations if desired, could be optimized for supporting deep space exploration in conjunction with the much larger Ares-V. The net result would be a robust launch infrastructure which virtually assured the U.S. would maintain uninterrupted access to low Earth orbit.

Well before its formal introduction, the COTS effort got underway with an "industry day" held at the Johnson Space Center on April 25th, 2005. At the day-long meeting, NASA outlined the basic requirements it was considering for the proposed program, seeking to determine the overall level of interest. Satisfied with the results, the agency began the challenging process of putting together the specific solicitation, including novel contracting terms which would define the program. NASA officially announced COTS program on January 18, 2006 with proposals due on March 3rd. The project would be handled by a newly established Commercial Crew and Cargo Program office, managed out of the Johnson Space Center by Associate Administrator for Exploration Systems, Scott Horowitz, and housed in the Exploration Systems Mission Directorate.

The COTS program office was established with three objectives. The first was to implement the official U.S. Space Exploration policy with an investment to stimulate commercial enterprises in space. Second, and more to the point, to facilitate U.S. private industry demonstration of cargo and crew transportation capabilities with the goal of achieving reliable, cost effective access to low Earth orbit. Finally, it was to "create a market environment in which commercial space transportation services are available to Government and private sector customers." [45] Though expressed in the neutral and seemingly harmless language typical of many government policy statements, those three statements effectively encapsulated what would become a transformative moment in the relationship between government and commercial space operations.

The overall effort was divided into two phases. Phase I called for technical development and demonstrations to prove the concepts which would be offered by industry. In this case, NASA would pay only for the achievement of certain agreed upon milestones on the path to developing the capability being solicited. If all the milestones, which were both technical and financial were met, then presumably the vendor would demonstrate its capabilities in a series of actual space flights, culminating in a trip to the International Space Station. Having done so, the vendor would presumably be in a position to provide the service NASA needed; cargo and possibly crew delivery to ISS on a regular basis. That service would be bid out in an entirely separate, firm fixed-price, Phase II contract, called Commercial Resupply and handled out of the International Space Station program office.

From the beginning, program officials and documents took pains to caution that particularly in regards to Phase I, the government was acting in the role of a venture capitalist pathfinder in which a reasonable, but not excessive amount of money was put forward for development with the understanding that the results would be a mix of both success and failure. The fundamental assumption was that on balance, the benefits accrued from those ventures which did succeed, yielding a high return on a limited investment, would outstrip the costs of those which failed. For once, failure actually was an option, just not the one preferred.

For NASA, COTS offered an alternative to the traditional large contractor, cost plus development model and a chance to test drive a new acquisition strategy. At the same time it still retained through the milestone payment arrangements, a high degree of control over the outcome, standing in marked contrast to other more questionable government/industry partnerships such as the highly controversial, and often ineffective loan guarantees like those which would become a point of contention in the solar power industry in the coming years. By structuring the payment schedule in concert with the achievement of performance milestones, NASA retained the ability to contain the costs of failures and to protect the program rationale and budget. It was a tool which would prove essential in fulfilling the agency's desire to create a competitive procurement process for the actual services, a goal which clearly required at least two success stories.

COTS was enabled through the use of a non-traditional form of contracting called Space Act Agreements, which allows the agency to enter into a relationship with a private company which is fundamentally different from the traditional contractor relationship. Rather than specifically directing industry activities, NASA and the company act more as partners, performing mutually agreed actions in which the agency has insight rather than direct oversight of the contractor, which is now referred to as an "industry partner." NASA describes the arrangement as "legally enforceable promises." Although the results can be the virtually the same, the chief advantage is that neither side has to adhere to expensive and constraining Federal Acquisition Regulations (FAR). Basically, the agency gives up direct control in exchange for allowing the "partner" the latitude to accomplish the mutually desired goal in the manner it deems most effective. Although fairly new in actual use, and certainly to the extent envisioned, the legal authority for Space Act Agreements extends all the way back to the organizing legislation that formed NASA in 1958, which allowed for the use of "other transactions" to achieve its ends.

Despite the novel arrangement, NASA would still maintain a close technical relationship with its new partners through the use of what it called "Partner Integration Teams" or PIT teams. With NASA personnel embedded at partner companies and working with them throughout the process, even on a sub-system basis, the agency would have close insight to the evolving systems, while also serving as an invaluable conduit for the agency's vast experience base in nearly every impinging field.

Phase I would proceed from an initial open solicitation, through a down select, to the selection of two finalists which would receive funding for achieving defined milestones, leading up to a planned 2008 decision as to which teams would win the series of competitively procured ISS re-supply contracts, provided the requisite capabilities were demonstrated in Phase I. For those bidders not selected in the first round for funded development, NASA issued an open invitation to enter into unfunded Space Act Agreements in which the teams and the agency could share data to refine other proposals in the hopes of being selected for the next stage, or supplying services at a later date.

In addition to carrying supplies to ISS, COTS also sought to address two other points of concern with maintaining the Station in the post Shuttle environment. First, the return of hardware and experiments was a critical capability which would be impaired with the retirement of the Shuttle. As the only other craft visiting the Station which returned to Earth intact, in addition to three tightly packed cosmonauts, the Russian Soyuz crew vessel could only carry an additional 1200 lbs. to what was not all that soft a landing to begin with. The Progress cargo freighter, based on the Soyuz, as well as the European and Japanese cargo freighters were each doomed to burn up in the atmosphere, ingloriously stuffed with trash. Absent a new form of return capability, the research capacity of the Station, significantly improved when it went to a six person crew, would be limited by the frustrating inability to return experiments to Earth other than in cinders.

To delineate the different capabilities it was seeking, NASA divided the COTS requirements into four overall categories; A, B, C and D. The first two called for the transport of external, unpressurized and internal, pressurized cargoes respectively. The third, COTS-C, was for a cargo return capacity, and the final, optional category, COTS-D, would examine the potential for crew transport services. In ending the Shuttle program, NASA was relegating access to the Station for its own astronauts, as well as those of its international partners to Russia, and to a single launch system. Although plans called for Project Constellation's Ares-1 to support the Station beginning in 2014, by 2006, that timetable was already slipping, and further delays appeared likely. On the off-chance that another American alternative was available, and the contractors showed sufficient progress to warrant it, COTS-D envisioned the transportation of astronauts as a way to leverage the investments by the private sector in delivering cargo. Presumably, if you could launch and safely return pressurized cargo in the volumes required, it was not a far stretch to extend that capability to crew transport as well. The key criteria would be the addition of an emergency detection system for the launch vehicle as well as a corresponding abort capability, manual control options for crew, and finally, a modestly upgraded life support capability beyond that already required for a visiting vehicle which while berthed to the station, would be hosting astronauts within its confines anyway.

COTS was a bold program, intelligently devised. Rich with the potential to both free NASA from Station re-supply as well as create an entirely new market for space transportation, it called for what could only be considered a very modest investment of public funds, initially $500 million, which would only be paid over several years as contractors passed specific milestones along the way. From the total amount, NASA estimated that its own program implementation costs would absorb roughly 3%, a remarkably low number for an acquisition program. While the immediate market in LEO was limited to a single customer, the government, COTS addressed the barrier of high launch costs, including a clear sense that the program was offering the possibility of doing something more, establishing a new method for how NASA does business, reflecting the Aldridge Commission's advice that the agency should seek to become a tenant rather than a landlord. In doing so, and in taking the role of anchor tenant, an agency often criticized as hide-bound to legacy systems was taking significant steps in creating a transportation infrastructure which could help foster a private market as costs came down. Indirectly, and depending on which vendors were ultimately selected, it also preserved some of the original intent of placing the precursor Alternate Access to Station Program within the Space Launch Initiative in hopes that it would lead to a bottom up approach to launch vehicle development.

For SpaceX, the official announcement of the COTS program was hardly a surprise; after all it was in pursuit of such an opportunity that the company had filed a protest with the GAO over Kistler's sole source contract. Nevertheless, as the program came into focus, it provided the company an opportunity to take a second look at its overall launch vehicle development plans.

COTS didn't simply offer the opportunity to develop a new launch vehicle, it was challenging industry to develop an entire ground to orbit and back to ground space transportation system, and more surprisingly, in calling for pressurized cargo to be delivered to the International Space Station, where it would be unloaded by astronauts in a shirt sleeve environment, it marked only a limited degree of separation from a crew transportation capability. Sitting right there in one government program was the opportunity to develop and fly several core pieces of the basic infrastructure SpaceX would require to fulfill its deepest ambitions.

From the outset, Elon Musk had made no secret of his ambition to progressively introduce larger and more capable launch vehicles, and well before COTS was introduced, SpaceX had already announced the development of a larger stable-mate to the Falcon 1, and the booster originally planned to support a Station resupply capacity; the Falcon 5. In the weeks following Griffin's speech, as the COTS program details began to emerge, the specific provisions suggested the need for a vehicle which stretched the projected capabilities of the Falcon 5, particularly if the company wanted to pursue the option of crew transport. At nearly the same time, "an unnamed government customer," also approached SpaceX with a request for a launch vehicle offering more performance than the Falcon 5. If the company wanted to fully pursue both opportunities, it would need to add another, larger rocket to a portfolio which still did not happen to include a demonstrably successful launch vehicle of any size.

On September 8th, 2005, just two months after Griffin's Space Transportation Association speech and well before the official introduction of COTS, SpaceX announced the development of the launch vehicle which would be the backbone of its COTS proposal, the Falcon 9. The new booster, intended to compliment, but not replace the Falcon 5, offered a performance range which brought it in to the EELV class of vehicles, opening up the potential to compete for both defense and commercial communications satellite launches. Compared to the limited and chronically unstable lower end of the launch market addressed with the Falcon I, the new Falcon 9 would not lack for opportunities, but capitalizing on them would present an entirely new array of problems. What COTS offered was the chance for

SpaceX to enter the market with NASA's help, and most critically, with a number of launch orders already in hand.

First though, they had to win. More than twenty teams submitted COTS proposals to NASA in March of 2006, a list which included companies both large and small, including a choice selection of the usual screwball entries which can be counted on for virtually any high profile NASA competition. After an initial culling, NASA invited six companies for further discussions. The competitors included Andrews Space of Seattle, WA; SPACEHAB, of Houston, TX, SpaceDev of Poway, CA, Transformational Space Corporation (t/Space) of Reston VA, and finally Kistler Aerospace. Notably absent among the first round semi-finalists, although present as suppliers for several entrants were the major aerospace contractors such as Boeing, Lockheed Martin and ATK. [46]

Among the competitors, SPACEHAB had a long history of integrating and flying component racks on the Shuttle and ISS, and could be expected to be on the most familiar ground in shaping the logistical aspects of their proposal. The company offered a craft originally referred to as Apex, but as the concept evolved, the name was subsequently changed to ARCTUS, an acronym for Advanced Research and Conventional Technology Utilization Spacecraft. Designed to be launched on a ULA Atlas V 401 configuration rocket, the ARCTUS craft was an amalgamation of components from other successful programs, including the hydrazine thruster assembly from the Centaur upper stage. Perhaps the most unusual feature was that the pressure vessel, fabricated from two domes of the Centaur upper tank, was designed to be recovered in midair by a helicopter, after being slowed and stabilized from descent velocity by a ballute. The core philosophy of the proposal was that it offered a minimum of technology development and thus would be a highly reliable entry. [47]

Andrews Space, building on previous work performed under the AAS program, offered a modular system consisting of separate pressurized, unpressurized and return segments, ultimately designed to launch on a proposed new RP-1/lox booster named Hercules, but with initial launches to be conducted aboard an existing EELV. [48] One of the most intriguing concepts was that offered by T/Space. Its proposal was based on an upgraded version of a rocket offered by team partner AirLaunch LLC, called Quickreach, a rocket concept it was developing for the DARPA/Air Force Falcon program (Force Application and Launch from Continental United States). The overall goal of the program, of which SpaceX was also one of 9 study contract winners in 2003, was the development of the capability to launch a strike on anywhere on the planet within 12 hours of an order to proceed. As Phase I of that program, DARPA was also interested in developing the capability to launch a small satellite into a Sun-synchronous orbit in a short time frame as well. If successful, the program would allow the Air Force to quickly place a satellite over any "hotspot" on demand, even as events were unfolding. The Quickreach vehicle, which was the winner, would be launched tail first out of the cargo hold of a C-17 aircraft in flight, utilizing an innovative deployment mechanism to eject the rocket and flip it clockwise into a launch configuration. The basic maneuver was tested successfully, but the program was never funded. The T/Space COTS proposal took this basic approach but expanded it to a much larger rocket to be carried aloft aboard a specially modified Boeing 747. As such it marked yet another in a long line of launch vehicle proposals to feature the jumbo jet as its first stage. [49] It would by no means be the last, and the next one would include a very familiar name.

Another significant entry, and one which would find a second chance in the follow on to COTS, NASA's Commercial Crew Program, was that offered by SpaceDev of Poway, Ca. SpaceDev, since acquired by Sierra Nevada Corporation was the brainchild of entrepreneur and lifelong space advocate Jim Benson. Benson developed one of the first functional text indexing systems for computers in the early 1980's and founded SpaceDev in 1997 to pursue his goal of affordable space transportation. In 2004, SpaceDev supplied the hybrid rocket motor for SpaceShipOne, playing a crucial role in helping it to win the Ansari X-Prize. Attempting to leverage that success, Benson announced in 2005 the development of a family of orbital and suborbital space tourism vehicles, appropriately named Dream Chaser. In pursuing the plan, Benson and partner Mark Sirangelo eventually settled on a design for a wingless lifting body, the HL-20 which had been proposed as a lifeboat for Space Station Freedom, and thoroughly studied by NASA Langley.

Although dating back to the HL-10, and the lifting body research which heavily influenced Space Shuttle design, and even provided the memorable opening sequence of The Six Million Dollar Man television show, the HL-20's unusual heritage actually came via the CIA and a dossier on a small scale Russian space plane, the BOR-4. The peculiar Soviet craft had shocked western analysts when it was photographed by an Australian reconnaissance plane being fished out of the Indian Ocean by a Russian trawler in 1982. The BOR-4 had originally been a sub-scale demonstrator of a losing entry in the Soviet Union's Buran shuttle program, but it eventually found a use as a test bed for the Buran's heat tiles. Intrigued by the photograph, the CIA approached NASA's Langley Research Center which made a wooden model based on the image, and was surprised to discover that compared to other lifting body concepts, it offered a remarkable degree of stability at different speeds without active flight control. [50]

As a result of the combination of Soviet flight history and subsequent refinement by NASA, the HL-20, which like so many other NASA concepts was never funded for development, presented something of a curiosity; a fully vetted concept for a space plane which was aerodynamically mature, despite languishing for many years. Securing the rights to use the design from NASA, SpaceDev entered the Dream Chaser in the COTS competition with a goal of making a fleet of the spacecraft available for other purposes as well, including satellite servicing. The Dream Chaser proposal milestones included suborbital test launches aboard a hybrid rocket SpaceDev planned to develop separately, with the orbital launch vehicle to be either the SRB segment of an Ares-1 with the same hybrid booster included as the second stage, or much more likely, an Atlas V.

In addition to SpaceX, the final competitor was Rocketplane-Kistler, a company formed in early 2006 when Rocketplane Limited of Oklahoma, a space start-up focused on sub-orbital space tourism, purchased the remaining assets of now defunct Kistler Aerospace, most notably the partially completed K-1 launch vehicle. Just as before, the RpK proposal focused on utilizing the two stage to orbit, reusable system, featuring interchangeable canister shaped crew or cargo modules.

NASA announced the results of the COTS competition on August 18, 2006. The winners were Rocketplane-Kistler (RpK) funded at $207 million and SpaceX, at $278 million. [51] Although SpaceX had failed in its first launch attempt only months before, the fact that the company was already developing and (test flying) much of the hardware which would find its way to the Falcon 9 weighed heavily in its favor. So too did the personal fortune and clear commitment of its founder. The selection of RpK as the other winner was somewhat more controversial. First, under new ownership after going through Chapter 11 bankruptcy, the prior management team led by Apollo veteran George Mueller had departed in 2004. Also, for its initial missions, the K-1 would not launch from the United States, but from a site in Woomera, Australia, with eventual plans to establish a second launch operation from a Nevada site north of Las Vegas.

The undeniable appeal of the RpK proposal was the extraordinary nature of the K-1 launch vehicle itself. Quite simply, if it worked as advertised, the K-1 would revolutionize the aerospace industry in an instant, drastically slashing the cost to orbit and opening up the commercial space frontier for real. Furthermore, there was nothing truly exotic in the design to warrant excessive concern over the ever elusive "unobtanium" which was often hidden deep in the specifications of many would be new space systems. It was also a fair assumption that although George Mueller had departed, during his six years at the helm, the CEO who among his many accolades was credited with the Apollo Program's critical adoption of all up testing, had refined the vehicle's design to the point that it was technically sound. The real question was obtaining the necessary financing, executing the design, and avoiding a catastrophic mistake on test flights.

Despite a promising design and the remaining presence of respected aerospace veterans among its upper management, the RpK approach carried a high level of risk. It was very different than either the Ares-1 or EELV approaches as well. Whereas SpaceX had built a business plan based on first developing and flying a much smaller and more affordable rocket, and in a fully expendable mode at that, the K-1 was a full up reusable vehicle with no prior history which would have to work nearly perfectly the first time. Even more improbably, it would have to prove the re-usability concept almost flawlessly from the outset to recover and re-use its launch vehicle for the multiple flights. RpK's effort would depend on raising outside investment, and that quickly proved its undoing. Despite the head start provided by working main engines which had already been successfully ground tested, RpK only managed to achieve the first three of 15 pre-set milestones.

After passing the third milestone, a Systems Requirements Review, the company faced an insurmountable barrier on achieving the fourth milestone, demonstrating its financial viability by successfully raising a pre-agreed level of external capital. Part of the problem was in the traditional approach the company's NASA dominated management team was taking. Rather than actually performing the development work themselves, RpK went about the project in a very conventional manner, operating as prime contractor and subbing out various components. It initially teamed with Orbital Sciences Corporation for vehicle integration, but after an early dispute, OSC pulled out, taking with it a $10 million cash contribution. RpK appeared to rebound quickly from this initial setback, signing instead with Alliant Techsystems for vehicle integration. The inclusion of ATK, which was in the midst of developing the 5 segment solid booster for the Ares-1, and Lockheed Martin which was contracted to build the fuel tank, brought legacy names, but at the cost of legacy overhead to the project. It was also difficult to see what real incentive either company had to make sure the K-1 was successful, considering the unpredictable tumult likely to occur in the launch industry if the K-1 actually worked. In the eyes of many, RpK had hired the fox to guard the hen house and would quickly be eaten.

As the May, 2007 funding deadline approached, and was subsequently extended to July 31st, it became apparent that RpK could not live up to its obligations. To many people's surprise and true to its word, as well as the guidelines of the COTS competition, the Commercial Crew and Cargo Program Office (C3PO) cut its losses and terminated the RpK Space Act Agreement. NASA quickly began preparing for a new interim round of competition, dubbed COTS 1.5, to select a replacement vendor. Among the $202 million originally allocated to RPK, only $31 million had been disbursed, still leaving 175 million incentives for entrants to submit proposals again, with a decision to be made in February 2008. This time the competitors were joined by Orbital Sciences Corporation, marking the entry of a full service, end to end launch and spacecraft provider to the competition.

OSC's entry consisted of a new expendable two stage launch vehicle which it already had under preliminary development, the Taurus 2, featuring a liquid fueled first stage, the first ever for the company, and a solid fueled second stage. OSC's new rocket did share one critical component with the now permanently defunct K-1 however—the engines. The Taurus 2 would be powered by two Russian NK-33 engines, of the same type which had been slated to boost the K-1. Reworked and modernized by Aerojet and renamed AJ-26, two engines would be integrated to a first stage built in Ukraine by the same factory building the Zenit and Sea-Launch rockets. The second stage would be an ATK supplied Castor solid rocket motor, the Castor-30A, with plans to upgrade to higher performing versions in subsequent flights. OSC offered a staged approach to developing its cargo craft, beginning with an open, unpressurized arrangement for early flights, with a pressurized vessel to follow. Continuing with the practice of integrating existing hardware rather than designing new, the pressurized cargo module would be a slightly reworked version of NASA's Multi-Purpose Logistics Module (MPLM) flown by the Shuttle and built by Thales Alenia Space of Turin, Italy. Unlike SpaceX's recoverable capsule, the OSC craft would burn up in the atmosphere upon re-entry, shirking COTS-C downmass capability, and rather obviously, COTS-D crew transfer capability as well.

Following the embarrassment of the RpK default, NASA followed a more conservative route and announced OSC as the winner of the intermediate round on February 19, 2008. [52] The decision was based in part on the fact that it would be adding a new medium lite class launch vehicle to the stable of potential NASA launchers, as well as pursuing a relatively safe approach in adapting mostly flight proven hardware. Off to a much later start than SpaceX, the experienced but definitely less agile Orbital Sciences would not be in a position to conduct its first launch until well after SpaceX had seen its entire system put to the test.

Viewed from the vantage point of hindsight, it perhaps seems rather obvious that SpaceX should have won at least a portion of the COTS competition. A contrarian might point out that it was bad enough that an unproven company would claim it could both build and operate a spacecraft sufficiently advanced that only a handful of national governments had accomplished the same thing, but to base such a bid on the ability to build an entirely new launch vehicle as well, bordered on delusion. The award decision can be judged in terms of the specific confluence of the times however. As of 2006, after two years of changes, Project Constellation was just getting underway, and although the warning signs would soon manifest themselves, appeared to be a very achievable endeavor, albeit a very expensive one as well. Consequently, COTS was still more of a convenience than a necessity, although if it ultimately failed to produce anything tangible, NASA would face a budgeting headache in trying to maintain its ISS obligations while pursuing the goal of a lunar return. Furthermore, the U.S. economy was roaring, and the Great Recession was still several years away. Creeping awareness of federal spending and accumulating national debt had yet to approach the levels it would soon see. The milestone based funding being put forward by the program was a gamble the country and NASA could afford. More cynically, until COTS actually showed signs of success, it did not yet pose a threat to the entrenched interests which would soon find themselves defending a massive government space program which was already well on the way to consuming nearly $17 billion dollars with nothing but a launch tower and a single suborbital test flight to show for it.

If NASA had wanted to go with the most conventional and safest technical approach possible, it might have been that offered by SPACEHAB, which featured the highly reliable Atlas V as its launch vehicle, and space proven components at every point. Nevertheless, in introducing the program, NASA Administrator Mike Griffin had made a compelling case for the necessity of lowering costs to LEO if NASA was going to go resume exploration beyond LEO. Also, both EELV rockets, the Atlas V and Delta IV, were already beginning a price climb which would reach almost unbelievable levels, and offered no chance for the type of breakthrough the agency was seeking. Such a decision would have also likely brought howls of protest from the entrepreneurial space community NASA was attempting to accommodate.

Consequently, even though neither original winner SpaceX or RpK was close to a first flight of the respective launch vehicle upon which its proposals was predicated, this likely worked in their favor. NASA wasn't merely seeking an alternative source of supply for ISS, the agency was at last seriously committed to both testing a new method of acquisition, and was looking for new launch solutions as well. On a more conspiratorial note, more than a few pundits suggested that COTS was meant to fail, and there was no way NASA would have chosen an existing EELV based proposal due to the potential of poor comparisons versus the Ares-1 launch vehicle the agency was developing for Constellation.

Within the world of 2006 however, compared to a number of other contestants, SpaceX was supported by the fact that it was working on actual hardware in a factory setting which, if successful in producing a launch vehicle, should by all reasonable assumptions be in a position to readily produce the spacecraft itself. Furthermore, the fact that the company was already well down the path of controlling its own production to an unusual degree suggested to the program office that it would be in a better position to control its cost. Combined with confidence in Elon Musk's commitment to his own venture and a small but growing list of commercial satellite launches which promised positive cash flow, SpaceX was an easy choice.

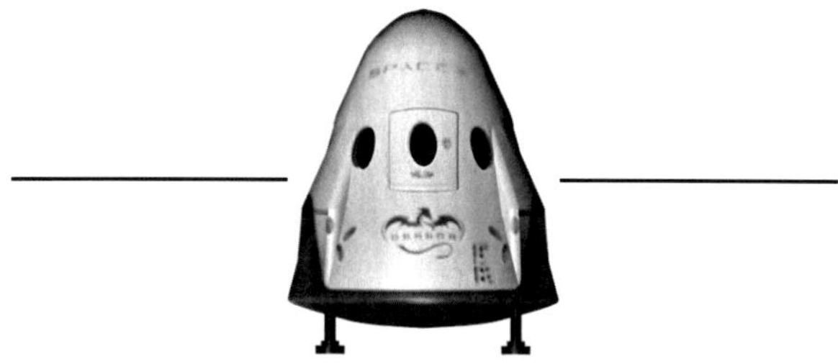

ISS as seen from Space Shuttle Endeavour on her final flight, STS-134. Credit: NASA

Chapter IX: Enter the Falcon 9

The formal announcement of the COTS program presented SpaceX with what can only be called a dream scenario, the opportunity to design and build not only a significantly larger launch vehicle which could open new markets, but what amounted to a complete end to end space transportation system readily adaptable to launching people and supplies to the International Space Station. Dreams though, sometimes have a way of turning into nightmares, and as SpaceX was discovering on Kwajalein, simple bad luck can undo the best laid plans. Emerging as one of two winners of the COTS award presented both an opportunity and a challenge. It also meant establishing a relationship with NASA which would fundamentally change the trajectory of the company, and to some extent, that of the agency as well. To a great degree the SpaceX which would successfully emerge as the first private provider to ISS and the leading contender to restore American manned spaceflight was born not in 2002, but on the day NASA announced the company as one of two winners of the COTS competition in August of 2006.

The SpaceX development plan approved by NASA originally consisted of 19 milestones, beginning with a Project Management Plan Review to be completed by September, leading to a projected first mission, labeled COTS 1, to take place in September of 2008. Prior to that point, SpaceX would conduct the maiden flight of the new Falcon 9 booster on separate terms, which the company advised NASA had already been sold to an (unnamed) U.S. Government Customer.

The set of milestones, with award amounts and target dates were:

1	Project Management Plan Review	09/06	$23,133,133
2	Demo 1 System Requirements Review	09/06	$5,000,000
3	Demo 1 Preliminary Design Review (PDR)	02/07	$18,133,133
4	Financing Round 1	02/07	$10,000,000
5	Demo 2 System Requirements Review	03/07	$31,133,333
6	Demo 1 System Critical Design Review	08/07	$8,133,333
7	Demo 3 System Requirements Review	10/07	$22,333,000
8	Demo 2 Preliminary Design Review	12/07	$21,133,333
9	Demo 1 Readiness Review	02/08	$5,633,333
10	Financing Round 2	03/08	$10,000,000
11	Demo 3 Preliminary Design Review	04/08	$22,333,333
12	Demo 2 System Critical Design Review	06/08	$21,133,333
13	Demo 1 Mission	09/08	$5,633,333
14	Demo 2 Readiness Review	12/08	$16,133,333
15	Demo 3 System Critical Design Review	01/09	$22,333,333
16	Financing Round 3	03/09	$10,000,000
17	Demo 2 Mission	06/09	$6,133,333
18	Demo 3 Readiness Review	07/09	$12,333,333
19	Demo 3 Mission	09/09	$7,333,333

In addition to the 19 primary milestones contained in the original COTS agreement, SpaceX also proposed seventeen optional milestones under the as yet unfunded COTS D element, Crew Transportation. If eventually funded, SpaceX envisioned conducting three additional launches consisting of an unmanned orbital test flight, a high altitude abort test and a crewed test flight to take place in April of 2012. [53]

Several notable items stand out. In the broadest sense, the milestones were divided into three categories; those relating to Design and Development, Test and Production, and finally Flight Demonstrations, the same approach which was subsequently applied to Orbital Sciences Corporation and its program for the Taurus 2/Antares booster and the Cygnus spacecraft. Also, the financial awards attached to each milestone were never intended to be matched to the level required to perform the respective tasks, a distinction which sometimes led to confusion regarding the milestone process. Instead, and in keeping with the goals Mike Griffin outlined in his introduction of the program, the pace of payments was meant to keep the company on track, but not extend the government's liability beyond a very limited horizon. The overall structure in some ways followed the aerospace industry's traditional design and development path. First, a Systems Requirement Review was conducted to determine if the proposed system met the program's overall objectives. Next, a Preliminary Design Review refined and matured the concept while identifying specific areas for attention. Finally, the Critical Design Review essentially checked each of the boxes before the actual production began. Prior to each of the scheduled reviews, NASA and SpaceX would agree on the specific deliverables, and then sit down for intense discussion in which NASA could either sign off, issue a request for more information, or a request for a change.

Reflecting NASA's overall level of concern about an all-new booster and capsule, the SpaceX milestones called for three separate flights, COTS 1, 2, and 3, as well as the independently funded maiden flight. Put another way, NASA was asking SpaceX to conduct 4 separate flights of an all new EELV class booster, as well as design, build, launch and recover three space capsules, at a total taxpayer investment of $278 million dollars, an amount so far below the historical thresholds that it defied comparison. For a startup company numbering less than 200 people at the time of the award, and struggling with a small and comparatively simple launch system, the odds of successfully building and operating a far larger vehicle capable of launching a full crew of astronauts to ISS appeared daunting to say the least. To many on the outside, they seemed absolutely insurmountable. In winning a share of the initial COTS award, SpaceX was somewhat in the position of the high school geek who scored an improbable date with the homecoming queen; it might turn out okay, but it could also be a life altering, embarrassing disaster, played out on a very public stage. The key to avoiding such an outcome is careful planning, leaving absolutely nothing to chance, and hoping that the luck which got you this far, doesn't suddenly run out.

One thing which did begin to run out was time, and as SpaceX struggled with the Falcon 1, it quickly became clear that the initial timetable was hopelessly optimistic. In February 2008, NASA and SpaceX negotiated a series of amendments to the original COTS contract, extending the timeline for Demo flights, combining the Demo 2 and Demo 3 System Critical Design Review into a single item. The amendment also added four new milestones, bringing the total to 22 while re-shuffling the payments among remaining milestones to reflect the same total of $278 million. [54]

NASA/SpaceX Milestones as amended February 2008

1	Project Management Plan Review	09/06	$23,133,133
2	Demo 1 System Requirements Review	09/06	$5,000,000
3	Demo 1 Preliminary Design Review (PDR)	02/07	$18,133,133
4	Financing Round 1	02/07	$10,000,000
5	Demo 2 System Requirements Review	03/07	$31,133,333
6	Demo 1 System Critical Design Review	08/07	$8,133,333
7	Demo 3 System Requirements Review	10/07	$22,333,000
8	Demo 2 Preliminary Design Review	12/07	$21,133,333
9	Draco Initial Hot-Fire	03/08	$6,000,000
10	Financing 2	03/08	$10,000,000
11	Demo 3 Preliminary Design Review	06/08	$22,000,000
12	Multi-Engine Test	09/08	$22,000,000
13	Demo 2/3 System Critical Design Review	12/08	$25,000,000
14	Financing 3	03/09	$10,000,000
15	Demo 1 Readiness Review (DRR)	03/09	$5,000,000
16	CUCU Flight Unit Design, Acceptance And Delivery	05/09	$9,000,000
17	Demo 1 Mission	06/09	$5,000,000
18	Demo 2 Readiness Review	09/09	$5,000,000
19	Demo 2 Mission	09/09	$5,000,000
20	Cargo Integration Demonstration	01/10	$5,000,000
21	Demo 3 Readiness Review	01/10	$5,000,000
22	Demo 3 Mission	03/10	$5,000,000

Although the revised milestone agenda allowed more time for SpaceX to begin its three flight test regime, the net slip of the COTS 1 mission from September 2008 to June 2009 still left the company facing a very tight schedule, and one which would prove difficult to meet.

One of the principle aspects of the COTS win was that it crystallized for the company the direction it would have to go in marshaling its limited resources in a program that appeared from the outside at least, to be somewhat scattershot. With the initial failure of one small rocket, two other rockets under development, and the company founder discussing much larger vehicles he wanted to develop in order to go to Mars, observers could be forgiven for concluding that a little focus was in order. The COTS award proved just that.

In particular, it narrowed launch vehicle development to the Falcon 9. If SpaceX was ultimately successful in winning the more lucrative Commercial Resupply Services contract for which it would still have to compete, then there was a clear path towards not only developing the new booster, but also in securing a substantial number of guaranteed flights which could in turn be used to leverage the more difficult science, commercial and defense launch markets.

It was not just a matter of building a bigger rocket than the Falcon 1 or the planned Falcon V. SpaceX would now be making its own entry into the size class at the heart of the international launch market and a category of vehicle capable of performing ambitious deep space missions. Critically, the Falcon 9 was intended from the outset to be SpaceX's version of a building block approach to launch vehicle development meant to enable very heavy payloads, the Common Core Booster. Though long established in conceptual plans dating back to the 1950's, it was not until 2004 that Boeing had brought the concept to life by strapping two additional, and nearly identical first stages onto the "core" stage of a basic Delta IV rocket in order to achieve what amounted to a poor man's heavy lift vehicle. If SpaceX could master the entry into this size class, the basic booster core, then larger and more capable vehicles, some even powerful enough to bring Mars into reach, were readily at hand. While Elon Musk was still talking about his long term plans for boosters on par with the Saturn V and beyond, there could be no doubt that the EELV class Common Core Booster was both the starting point, and the lowest common denominator, in terms of the capacity to reach the rest of the solar system.

If the Falcon 9 is in fact the launch vehicle which ultimately transforms the space age, it will forever remain somewhat of an irony that it started out life as a smaller rocket, the Falcon 5, which was upgraded when a single customer requested a booster of modestly larger lift capacity. Historically speaking, it will be in pretty good company. The Douglas DC-3, the iconic twin engine aircraft credited with giving birth to civil aviation as we know it today, was built only with great reluctance on the part of Douglas Aircraft, after American Airlines ordered a sleeper aircraft, the Douglas Sleeper Transport (DST). The DST was quite an anachronism, an aircraft which appealed both to upscale rail passengers who were accustomed to sleeper cars, as well as the small percentage of early ultra-wealthy airline customers already flying in slow, lumbering aircraft which made their way across the country in short hops aboard similarly equipped aircraft. It was only after the first seven were ordered to cater to luxury clients that American asked for a "daytime only" version to take advantage of the additional seating capacity inherent in the DST design. What started out in response to filling a customer requirement for a niche market which would soon disappear in any event, sleeper aircraft, resulted in changes to the already successful DC-2, but it was just enough to alter the dollar per passenger mile formula and establish commercial aviation for good.

The story of the Falcon 5 is a strange one, which began almost 18 months earlier, on December 4, 2003, at the official unveiling of the Falcon 1 launch vehicle in Washington D. C. where it was being displayed on its transporter truck parked on Independence Avenue across the street from the National Air and Space Museum. Following a reception to showcase the company and its rocket to policymakers and Congressional staff which had just taken place inside the museum, Elon Musk walked outside to the Falcon 1, and casually announced that it would soon be joined by a considerably more powerful vehicle called the Falcon 5, a rocket capable of lofting a crew capsule. A month later, on January 7th, 2004, SpaceX officially introduced the Falcon 5 as a true medium class vehicle, offering a payload capacity of 4,200 kg to LEO. The new rocket would essentially be a super-sized version of the Falcon 1, powered by 5 first stage Merlin engines and designed as a competitor to the highly popular Delta II. Like the Falcon 1, its introduction was highlighted by an extraordinarily low launch price, in this case $12 million, resulting in an estimated cost per pound to LEO of $1300 in 2004 dollars.

Prior to the formal introduction of the Falcon 5, Elon Musk mentioned on several occasions the possibility of building a "heavy" version of the Falcon 1, which would have consisted of a stretched Falcon 1, flanked by two standard Falcon 1 first stages, serving as strap-on liquid fueled boosters. It was an interesting concept, and would have represented a scale model of a CCB, and a rather unique approach for developing what would still be a small launch vehicle offering a capacity of only 2,000 kg to LEO and 750 kg to GTO. Although it might have proved beneficial in laying the groundwork for what would come later, it is probably fortunate the company went in a different direction. To gain any efficiency of production, SpaceX needed both a bigger market and bigger rocket.

In approaching the situation tactically, SpaceX faced a problem. Based on current market numbers, the next size class was that of Medium lift vehicles, defined by the capability of launching a minimum of 5,000 lbs. to a 100 mile orbit, a category dominated by the Boeing Delta II rocket, a craft whose origins dated back to the Thor IBM and the early days of the space age. One of the most popular and successful launch vehicles in history, the Delta II was also versatile, coming in a number of configurations with a payload capacity ranging from 6,000 – 12,000 lb. depending on the number of solid rocket strap-on boosters to be utilized.

Unlike smaller rockets such as the Falcon 1 and its designated target the OSC Pegasus, the Delta II also sported a respectable payload capacity to geosynchronous and interplanetary transfer orbits, having boosted a number of groundbreaking science missions to interplanetary space. Compared to the much more expensive Atlas V and Delta IV launch vehicles just coming on line, the Delta II historically offered a good balance between cost and capacity and was a NASA favorite for this reason. In short, it was a highly popular and respected launch vehicle which would be difficult to dislodge from its position as NASA's preferred science launcher by any competitor which had not accumulated a very substantial track record on its own. Furthermore for the science missions for which it was generally required, as long as the pricing remained relatively stable, and consumed no more than 15-20% of any given mission's budget, project scientists who were understandably much more interested in their respective spacecraft than the vehicles which launched them, had no interest in substituting an unproven launch vehicle for a known and reliable rocket like the Delta II. Underscoring this point, in 2002, NASA Launch Services committed to a single purchase of 19 Delta II launch vehicles intended to support a spate of Science Mission Directorate missions, a transaction which came to be referred to as the 19 pack. With the remaining inventory of Delta II's scheduled to last through 2011, it would be some time before this market opened up.

Still, there were some interesting limitations to the Delta II vehicle. In the first place, even with the large block buy, prices were on the rise, at approximately $60 million per vehicle, nearly 5 times the advertised price of the Falcon 5. Also, the Air Force was beginning to demonstrate an increased commitment to the EELV class of rockets as a one size fits all launch solution, suggesting that the Delta II's long term future was not as solid as it might appear. If the Air Force backed away from the Delta II entirely, then NASA would likely be hung with the substantial maintenance costs of the booster's dedicated launch pads on both coasts, a prospect which promised to drive the agency's costs even higher. Also, the Delta II had never been man-rated, and depended on solid rocket boosters for all but the most basic configurations. While otherwise insignificant, this fact took on new importance in the chaotic post

Columbia environment which included the growing possibility that NASA might have no choice but to consider a proposed course of commercial re-supply of ISS, including possibly, crew transfer. If that happened, there would be a clear opportunity for a rocket designed from the outset to be crew rated, a fact which Elon Musk mentioned when introducing the Falcon 5. Pursuing this possibility, SpaceX signed an unfunded Space Act Agreement with NASA's Johnson Space Center on June 2nd 2005, allowing the two organizations to legally exchange information, the beginning of a process which would soon accelerate much more quickly than anyone imagined.

The distinguishing feature for the new booster was that it was to be powered by 5 first stage engines. Although born of necessity, the multi-engine approach was hardly controversial, after all the Saturn 1 and 1B boosters had been powered by 8 first stage engines and the Saturn V had featured five. Furthermore, the Russian Soyuz, little changed since the dawn of the space age, and now carrying American astronauts to the International Space Station was well on its way to racking up more than 1800 launches, each one lifting off with 5 main power heads, 20 main combustion chambers and 12 more vernier chambers, used for steering, resulting in a grand total of 32 combustion chambers blazing. Nevertheless, more recent trends were toward a single large engine per stage. The Delta II, Delta IV, Atlas V, Ariane V and Ukraine Zenit and Sea Launch all featured a single engine powering their respective cores, though in many cases the cores were supplemented by strap-on solid rocket boosters.

A five engine core without strap-ons was definitely a throwback, but the new Falcon would offer a feature none of the aforementioned rockets did, engine out capability all the way to space. If one engine failed, then the other four could take up the slack through conducting a longer burn, and still impart the thrust needed to reach the desired orbit. Depending on the phase of flight in which things started going badly, and the payload aboard, a Falcon 5 could lose as many as three engines and still achieve its objective, an interesting if somewhat disturbing prospect.

In most respects the Falcon 5 was designed to be a bigger version of the Falcon 1, sharing a common architecture but on a larger scale. The intended lift capacity was 9,200 lbs. to LEO, with both GEO and Earth departure mission capability as well, supported by a 4 meter payload fairing. In order to achieve the necessary performance, the Merlin 1A engine which powered the first Falcon 1 attempt would be upgraded to a new model, the Merlin 1B, featuring a sea level thrust of 85,000 lbs. and a specific impulse of 261 seconds. The extra power would be delivered by an upgraded turbo-pump generating 2,500 as opposed to 2,000 horsepower, with an increased turbine nozzle count, and increased fuel flow capacity to an engine which would still be based on an ablatively cooled thrust chamber. Clustering 5 engines together offered the opportunity for simplifying and reducing the costs of the engine by removing the roll control actuators and allowing carefully programmed angling of the turbine exhaust nozzle to counter balance the natural tendency to rotate on takeoff. Another significant change would be the removal of torch igniters in favor of industry standard tea-tab ignition fluid.

SpaceX took a similar approach in deciding how to perform what would necessarily be a significant upgrade to the second stage powerplant. After briefly flirting with acquiring the cryogenic Rocketdyne RL-10 engine for the new rocket's upper stage, an approach which would demand a radical redesign of the stage architecture to accommodate the RL-10's hydrogen fuel, engineers settled on the much easier solution of adding a second Kestrel engine from the Falcon 1, to form a twin engine upper stage. Fuel tanks would be enlarged and stretched to accommodate the significantly larger volumes, but would still be manufactured along the same lines out of sheet aluminum with welded stiffeners and fabricated into a pressure stabilized structure, with the second stage now merely being a shorter version of the first.

When the Falcon 5 was announced in January 2004, SpaceX fully expected to conduct a first launch of the Falcon 1 later that same year, and anticipated applying flight experience with the smaller rocket to the development process of the new booster. Instead, the harsh realities of developing an entirely new launch system from the ground up rendered the plan unworkable. As the Falcon 1 launch schedule steadily stretched out, much of the development work for the Falcon 5 was effectively delayed into the 2005 time frame, placing it into conflict with the new demands placed by a COTS program which appeared increasingly certain to be introduced. All of which, most likely turned out for the better.

Even in cases where there is a history of flight experience, developing a new launch vehicle designed around heritage components is still no small task, a fact aerospace giant Boeing demonstrated the hard way when it sought to introduce the ill-fated Delta III rocket in the late 1990's. Responding to the growing demand for larger satellites than could be lofted by the Delta II, Boeing endeavored to introduce a new member of the family, the Delta III, which would be powered by the same RS-27 keralox first stage engine as the Delta II, but with a new high performance upper stage based on a variation of the cryogenic RL-10 which also powered the Atlas II booster. To accommodate what was now a longer rocket and keep it controllable in upper level winds, Boeing redesigned the fuel tank and main body to a shorter, wider dimension. The first launch attempt came on August 26, 1998, with a paying customer in the form of a Hughes Galaxy X communications satellite. The result was a dramatic failure, which ensued when a problem in the guidance control software, adapted from the Delta II, caused a repeated series of rapid course corrections, resulting in the rocket running out of a finite reservoir of hydraulic steering fluid. The range safety officer dutifully destroyed the rocket, sending the $225 million Galaxy X satellite into the Atlantic Ocean.

The second launch, in May 1999 proved only marginally more successful, and was brought to a premature conclusion when the second stage engine ruptured and shut down, placing yet another Hughes satellite, the Orion-3, into a useless orbit. The third and final launch attempt of the Delta III took place on August 23rd of 2000, carrying only a dummy payload and an array of sensors. For once, the Delta III finally came within close enough range of the desired orbit to be labeled a success, but by this point the commercial satellite market was already beginning to collapse and Boeing mercifully terminated the unhappy rocket, proceeding with the far more successful Delta IV, a clean sheet design which shared little in common with its predecessor other than an RL-10 based upper stage. The poor launch campaign for the Delta III was a major source of concern for the Air Force, which was in the middle of supporting two new launch vehicles as part of the EELV program, and its problems led to significantly increased scrutiny of both Boeing and Lockheed Martin's development efforts. For SpaceX, the saga of the Delta III, which by all rights should have been an easy success, proved a cautionary tale, highlighting the need for exhaustive testing and leading directly to the decision to use the on board kerosene fuel supply as the source of hydraulic fluid for the Merlin engine, effectively ensuring that it could not run out.

SpaceX secured a limited number of launch contracts for the Falcon 5, including that of an inflatable test article for aspiring private space station company Bigelow Aerospace, but mid-way into 2005, ongoing discussions with NASA officials suggested that if the COTS program Mike Griffin had outlined several months prior were fully implemented, that under the COTS D category, the agency would be particularly interested in a capsule which could fulfill the long sought goal of accommodating the Station's planned full complement of seven astronauts in the event of an emergency. If it wanted to pursue this possibility, SpaceX would need to field a rocket with more capacity than that allowed by the Falcon 5.

In attempting to meet the requirement, Musk's team did not immediately jump to the solution which it ultimately developed into the Falcon 9. At first, engineers considered a number of approaches to achieve the greater liftoff thrust, including adding either liquid or solid strap-on boosters, or significantly upgrading the Merlin main engine while maintaining the 5 engine arrangement. Eventually however, with initial plans calling for the simultaneous production of both the Falcon 5 and the new Falcon 9, the more expedient course proved to be adding four more engines to the thrust structure designed for the Falcon 5, and stretching the tanks in order to produce the larger, EELV class rocket. What had been a five engine thrust structure would now become a 9 engine design with the engines arranged in three rows of three engines each. This solution would allow manufacturing of both vehicles with the same tooling and equipment. In a curious turn of events, the Falcon 5, whose dimensions and capabilities had changed several times since its original announcement, would for a time become a depowered version of the flagship Falcon 9 with 4 engines removed.

Once COTS was formally introduced in January 2006, SpaceX, still grappling with the Falcon 1, faced a decision as to whether or not to maintain the Falcon 5 in the light of changing requirements, particularly given the fact that despite the commonality between the two vehicles, as a small company with limited resources, it was not in a position to fully proceed with both. In the end, the company had little practical choice but to delay work on the intermediate booster in favor of the more powerful Falcon 9, the schedule for which was now being driven by the COTS program. Ultimately, the Falcon 5 vanished from the SpaceX lineup, seemingly for good.

If the Falcon 5 went quietly into the night, its successor stepped squarely into the spotlight, promptly attracting a healthy dose of criticism, much of it stemming from the 9 engine first stage arrangement. Even its immediate COTS competitor, the Kistler K-1 was powered by only three first stage engines. As such, it was met with more than a little skepticism, and even ridicule from a traditional aerospace establishment which viewed the plan as borderline sophomoric, particularly for a company which had yet to demonstrate any actual flight experience.

SpaceX however, had both history and data on their side. While it stands to reason that the risk of failure of any one engine increases as more engines are added to a given rocket, redundancy also rises. Just as was the case with the Falcon 5, probability analysis indicated that increasing the number of engines would improve the odds of a successful launch, and in the event of an engine shutdown, the eight remaining engines could reliably enable overall mission success by operating for a longer burn. This is a point Musk would repeatedly make, asking skeptics whether they would like to fly on a commercial airliner which was obligated to crash in the event a single engine failed. As for historical precedent, it came from a source regarded as the pinnacle of American achievement. The Saturn V booster featured multiple engines on both the first and second stages, and overcame engine failures on two occasions, the unmanned Apollo 6 test flight, in which two of five second stage J-2 engines shut down prematurely, as well as on the ill-starred Apollo 13 flight in which the second stage middle J-2 engine failed as well. In the Apollo 6 incident, the failure, caused by vibrations along the length of the vehicle as it accelerated, fractured an engine igniter feed line in one of the second stage's 5 J-2 engines, prompting the flight control computer to issue a shut-down command. Two seconds later, another J-2 shutdown as well, caused by a cross wiring issue when the initial engine chamber failure occurred. Nevertheless, now running on three rather than five engines, the second stage successfully completed its mission by extending its burn time. The vibration problem incidentally transferred into the third stage as well, which failed to re-ignite on command after its single J-2 engine placed it into a preliminary orbit. Occurring as it did on the morning of the Martin Luther King Jr. assassination and only four days after President Lyndon Johnson's surprise decision not to seek re-election, the trouble plagued test flight escaped substantial attention from a national press following bigger stories.

Both the Apollo 6 and 13 engine incidents occurred well after the boosters had developed sufficient velocity to maintain trajectory while allowing the remaining engines to assume the load. Much like an airplane, the most dangerous time for a loss of thrust in a multi-engine rocket is in the first moments after ignition, when the sheer weight of unused fuel would overwhelm the reduced thrust resulting from a failed engine and cause a catastrophic loss of momentum. As with the Falcon 1, SpaceX planned to minimize this potential failure mode for the Falcon 9 by physically restraining the rocket on the pad until computers verified that the engines were operating at full thrust and demonstrating no undesirable trends, much in the same way aircraft pilots used to begin the takeoff roll by revving the engines up to full throttle and standing on the brakes momentarily before committing to takeoff. In either case, while a catastrophic engine failure is always still an uncomfortable possibility, every practical precaution has been taken to make sure that it does not occur in the initial moment of flight when the balance of forces are at their most precarious.

For a company publicly outspoken about its goal of enabling missions to Mars, the reliance on a clustered small engine solution for what would now be an EELV class vehicle seemed to highlight the disconnect between its lofty goals and less than impressive capabilities, particularly when contrasted with the same established EELV boosters against which it would now have to compete. Compared to the Merlin's paltry 85,000 lbs. of sea level thrust, the Boeing Delta IV's single RS-68 main engine generated 663,000 lbs. of sea level thrust, subsequently upgraded to 705,000 lbs., and the Atlas V's RD-180 sported an even more powerful 860,000 lbs. sea level thrust, more than 10 times the power generated by the Merlin 1B. Furthermore, both competitive engines were comparatively sophisticated, particularly the staged combustion RD-180, whereas the gas generator cycle, ablatively cooled Merlin 1B was anything but.

The comparison grew even more strained when one considered the original "heavy" version of the Falcon 9 which SpaceX announced at the same time. It would feature a single core Falcon 9, flanked by either twin Falcon 5 or 9 first stages acting as liquid strap-on boosters, much like the configuration of the Delta IV heavy. The difference here was that the SpaceX version would feature not three, but either 19 or 27 engines all lighting at once, a proposition which could not help but raise the specter of perhaps the single worst launch vehicle in history, the giant Russian N-1 Moon booster, which was powered by 30 first stage engines and scored a perfect flight record of zero successes in four launch attempts, never even reaching the point of first stage separation.

The 9 engine configuration was in the end, a decision born of necessity, as SpaceX lacked both the time and the financial resources necessary to develop a larger engine similar to its competitors. There was however, an intriguing long term angle to the plan as well. Elon Musk had never been shy about his ultimate ambition to build a truly gargantuan booster, sometimes referred to as the Falcon XX, exceeding the scale of the Saturn V. To power such a booster, SpaceX would need a much larger main engine, something on the order of the Saturn's mighty F-1. Rather than follow recent precedent and move on to staged combustion for a much larger engine, SpaceX indicated that it would remain with an open-cycle architecture. In that light, the Merlin was specifically intended to serve as a working scale model of the engine Musk really wanted to build, offering invaluable experience into manufacturing and operations well before it was time to "super-size" the effort.

Such grand ambitions were a long way into the future however, and at the time SpaceX announced the Falcon 9 in September 2005 it still had no flight experience at all. The last West Coast Titan IV was still stuck on the pad at Vandenberg, tying up the range, and SpaceX was just beginning its ordeals on Kwajalein. Although the first flight attempt would eventually occur the following May, a successful orbital launch, was in fact still three years away. Given that any problem which arose in proving the Falcon 1 could similarly affect the Falcon 9, the larger vehicle would develop only at the pace allowed by progress on the smaller rocket. On the other hand, the Falcon 1 offered a low cost alternative for developing much of the capability of the larger and far more expensive Falcon 9, a path designed to protect the company from financial disaster should early attempts prove less than successful.

As originally conceived, the initial version of the Falcon 9 was to be powered by the same upgraded Merlin 1B engines which were already under development for the Falcon 5. Although sufficient for the requirements of the COTS program, a Merlin 1B powered Falcon 9 simply didn't offer the performance required if the company wanted to make a serious entry into a geostationary commercial launch market witnessing steady growth in the mass of satellites being launched. Consequently, according to its original plans for the Falcon 9, while the first version of the booster would be powered by the Merlin 1B, subsequent versions would be equipped with a more powerful, regeneratively cooled version of the Merlin engine. In the wake of the first Falcon 1 failure in May 2006, and the ensuing delay in conducting the next attempt, SpaceX elected to shuffle its development schedule and incorporate the new engine, to be named Merlin 1C into the Falcon 9 from the first flight.

There was also a more practical reason for switching to the Merlin 1C as well. In qualifying its first engine, the Merlin 1A, SpaceX struggled with difficulties in containing burn-throughs of the ablative thrust chamber. Although the problems were eventually resolved, any effort to increase overall engine performance without upgrading the thrust chamber promised renewed problems. That was not the only challenge, as SpaceX now needed a new upper stage engine as well. While the second stage of Falcon V was intended to be powered by two upgraded Kestrel engines, this option was insufficient for the Falcon 9. Having already considered and discarded the possibility of an RL-10 based upperstage, the obvious solution was a vacuum rated version of the first stage engine. This however, entailed adding two critical capabilities common to upper stage engines which were not present on the first stage. First, it would need to be re-startable in space, and second, the engine would require a throttling capability. Finally, like any upperstage engine, it would need to be able to withstand extended burn times, a requirement which further argued against the 1B.

The bottom line was that everything depended on introducing a new engine, the Merlin 1C, which would be much more an evolution than an upgrade. In this case, with the first flight of the Falcon 9 planned for 2008, evolution would need to take place rather quickly. As a result of the decision to proceed directly to the new engine, SpaceX dropped what had been its plan to introduce the Merlin 1B engine on the Falcon 1's next flight, and desperately needing to get a success under its belt, substituted a first generation 1A instead. The 1B would never fly.

Designing the Merlin 1C

There are two basic approaches to designing the regenerative cooling mechanism of a rocket engine. The first, and historically the most common, is to form the thrust chamber walls out of hundreds of small tubes brazed together and constrained by expansion bands on the outer perimeter much like hoops on a barrel. The tubes extend to the throat area, which experiences the greatest degree of heating, and further down to the expansion skirt, where fuel is collected in a collar around the perimeter, and after sometimes being routed back through the passages again, is injected into the engine for combustion. Manufacturing a tube wall engine is a very time consuming and tedious process, particularly when accommodating the changing contours of constriction and expansion, and developing reliable methods for doing so was one of the principal challenges in the early days of the space age.

The search for a better method led to the second design, a combustion chamber milled out of a solid billet of metal, generally steel or copper, with grooves then milled along the outer surface to function as channels for the cooling fuel. The challenge here was how best to place a jacket on the outside of the engine to seal off the individual channels and avoid leaks, a term called "closing out." The solution was electro-plating. First the individual channels would be filled with a sacrificial wax, and then the entire thrust assembly would be submerged in a vat containing a suspended solution of the desired material, and with electrodes applied and a current generated, one by one ions would attach themselves to the exposed face of the grooves, eventually building up a coating which spanned the channels, forming a solid metal jacket. After that, it was a simple matter to heat the wax and empty the channels. This was the method SpaceX chose for the Merlin 1C, utilizing a copper combustion chamber with milled walls for coolant flow, surrounded by an electroplated nickel-cobalt jacket, which was then attached to a brazed tube wall nozzle chamber. [55]

The implementation of active cooling and much higher thrust levels intended for the Merlin 1C placed demands on the fuel delivery system which went well beyond the capacity of the Merlin 1B improved turbo-pump, resulting in the need for an all new turbopump. Like the original version, it was also a single shaft model manufactured by Barber Nichols, but its principal design was by SpaceX and its introduction marked a complete break from the historical tie to the NASA Fastrac engine. At full power it would produce an output nearly twice that of the Fastrac and the original Merlin. Following the decision to proceed with the Merlin 1C, initial development proceeded quickly and by August 2007, a demonstrator unit was ready to begin the first trials of a marathon testing regime unfolding at the McGregor, Texas facility. What the company needed most, however, was a successful test flight, an accomplishment which was proving very hard to come by.

Merlin 1C engine test. Credit: SpaceX

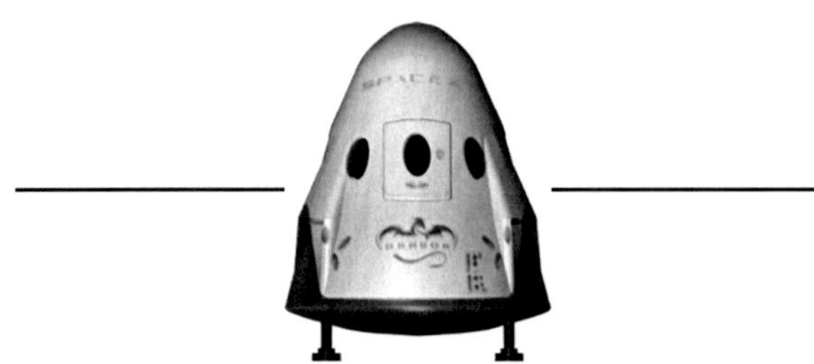

Chapter X: More Setbacks; Falcon 1, Flight Number 2

The failure of the Falcon 1 on its maiden flight only months before the COTS award was the first in what would be a series of three hard won learning experiences in anticipating the small details that spell the difference between success and failure in launch vehicle development. Fully aware that it needed to rapidly translate experience with the Falcon 1 into a successful introduction for the Falcon 9, SpaceX returned to Kwajalein in the late fall of 2006 with a new Falcon 1 rocket significantly updated from its first flight vehicle.

Following the flight 1 failure on March 24, 2006, SpaceX began the necessary process of reviewing all available data from the launch in order to determine what had gone wrong. As is common in all launch vehicle failures, an independent review board was established to pore over the data and conduct its own investigation, one which would look not only at the primary cause of failure, but into any other anomalies as well, even if they had no bearing on the final result. In this case, the review was conducted by the flight's sponsor, DARPA, and led by Air Force Lt. Col. Simon "Pete" Worden, who had previously overseen the Strategic Defense Initiative's groundbreaking DC-X Single Stage to Orbit suborbital test project, and was a long-time advocate of innovative space launch solutions.

With the wreckage generally scattered over hundreds of miles of open ocean, or in upper stage failures incinerated in the atmosphere or even floating uselessly in space, it is often not possible to establish with 100% certainty the precise cause of a launch accident, and many times the result of a long and tedious review is still only a "best guess" approximation which leaves everyone involved somewhat unsettled until the next flight seemingly confirms or contradicts the analysis.

Fortunately for SpaceX, a combination of factors led to a quick and comparatively decisive review, released on June 25 which identified the most likely cause of the failure, as well as noting several other areas of concern. [56] Although an initial examination of flight photos in the hours following the launch clearly showed a fire emanating from the well at the top of the Merlin engine 25 seconds into the flight, closer resolution images revealed it was already underway only seconds after the rocket cleared the tower. The most reasonable explanation was a fuel leak, but engineers were at a loss to identify the possible source. The Merlin engine program had endured its share of problems over literally hundreds of test firings, but fuel leaks had not been an issue.

Based on careful analysis of available telemetry however, the most likely source was eventually traced to a failed "B nut" on the turbo pump inlet pressure transducer which allowed fuel to begin leaking approximately 400 seconds before liftoff. Upon engine ignition, the leak was transformed into a steadily burning conflagration which eventually severed a pneumatic line, triggering an engine shutdown. Flight 1 had been doomed before it ever left the pad.

The nut in question, which was galvanized aluminum, was designed to protect against the corrosive salt air, and had not failed in the course of repeated test firings including the static fire two days before the launch. Yet despite all this, somehow it had failed on launch day. A possible contributing factor was that the nut, being aluminum, was subject to the effect of galvanic corrosion from contact with the stainless steel feed lines. As Elon Musk ruefully observed, countering some suggestions that SpaceX used "cheap" components in a rash effort to build a low cost rocket, had the company actually utilized a less expensive but ever so slightly heavier stainless steel nut instead, the problem might likely have never occurred. Instead, following multiple test firings, including a final one only days before, during any one of which a fuel leak would have been quickly detected and prevented from re-occurring, the problem had manifested itself on launch day, most likely as a result of extended exposure to the elements as the Falcon endured the various travails of the exasperating launch campaign. The cloud of bad luck which had been hovering over the western Pacific since the day SpaceX had arrived showed no signs of going anywhere fast.

On the positive side however, the mission had been classified as a test launch, and the loss of the payload, built by cadets at the U.S. Air Force Academy, while disappointing to those who worked on it, at least spared SpaceX the additional publicity of losing a more expensive piece of hardware. Furthermore, despite the fire and resulting first stage failure, the flight had met a number of test objectives, including vehicle boost past the zone of maximum dynamic pressure. Unfortunately, the loss ensured that the next effort would also have to be a test flight, one which would carry only a mass simulator instead of an operational satellite. Making the most of the forced stand down, and keenly aware of its new, much higher profile as a result of the COTS award, SpaceX undertook a series of improvements to the entire F-1 system, ranging from modest upgrades to the engine, to a major overhaul of the vehicle monitoring system. Though physically indistinguishable from a distance, the next booster to leave the pad would be considerably more sophisticated than its predecessor.

As to the immediate cause of the failure on flight one, engineers threw everything but the kitchen sink at the problem in order to make sure it never happened again. First up were a number of upgrades to the physical infrastructure on Omelek itself designed to counter the tropical conditions which had affected the booster, including building a new clean room, and doubling the insulation in the rocket hanger. In the days and weeks leading up to subsequent launches, the Falcon 1 would be kept in temperature and humidity controlled conditions. Galvanized nuts throughout the engine assembly were replaced with either stainless steel nuts or welded into place. In case an onboard fuel system fire started anyway, vents would be directed into the exhaust stream. Also, the engine bay doors would now be covered with fireproof blankets, and the enclosed spaces bathed in a steady wash of inert nitrogen. Going a step further, the pneumatic hoses which had suffered a burn through were replaced instead by metal overwrapped hose. The Falcon might surely fail again on another day, but it would not be for the same reason.

The stand down caused by the DARPA investigation also provided an opportunity to make broader changes to the Falcon vehicle and flight preparation procedures as well. If the Merlin 1A engine was given a mild revision, characterized in software terms as a 1.1, the flight control system received what Elon Musk termed a 2.0 plus upgrade. Whereas the health monitoring system for the initial Falcon was configured to cover known or otherwise obvious potential failure points, the upgraded system would now track virtually everything, expanding from a few dozen checks to over

700 inputs in a vastly improved network of sensors. The upgraded system also made heavy use of trend analysis, issuing console warnings in the event of developing deviation from acceptable limits, and automatically halting the countdown and safing the rocket in the case of a critical problem. To aid in the monitoring process, and to put more eyes on the system, El Segundo headquarters would receive one-way, real time direct feed of system telemetry, including voice access to the launch control center on Kwajalein Island. At the same time, the launch control team stationed on-site, now aided by the home office, was pared back to reduce cost and wasted man-hours.

In the immediate aftermath of the first flight, as analysis began to center on the failed B nut which caused the fuel leak, the investigation team originally suspected that the cause might have been human error, resulting from the failure to properly secure the nut with a retaining wire. Although that theory was ruled out when recovery efforts discovered clear evidence that the lock wire was in place, SpaceX also instituted a significant change designed to prevent human error in all aspects of the ground operations procedure. Going forward, any correction to the launch vehicle performed by a technician would be signed off by both the tech and their responsible engineer, and then entered into an evolving database by the newly created position of quality assurance officer, resulting in a three way cross check similar to the way controlled narcotics are dispensed in major hospitals. SpaceX also instituted a series of upgrades aimed at reducing the potential for human error in the countdown process as well. Building on experience gained through multiple test fires and the flight 1 launch, engineers further automated the process, removing manual checkpoints which held the potential for diverting attention from the more important task of data analysis.

Anticipating that breaking in the new system, with its greater number of sensors and conservatively established parameters would likely trigger more than a few false alarms, Musk warned reporters as well as followers in his regular internet postings that after increasing the scope of potential system aborts by a factor of 30, it was reasonable to expect a new series of delays before the Falcon finally took to the air again. This was one prediction which held true.

With the release of the Demoflight 1 accident report, the countdown clock started for the next launch, Demoflight 2, which although not carrying a deployable satellite, was equipped with two experiments attached to the vehicle's second stage. The first, provided by the flight's primary sponsor, which was again DARPA, was the Autonomous Flight Safety System (AFSS) and a NASA payload, the Low Cost Tracking and Data Relay Satellite System Transmitter (LCT2) as well as an inert payload simulator which would be deployed into the target orbit at 685km at 9 degrees inclination.

Before returning to Kwajalein, engineers wanted to conduct a full mission cycle acceptance test of the entire upper stage, which for obvious reasons had not seen any benefit at all from the first launch. Following a successful test in Texas, the 1200 lb. structure was loaded aboard a commercial aircraft for delivery to Kwajalein. By December 8, all of the components necessary for the next launch were en route to the island, and a possible launch attempt in January.

After integrating the Flight 2 booster, the launch team conducted a Flight Readiness Review and wet dress rehearsal in mid-January, opening a window for an end of the month launch attempt. Just prior to the next step, a first stage static fire, further pre-flight testing revealed a problem with the pitch actuator for the second stage Kestrel engine's thrust vector control. Cancelling the test fire, SpaceX elected to stand down and proceed with a cautious series of checks back in the hanger while the Reagan Test Range was configured for a Minuteman missile test. Following another range delay, this time due to the absence of range safety personnel, the launch team conducted a flawless static fire on Saturday March 17th. With no issues to resolve other than a minor problem with a backup GPS receiver needed for the stage 1 recovery effort, a flight readiness review conducted on Sunday set the launch attempt for 4:00 p.m. PT the following day, Monday the 19th.

A full year to the week after the original flight attempt, and with numerous new procedures in place, SpaceX was ready to try again. With the company anticipating a first flight for the Falcon 9 in late 2008, keeping that timetable meant the second flight of the Falcon 1 would need to be successful. Unsurprisingly however, and true to Musk's predictions, after initial delay due to difficulty in routing the rocket's telemetry feed to SpaceX headquarters in El Segundo, the countdown resumed only to come to a halt less than two minutes before liftoff.

As is typically the case, at some point the various connections which have been routing data between ground control and the vehicle must switch from landlines to radio transmissions which last throughout the flight. In this instance, the switch, at approximately T-90 seconds, caused a tiny gap in communications, amounting to only a few hundred milliseconds, but it was enough to trigger a hold in the ground control software, resulting in an abort. After briefly considering updating the software on the spot to account for the hardware driven delay, which had not shown up in simulations, controllers instead scrubbed the launch and safed the vehicle. Back in California, software engineers worked through the evening to program and test the necessary changes. With the new software tested, downloaded and installed, the next launch was scheduled for the same time, 4:00 pm (PST) the following day.

Tuesday, March 20th, found SpaceX once again counting down to what all hoped would be its first successful mission to orbit. As it had the day before, after a few brief delays, this time due to a potential radio frequency compatibility issue with the payload, which pushed the estimated launch time back to 5:05 pm PT, the countdown got underway in earnest, and proceeded smoothly at first, and as the final seconds ticked off, everything looked go for launch. Right on cue, the Merlin 1A engine roared to life, and then promptly shut back down.

This time the analysis and response time once again challenged existing records for a launch scrub and recycling. The problem had arisen from the fact that following the previous day's launch attempt, the kerosene fuel drained from the now safed rocket, returned to its insulated storage tank much colder than it otherwise would have been, and now pumped back aboard the booster, was further chilled by the adjacent, super cold liquid oxygen. Consequently, as the engine began the ignition sequence, the initial onslaught of fuel delivered to the engine was still slightly colder than parameters allowed, only half a degree, but it was enough to trigger an abort due to a minute decrease in main engine pressure. The solution was both simple and quick. Engineers cycled the system, draining 50% of the fuel from the first stage fuel tank and refilling it with warmer fuel from the ground supply. Less than an hour later, the countdown was underway again.

At 6:10 PST, the Merlin engine came to life for the second time that day, and for once it kept right on burning. With a flawless liftoff, the Falcon 1 climbed its way into the cloudy tropical sky, sending expectations soaring. With improved telemetry now coming from separately grounded systems on both the first and second stages, the Falcon effortlessly sailed thought the zone of maximum dynamic loading without a hitch. The next milestone, first engine cutoff and stage separation took place on schedule, but not without cameras showing a brief bump and course correction as the second stage engine extension brushed the interstage. Undeterred, the Kestrel engine flared to life, and as the second stage began a steadily accelerating climb to orbit, it appeared that maybe SpaceX had outlasted the gremlins which haunted it on the Kwajalein atoll.

The fateful combination of inexperience and Murphy's Law was not done with SpaceX yet however. Unknown to the launch team, the brief bump during stage separation was more serious than it initially appeared. The first sign of a problem came midway through the second stage burn, when on board video showed the rocket developing a slight wobble. By itself, the motion was not a critical problem, as long as the trajectory remained nominal. The Falcon could still make orbit, even if those watching it became somewhat dizzy. Instead, the wobble began to intensify, growing steadily more pronounced. Less than three minutes before a historic second stage cutoff, and orbital insertion, the second stage.... cut off. This time there was no fuel induced fire, rather the problem was a lack of any fuel at all making it to the Kestrel engine.

The design of the Falcon 1 was heavily influenced by an Aerospace Corporation study which clearly suggested that stage separation events were one of the trickiest elements of any launch system, and one of the most insidious elements was the simple fact that unlike a first stage hold down test, there is simple no way to completely simulate a stage separation on the ground. Hopefully, with the use of good computers, great software and even better analysis, born of experience, you identify and correct for trouble spots before a vehicle is lost, and there are a whole lot of people replaying it on YouTube and wanting to discuss it. Unfortunately, that was just what happened.

As the second failure review board would ultimately conclude, the first several seconds of stage separation went pretty much as planned, but that was the problem. Because the first stage was intended to be recovered and reused, the stage separation procedure required that the Merlin engine be programmed with a quick shutdown which would prevent damage to the turbo pump from operating in a fuel starved mode in the final seconds of flight. That happened, but with an unanticipated side effect. As result of the quick shutdown, and even before separation, the complete vehicle began a slight pitch over. Engineers had anticipated and planned for the motion, but not to the degree it took place. Seconds later upon stage separation, the first stage, now freed from the mass of the second, accelerated its pitch, again only slightly, but it was just enough to cause the interstage adapter to strike a glancing blow to the extreme end of the second stage Kestrel engine's bell shaped expansion nozzle. The strike, clearly visible in the rocketcam mounted to the second stage, caused little in the way of physical damage, but it was sufficient to knock the upper stage significantly off course. To the design team's credit however, the flight computer commanded a hard slew which immediately put it back on course accelerating towards space. Regrettably, the slew induced a slight sloshing motion to the mass of fuel contained in the second stage tank.

As anyone who has tried to drive with an open cup of coffee in their car can attest, once a motion is induced, say from that pothole which you know is there but you always forget until you are swearing in disbelief while looking for a paper towel, once a liquid in a container starts moving, it is difficult to calm it down. In this case, following the hard slew, the second stage appeared to begin a normal acceleration with no noticeable oscillation. It was not until several minutes into the burn, after successful fairing separation, that the tiny motion began to manifest itself, revealed by a slight wavering in the video feed showing the bell nozzle moving against the background of the Marshall Islands, still visible below. Then inexorably, even as flight control continued to verify that engine burn and trajectory were normal, the movement began to steadily accelerate in a growing corkscrew motion that indicated that something was badly amiss and getting worse.

The sloshing effect on the liquid oxygen caused the rocket to attempt to compensate, which in turn accelerated the problem, causing the second stage to compensate even further, The net result was a steady but gradual increase in the slewing effect causing the rocket to fly in an ever expanding spiral, albeit one which was still pointing towards space, and a qualified success.

As the second stage continued to accelerate, rapidly consuming the remaining LOX in the tank, the increased sloshing effect eventually threw all of the fluid away from the pickup sump, resulting in what amounted to an engine flame out. Even so, the now inert second stage continued on a parabolic trajectory which took it to a maximum altitude of nearly 180 miles, three times the recognized barrier of space, but 6,000 thousand miles per hour short of the velocity required to remain there. Just for good measure, not only did the payload fairing separate successfully, but the mass ejector ring, designed to deploy satellites, successfully ejected its mass simulator to begin its plunge back to the planet waiting below. Initially it was not clear that the Falcon had failed to reach orbit, and in immediate comments after the launch, Musk entertained the possibility that the second stage might have achieved a low orbit after all, but any lingering hope was quickly dispelled when flight telemetry confirmed that final velocity was well below that required to stay in space.

Although this second launch attempt had failed to reach orbit, it did in fact achieve a large number of the flight's objectives, leading Elon Musk to accurately assess it as a qualified success, within the context of being a test flight. Among the objectives met was a successful demonstration of the changes made after the first flight, as well as a validation of the second stage control systems which fought valiantly to keep the vehicle on track. Prior to that, and after the two previous aborts, the main engine performed flawlessly, smoothly driving the rocket skyward, and through Max-Q. Furthermore, with the payload fairing separation apparently a success and the payload ejector functioning as planned, the remaining area of risk was sufficiently small and understood that SpaceX felt justified in declaring the test flight phase was completed, and pending final analysis, the company could begin planning for its first operational launch, that of the long delayed TacSat-1 now scheduled for later in the year. Commercial operations could now begin.

Reactions to the second launch attempt varied widely, with some critics condemning the claim of a qualified success as spin, and even dishonest. Others took obvious glee in Musk's' apparent comeuppance. And to some extent, the way other observers regarded the second flight changed when following a lengthy delay, the next attempt fared even worse. Still, one of the key tenets of both the scientific method and engineering development is that there is no such thing as a bad experiment, because you can often learn more from failures than from success. The problem in many endeavors of course is the high cost of failure and what can be a very expensive education. In the early years of space exploration, and in particular before the advent of advanced software analysis, success was often attained by doggedly eliminating all the known modes of failure until in the end, there was nothing left to fail, leaving success as the only option. There is perhaps no better example than that of NASA's Ranger program, which suffered six consecutive failures in its efforts to provide close up imagery of the Moon in advance of the Apollo landings. Success did not come until July 31, 1964 when Ranger 7 finally managed to transmit more than 4,000 images in the final 17 minutes of flight, before it impacted the lunar surface. Backed by the full resources of the United States government, or the Soviet government for that matter, this was a viable strategy for a fledgling industry, but not so much for SpaceX. Originally asked how many attempts he could afford to make without success, Musk's answer; three. One thing was clear though, Musk wanted to get a new rocket on the pad as quickly as possible, assuring supporters that the newest delay would not stretch on for a year, like the previous one. It didn't; instead it lasted even longer, 17 months.

With supporters and detractors left to debate the relative success of the second Falcon 1 flight, SpaceX, declaring the Falcon 1 operational, was moving on. Next up was the NRL launch originally scheduled for Vandenberg, and shortly after that, the company's first purely commercial mission, a remote sensing satellite for Malaysia. Before either could take place however, the company would need to complete the development and qualification for the new Merlin 1C engine upon which everything depended.

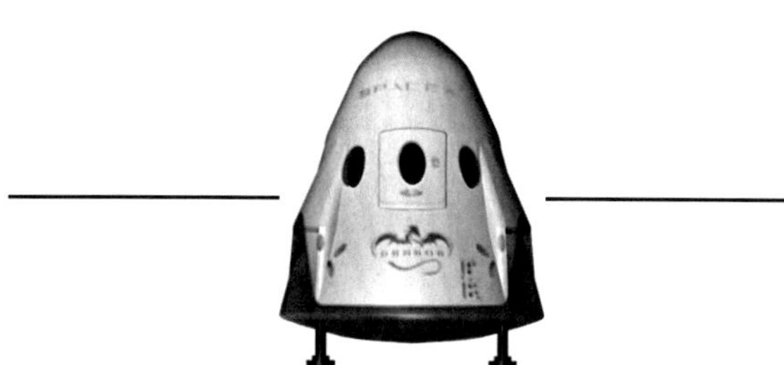

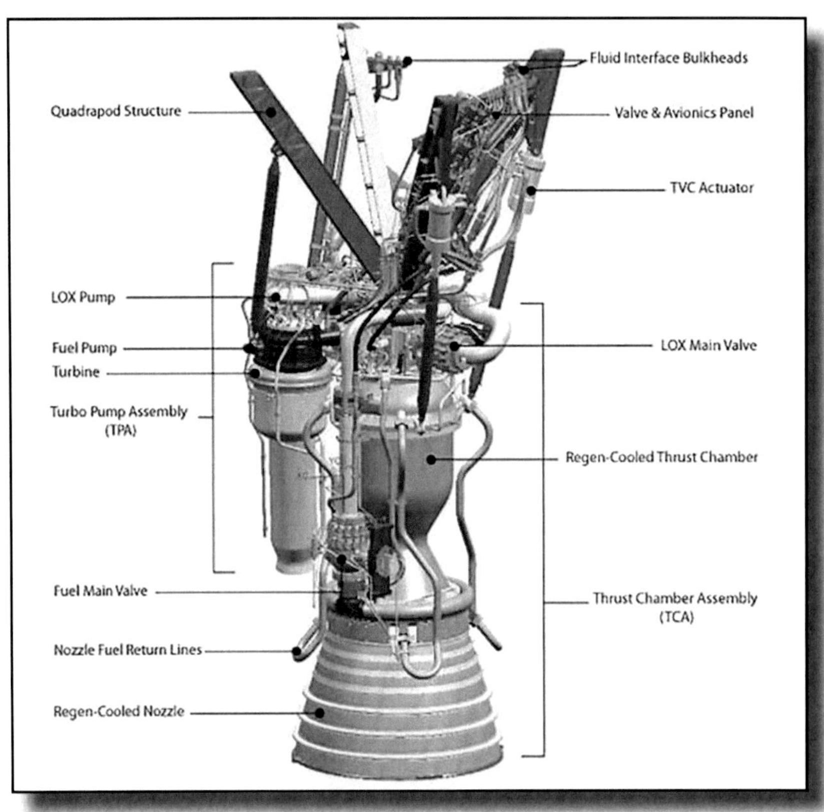

Graphic of Merlin 1-C. Credit SpaceX

Chapter XI: Designing for Manufacturability

Just a day prior to his assassination in Dallas, President Kennedy attended the dedication of the new Aerospace Medical Health Center at Brooks Air Force Base in Houston, TX. Speaking to the assembled crowd in what would be his last formal public address, the man who had set the nation on its first journey to another world channeled his Irish routes in assessing the challenge he set before the country, observing that "this nation has thrown its cap over the wall of space." In establishing a definitive time frame for achieving the lunar landing, "before this decade is out," NASA found itself in a similar circumstance to the suddenly bare-headed youth of the Emerald Isle, once the cap was over the wall, there was little choice but to follow it. Even though plagued by doubts, and concerned over the expense of the Apollo program, and its potential political consequences in the upcoming 1964 election, President Kennedy had in effect thrown his own hat over as well, and could not go back. While Kennedy would not live to see the accomplishment, NASA marshaled the best of American technical talent, and made it over the wall.

Following Apollo, and a retreat from a peak of 4.41% of the total federal budget in 1966, a number it would never remotely approach again, the concept of hard and fast goals steadily slipped away from the American space program. Planetary missions continued to break new ground, but the manned space program, absent a destination driven architecture, began to drift, concentrating on operations in low Earth orbit. With the Vision for Space Exploration introduced in 2004, the White House once again directed the agency's ambitions towards the Moon, this time with the explicit and ultimate goal of reaching Mars as well. With a projected annual budget which never approached the level of commitment seen in the Apollo program, there was little choice but to adopt a very different "go as you pay" approach. By 2007, NASA's plan to return to the Moon, Project Constellation, optimistically described as "Apollo on Steroids" was already exhibiting some of the early signs which would lead to its eventual demise. Without the leadership, the talent, or the budget of the iconic program it sought to surpass, Constellation quickly devolved into yet another in a long line of large NASA projects with slipping timetables and ballooning costs. [57]

At the same time, even as it was struggling with the Falcon 1, SpaceX was at least making progress with its COTS objectives, and by the same time NASA pulled the contract on RpK, SpaceX had completed six milestone objectives, including a Critical Design Review, earning $95 million out of the allotted $278 million total. Where Kistler had failed in meeting financial requirements, SpaceX, personally backed by Elon Musk, and following a completely different approach than RpK, was on much firmer ground, provided the founder still had money, was willing to support his dream, and things went as planned. Even if Elon Musk spent every penny he had on the project, he still had nowhere near the funds to make an investment on par with what had been cumulatively spent on the K-1 throughout its long, strange history. With concurrent projects in Tesla Motors and Solar City all coming from an estimated $200 million gain on the sale of PayPal to EBay, Musk ultimately invested nearly the entire sum into the three projects, with the lion's share going to SpaceX. [58] Quite simply, although his own contribution was a very substantial one, SpaceX needed both positive cash flow from launch contracts, as well as the ongoing stream of COTS milestone payments to bring the Falcon 9 to the launch pad.

Still, NASA's commitment of $278 million dollars was in historical terms a pittance compared to what governments generally spent on developing a new launch vehicle, as evidenced by the fact that the agency was simultaneously burning its way through that amount almost every 90 days on the Ares-1 program. What was more surprising was the fact that unlike Boeing, Lockheed Martin and ATK, each of whom began development and manufacturing efforts with proven systems and infrastructure already in hand, SpaceX was not only starting from a limited experience base with the small and unproven Falcon 1, it was also expected to develop a fully functional, human adaptable, Dragon spacecraft out of those same funds. With roughly a 50/50 resource allocation between Falcon 9 and Dragon development, NASA was paying SpaceX less than $150 million to develop a spacecraft which was functionally similar to the Orion spacecraft the agency was on its way to paying Lockheed Martin $7 billion to develop at the same time.

Elon Musk and SpaceX had fought hard to win this opportunity however, and in signing the Space Act Agreement which made it possible, had thrown their own caps over the wall of space as well. It didn't stop there. Elon Musk wasn't simply promising to develop a launch system for far less than any previous development effort, he was also publicly promising to offer a rocket which would cost only a fraction of what other systems did to both build and launch on a per unit basis. The cost estimates, and specifically the stated prices which accompanied the introduction of each of its three products; Falcon 1, Falcon 5 and Falcon 9, became the equivalent of Apollo's "before the decade is out challenge." Even as time tables slipped, bringing the company under increasing criticism, SpaceX maintained a focus on meeting the price goals which the rest of the industry thought at best were naively optimistic, and more likely evidence of a cynically designed marketing campaign. For SpaceX to be even discussing retail launch prices at all seemed an affront to established tradition, which held that launch costs, like a lady's age, were matters one simply didn't speak of in polite company.

Both before and after the merger which created United Launch Alliance, partners Boeing and Lockheed Martin simply refused to disclose prices, and in some cases went to considerable lengths to keep the numbers concealed. It was a tactic Elon Musk characterized as being somewhat like a Middle Eastern rug bazaar, where the price changed based on the buyer's ability to pay. NASA, for its part, was hardly any better where the Shuttle was concerned, frequently offering confusing or conflicting assessments of its costs. Any accounting of the system depended on one's perspective, and decisions whether to include amortized development costs, return to flight costs after two fatal accidents, marginal costs or overhead covered under other government accounts, all resulted in a system which was so convoluted that it is questionable whether the agency could have ever come up with a solid cost estimate of a single Shuttle flight even if it wanted to. (It did not.)

SpaceX took the opposite approach, stating prices publicly on its website, and even refusing to play the "list price" game beloved of car dealers and retail clothing shops. Prices were fixed, allowing only "modest" discounts for multiple unit orders. The price for the original Falcon 9, equipped with Merlin 1B engines was set at $27 million, less than one fourth the price of competitive vehicles. Elon Musk hadn't merely thrown his hat over the wall, it was already in orbit, and he would have to go there to retrieve it.

While there are theoretically any number of ways to reduce launch costs, on a spectrum of technologically advanced fully reusable systems on one end, to crude but inexpensive Big Dumb Boosters on the other, SpaceX had already chosen many of the key elements of its particular approach when it defined the primary characteristics of the Falcon 1 and then Falcon 5 launch vehicles. In extending those decisions to the Falcon 9, without predicating it on recovering and reusing any elements of the booster, the company was betting that the answers lay as much in manufacturing and operations as they did in design and technology. These would then become the fronts on which SpaceX would conduct its campaign.

The most damning aspect of targeting such as low price to orbit was the fact that there wasn't anything truly unusual about the launch vehicle itself. Rather than pinning hopes on a miracle breakthrough, a radical new building process, or on an irrationally optimistic production and flight forecast, SpaceX was instead counting on applying sound manufacturing principles common to other industries to a moribund aerospace manufacturing sector perennially protected from the pressure of competition by its privileged government relationships and the insular world of cost plus contracting. This, even more than the failure to develop reusable launch technology, was the single greatest reason the United States was no closer to Mars, and the average citizen was no closer to space, than when Explorer 1 first made it into orbit in 1958.

By any standard, expendable launch vehicles are an outrageously wasteful mode of transportation, but even with each carefully crafted system sharing the same ignominious fate, some are more wasteful than others. The difference lies in how costs are managed in five overall areas; the design choices made, the type of contracting body, the method of integration, the manner of construction, and finally the nature of launch operations. For SpaceX, achieving its target meant driving the costs out of all five. Only then, after having established a permanent presence in the launch industry, could it realistically pursue a quest to permanently sever the term "expendable" from "launch vehicle" by pursuing reusability.

In the most elemental sense, efficiency begins before any discussion of the technical details of the launch vehicle, with the nature of the organization performing the work, and the degree of corporate cost load, or management overhead, which is brought to every product it touches. Here at least, SpaceX had a head start, occupying the lowest level of what could be considered a three tiered layer of expense.

The most expensive, was the historical approach and the one being taken concurrently by NASA in the development of the Ares-1 launch vehicle. NASA, as a government agency is generally not in the business of building the space hardware it uses. Instead, depending on the size or complexity of the project, one company serves as the integrating contractor for a project which can involve dozens or even hundreds of sub-contractors. For larger projects with distinct components such as the Saturn V or Space Shuttle systems, there was a different integrating contractor for each major system. The relative success of NASA's job, which was to effectively manage the entire effort, depended in large part on the technical proficiency of teams in charge of oversight.

By the time the Ares-1 was begun, after 25 years of Shuttle operations, more than

35 years since it had last designed and fielded a launch vehicle, time and inactivity had taken its toll, and lacking the technical skill it once had, NASA faltered badly. The effort began with a seemingly simple, yet flawed initial design which soon led to a lengthy and expensive series of changes. On the receiving end of each of the changes were the only modestly more efficient bureaucracies of the aerospace/defense contractors holding the contracts, Boeing, Lockheed Martin and ATK. The inevitable result was layers and layers of management, oversight and the endless documentation and verification procedures involved to make sure everybody's risk was covered as the effort groped towards what ultimately would have probably been a functional, but very expensive launch system.

Provided the funds keep flowing, there is no denying that the cost plus system can work brilliantly as it did with Apollo, or reasonably well as it did with the Shuttle. It can also produce a complete train wreck, but even with the best of outcomes, it is rarely cost efficient, and frequently a formula for financial largesse, which only grows worse when one factors in the time wasted, and labor costs spent as the smallest of decisions are routed up and down the line, and around the nation. In the case of an international development effort such as ISS, there is even another layer of inefficiency as national government agencies must now negotiate decisions around the planet.

Time and again, when major contractors are confronted with glaring examples of outrageous costs, they point the finger back at the government, citing ever changing design requirements and a stop start process driven by funding uncertainties. To some extent, they are quite justified in doing so. As often as not, NASA in turn points back to Congress and to a highly erratic budgeting process which aggravates program planning and leads to frequent and ultimately pointless delays. At every stop along the way, times stretches out and costs go up.

For example, every time NASA changed its requirements for the Orion spacecraft as a result of problems or changes in the concurrent Ares-1 program, Lockheed Martin could legitimately charge an increased cost, and thus an increased total profit as a result, even if the margins remained the same. While the contractor and in particular the employees working on a given project want to see the results of their labor perform, there is no realistic incentive for keeping prices in check, and absolutely no chance of reaching the lowest possible cost solution. The bottom line is that when a program is developed under a cost plus contract, price efficiency is precluded from the outset. Sometimes it is unavoidable, as is most often the case when there is no corresponding commercial market to provide a sanity check. The question being pursued by COTS was whether or not, as of 2006, medium class launch vehicles and automated LEO delivery systems were still one of those circumstances.

A somewhat better outcome can ensue when a traditional aerospace corporation instead acts as its own general contractor on behalf of, but without direct oversight and control from an external agency. This was the case with the Evolved Expendable Launch Vehicle Program, which at least got off to a reasonable start, reducing launch costs by 25% over the heritage systems it was commissioned to replace. In that case, the Air Force initially funded Boeing and Lockheed Martin approximately $500 million apiece to design and build their respective systems, with both contractors required to supply the remainder of the development expense out of internal funds.

By the time of the first launch in 2002, both companies reported that they each made internal contributions several times the investment made by the Air Force. Taking both at their word, the total combined cost of roughly $2 billion per launch vehicle was a relative bargain compared to the billions NASA began plowing into the Ares-1 only a few years later. Unfortunately, due to a combination of factors, the original savings proved illusory, and with the loss of competition resulting from the ULA merger, and steadily increasing annual launch infrastructure subsidies, the EELV program's operational, as opposed to developmental costs, soon ballooned far beyond original predictions.

The third approach and the one yet to be fully proven but perhaps most closely approached by the OSC Pegasus, was that of a private contractor, working on their own design, and predominately at their own expense. In this case, the contractor suffers no lack of incentive in finding ways to reduce costs and increase efficiencies. Failing to do so is the difference between success and failure, profit and loss, and sometimes even bankruptcy. By contrast, the consequences of failure are rarely as severe when the taxpayer is footing the bill.

Even when a company is developing a proprietary design, working off its own capital, including that from creditors, there are still meaningful differences in terms of how costs are built into the system, based on the degree to which it performs work in house, or subcontracts out large segments, and the history of the aerospace industry contains a strong bias towards subcontracting, which can bring in valuable expertise which may be difficult and expensive to develop internally. The counterargument is that it is also a very expensive method. Each new contractor brought in, arrives with its own overhead expense, which is added to the integrating contractor's expense in managing the subcontractor. If you are being paid on a cost plus basis, this is another cost you can pass along. If you are not, it is a way to burn through resources very quickly, as Rocketplane-Kistler discovered with the K-1 launch vehicle.

In choosing an inherently simple format for the Falcon 1, and then in upgrading this design for the Falcon 9, SpaceX charted a conservative course, avoiding the necessity of bringing in external expertise to manage the demands of a technology outside of its core capabilities. Instead, it built an internal knowledge base by steadily expanding its workforce and remaining focused on in-house development of the systems it knew. Learning by on the job training has its own costs, as witnessed by Falcon 1's two failed flights, at least one of which could have likely been avoided with more and different eyes on the system. Still, if you can absorb the costs of your own mistakes, there is no substitute for the experience gained.

Even before the first piece of metal was bent, SpaceX possessed three crucial advantages either neglected or unavailable to other launch vehicle efforts. A concept with direct control over its Design, Development, Testing and Evaluation (DDT&E) effort, and low corporate overhead all set the stage to achieve a desirable outcome with the Falcon 9, but those factors alone were not nearly enough. What remained was the biggest challenge of all, to engineer the rocket for manufacturability. The process began with manufacturing the site itself.

SpaceX began operations in 2002 at its first factory in El Segundo, Ca. It was sufficient for the Falcon 1, but woefully inadequate to handle the Falcon 9, which measuring 157 feet with a Dragon Capsule, was longer than a Boeing 737, absorbing nearly the entire width of the building. Winning COTS meant that SpaceX needed a much larger facility which could handle not only the assembly of Falcon 9 stages, including a production of 10 engines per flight, but also the Dragon spacecraft, all its associated systems, subsystems, testing and a rapidly growing staff. It also needed the space onsite for a full-fledged mission control center which would be called on not only to support launches, but to assert control of the Dragon spacecraft which would be handed off as soon as it entered orbit on its way to approaching and berthing with the International Space Station. None of this would be cheap, and yet it would have to be developed within the funding limit of the COTS program, plus whatever Elon Musk chose to invest of his personal fortune, the company's meager cash flow or from outside investments.

Preferably, a new facility would be as close as possible to the current one, and to the existing aerospace talent which had drawn it to Los Angeles, and to El Segundo in the first place. Just such a facility came on the market in 2005, when Vought Aviation significantly reduced its manufacturing footprint for Boeing 747 fuselage panels at a sprawling factory in Hawthorne, California only a few miles away. As part of the process, Vought sold the property, and then leased a section of it back. The other parcel, now vacated, and located at the opposite end of the Hawthorne Municipal Airport was originally leased to an R.V. company, but came back on the market in 2006. Despite the building's size and history, and the presence of an airport runway just outside the back door, it is almost stereotypically, and uniquely American. Across the street stands a Lowes Home Improvement store, and next to that, a Target, with the ubiquitous Starbucks across the opposite street.

SpaceX occupied the 550,000 square foot facility in fall 2007, and quickly went about the business of establishing a manufacturing operation which combined distinct elements of Silicon Valley software development and modern automotive manufacturing environments into an approach as radically different for building engines, launch vehicles, and spacecraft from the norm as Elon Musk was from the typical corporate CEO. SpaceX was not simply acquiring a larger facility to accommodate the needs of building a physically larger product, just as importantly it needed a space of sufficient size to allow it to fully implement a different approach to the DDT&E process, and then integrate the results into the manufacturing operation. Reflecting Musk's personality and experience, the goal was to develop a type of corporate campus not dissimilar from that which Apple, Google, and Microsoft had established as hallmarks of the computer and software industries.

Operating with a reduced corporate cost load was an important start, but in the absence of reusability, cutting launch costs by more than half could only be achieved on the manufacturing floor. And SpaceX would need a lot of that. Following the template set forth with the Falcon 1, the Falcon 9 was intended to be manufactured in house, and under one roof rather than contracted out and built in pieces from multiple locations. This marked a departure not only from most of the rest of the launch vehicle industry, but even from the aircraft manufacturing industry as well. The Hawthorne facility was built in 1966 specifically to accommodate the manufacture of fuselage sections for the Boeing 747 jumbo jet, and as such needed enormous square footage, and large overhead cranes with unusual lengthy spans and a roof height of 60 feet to handle the fuselage sections. And yet, for all this, and despite their enormous size, this aircraft itself wasn't built on the site, but instead at Boeing's even more massive site in Everett Washington, with the panels being shipped north via oversized railcars, where they would be joined by tail sections built in Texas, rudders built in Florida, doors from Georgia and engines built by three different manufacturers all over the world.

SpaceX by contrast, envisioned a turnkey factory with raw materials coming in one door, and complete launch vehicle components coming out the other. This was in some ways a throwback to an earlier era in which with the exception of the engines, aircraft were all built on site and flown off the adjacent runway to meet waiting customers. Launch vehicles however, are not aircraft. With the total annual worldwide launch rate of commercial satellite missions rarely exceeding 20 per year, and in recent years averaging 14 launches shared between multiple providers, with some launches being dual manifested, the launch vehicle industry never approaches the type of output which lends itself to airline style assembly line manufacturing. Commercial launches are only part of the overall market, but with almost all military and most national science launches restricted to domestic providers, the commercial market is the only segment which is remotely open for international competition. That market however, is comparatively price sensitive. Boeing for instance, built its enormous 1.5 million square foot Decatur Alabama Delta IV manufacturing plant to handle an eventual capacity of as many as 40 units per year, at an estimated cost of $450 million, betting that it would capture a significant share of launch market which was peaking in 1998. Four years later, when the Delta IV made its inaugural launch, the market had crashed, and production costs for the all cryogenic rocket soared. Suddenly uncompetitive, the defense and aerospace giant withdrew the rocket from the commercial market after only one launch. Prior to the merger with the Lockheed Martin Atlas V assembly line in 2007, the factory's annual output never exceeded 6 rockets. Even with joint production, the total averaged 8 per year through 2011. [59]

Although SpaceX came into the market offering substantially lower launch prices, there was no rational expectation that doing so would result in an increase in the overall size of the launch market itself, at least not for a number of years. Consequently the best that could be hoped for was to gain a significant portion of the existing commercial segment, along with a commensurate portion of NASA and possible DOD launches, all of which added up to a manufacturing forecast beginning at less than 4 units per year and increasing to a foreseeable short term peak of 12 cores. Achieving the lowest possible cost at the expected production rate meant building a facility which allowed a direct interchange between engineering and production to develop solutions which took production costs and flow into account from the outset.

Regardless of the specific industry, there is a natural tension between engineering and manufacturing, one which can periodically erupt in outright hostility between the two, and is never easy to manage. Designing the Falcon 9, (as well as Dragon) for manufacturability required blurring the lines and fostering a sense of teamwork between the two historically different cultures, which in many cases are housed in separate buildings, and often located hundreds or even thousands of miles apart.

The answer was an open production facility with both elements under the same roof, and only footsteps away from each other. This extended to the office areas as well, with an absence of walls except for Human Resources and Accounting. Here, Elon Musk set the example, occupying an open cubicle readily accessible to other personnel. The only indulgence was an iconic photograph of a young Muhammad

Ali standing over the vanquished Sonny Liston, seemingly daring the world to take him on. On the expansive factory floor itself, SpaceX established a flexible, work station style layout where virtually everything could be relocated to accommodate changing demands. Linking it all together was a comprehensive off the shelf commercial software package from Siemens, which including NX CAD and an overall document management tool, Teamcenter served to connect everything and reduce the rippling effects of design changes in independent programs.

As the design of the Falcon 9 progressed, close coordination between R&D and production allowed the inevitable changes which occur when theoretical design solutions run into operational assembly techniques to be resolved quickly and without internal restrictions or impediments common to multi-tiered, multi-site, and sometimes multinational operations. Boeing for example, took dispersed manufacturing to a new level in developing its 787 Dreamliner. As compared to the 747, which featured 95% domestic content, the Dreamliner development team farmed out 30% of the new airplane's manufacturing, with costly consequences. Some of the pieces arriving in Everett Washington for finally assembly literally did not fit, requiring extensive modifications and lengthy delays as Boeing sought solutions with vendors ranging from Sweden to China, some of whom were reportedly never provided final drawings. [60]

Separated by a few steps rather than oceans, SpaceX established an open access factory, effectively creating an environment which supported talented personnel in finding new ways to drive the rate of progress with an absolute minimum number of unnecessary expenses or roadblocks. It allowed the SpaceX team to learn how to evolve the design for both reliability and ease of production at the same time with a quick fire decision making process which came as a major surprise to the NASA personnel embedded with the company. Meetings were short and to the point, with a strong emphasis on getting back to the floor and physically working through solutions. Together with Silicon Valley style amenities such as a free cafeteria with dedicated chef, and company provided transportation on buses equipped with high speed internet, the Hawthorne facility provided a unique opportunity to push the envelope of vertical integration.

A factory, no matter how cleverly laid out, is much more than a building and tooling. The most important component is the people it hosts, and as SpaceX began working on the Falcon 9 and the Dragon capsule it would carry, hiring picked up dramatically. As it had from the outset, SpaceX recruited heavily from the surrounding talent pool in the Southern California aerospace environment. Many new employees were "fresh outs" coming straight out of college where SpaceX was beginning to emerge as one of the most desired places in the world for aspiring young engineers to work. Compared to the stiff, buttoned down atmosphere of other companies such as Boeing, Raytheon, Northrup Grumman, and Lockheed Martin, the almost shockingly casual atmosphere and attire, where shorts and tee-shirts are as common as suits and ties, belied an engineering culture which demanded the very best. It was not only a matter of hiring enthusiastic whiz kids who didn't know that what they were doing was supposed to be impossible. Along with younger personnel, the company was also gradually filling its ranks with experienced senior engineers and managers from every major "established" contractor, with histories which went long and deep with Atlas, Delta and Shuttle. The manufacturing philosophy though, came as much from other sources as it did from the aerospace establishment and as a result, it was not always a smooth fit. More, than a few high profile hires departed almost as soon as they arrived. Among them was Jim Maser, hired in as SpaceX President (Elon Musk was CEO and CTO, Chief Technology Officer) away from the same position at Sea Launch. Maser left the company after little more than year to take on the reigns at engine builder Pratt and Whitney Rocketdyne.

Job hopping among corporate executives is not an uncommon occurrence, and while it can be disruptive for a company heavily engaged in technology development, consistent leadership in key engineering positions is even more important, and in this regard SpaceX benefitted from the continued influence and relentless energy of its founder as well as three critical leaders, Tom Mueller, the driving force behind the Merlin engine, Chris Thompson, Vice-President of Product Development and Gwynne Shotwell, Vice-President of Business Development, and the seventh person hired. Thompson, like Mueller had joined the company in 2002, bringing in invaluable experience from Boeing in managing development and production of the Delta family of launch vehicles. After winning some of the company's earliest and most critical sales contracts, Gwynne Shotwell, an engineer who had previously worked at the Aerospace Corporation and space start-up Microcosm, would be promoted to SpaceX President and Chief Operating Officer in 2008, as Elon Musk maintained the title of CEO and Chief Designer. As SpaceX began to enter the public awareness as a result of the COTS award, Shotwell would soon become nearly as well known as Elon Musk as the face of the company, and the one individual most directly responsible for managing the endless demands of a business experiencing rapid expansion in its products, personnel, markets and geographic footprint. The key, as always, was containing costs.

What SpaceX first learned from early efforts at outside supply with the Falcon 1 was that as a smaller company, they were completely at the mercy of vendors who were both slow to produce, and operating under the high cost infrastructure of the conventional aerospace industries, were not particularly responsive to the idea that being asked to provide an estimate meant sharpening your pencil the very first time. As much as anything else, this brought home the real effect of the entrenched costs imbedded in the arsenal system. With no real competitive pressure anywhere within the supply chain to produce a quality product at the lowest possible cost, the results were entirely predictable, a condition Musk assessed as being "wide, but not deep." Even in cases where "off the shelf" products were available, the pricing was still "top shelf," reflecting the degree to which legacy costs were embedded into almost every aspect of the aerospace environment. Although never referenced directly, looming everywhere was an implied perception that for vendors who also serviced much larger ULA, there was interest in remembering who their friends were, a reality which opened the door, whether by design or not, to slow walking SpaceX to death, an unacceptable situation for a company which was in a hurry to begin with.

Cost averse and very wary of being saddled with a sole source vendor in any event, the response was to invest in the tooling necessary to allow Hawthorne personnel to design, test and fabricate for themselves items which any other manufacturer would not think twice about outsourcing. Nowhere was this more evident than in the manufacturing process being devised for the new Merlin 1C engine. In addition to thrust chambers and turbo pumps, rocket engines are nothing if not an assembly of highly complicated metal tubing and hoses, tasked with routing super cold fluids under tremendous pressure and flow into a blistering hot environment, and often in a seemingly Byzantine manner. Even with the aid of the very best in computer assisted design, there is often still no way to really know whether a particular solution will work in all aspects; resistance to fatigue, vibration and acoustics, not to mention ease of assembly, inspection, repair and replacement until they are thoroughly tested. One small change can easily require another, and then another, and then yet another.

The cost of managing design changes in such an environment can be astronomical, unless you develop the capability to make the changes yourself. SpaceX's rapid progress in taking a one-engine solution, upgrading the engine and then developing a thrust structure which accommodated the needs of 9 engines operating in close proximity was made possible by the fact that it invested in the tooling and personnel to produce whatever assembly was required right on the spot, and then integrate the changes into production.

As anyone who has worked on a modern automobile can attest, decisions made to simplify the production process often lead to infuriating and seemingly pointless procedures when it comes to performing basic maintenance. Such tradeoffs are considered acceptable because the greater priority is to increase sales through a lower production cost, and in the case of buying a new car, the deal is long over before the owner discovers that changing the oil filter requires removing half the engine to get to it. Of the numerous and extended delays in the history of the Shuttle program, the length of the delay was often driven by the difficulty in simply reaching a component or even a sensor on a component which was suspect. With extensive documentation and verification required every step of the way, in both the dis-assembly and re-assembly process, delays were fantastically expensive in terms of the man power and labor hours involved.

Reducing the overall cost of building and launching the Falcon 9 meant ensuring the design worked in all aspects and that whether testing in Texas or in the split second before launch in Florida, whenever the inevitable failure occurred, the effort to repair or replace a faulty component could take place with a minimum amount of time and complexity. With 9 engines crowded into a thrust structure under the 12' diameter first stage, this was a challenge which could only be solved in the interchange between design and production, the ultimate result being a rapid turnaround in removing and re-installing a questionable engine.

The outcome was a hands-on process which resembled a NASCAR race shop as much as it did anything else, and that is not an insult. NASCAR operations possess some of the best tooling and equipment in the world, and pride themselves on being able to tear down a car after a race on Sunday, manufacture literarily any part required by the car, and have it ready and installed in time for qualifying the following Friday. It is a can-do, results oriented culture which both depends on and actively fosters a teamwork environment in which the product is paramount, results are instantaneous and the level of overall technical competence is remarkable. Applying the same ethos to the limited production environment of the SpaceX factory as the Falcon 9 production line was beginning to take shape, it enabled the company to establish an iterative, constantly improving product and process, which starting with a brilliantly simple basic design, made it possible to shatter the prevailing cost structure.

The Stage Structure

For those who have had the chance to visit a "rocket garden" like that at the Kennedy Space Center Visitors Complex or the U.S. Space and Rocket Center in Huntsville, Al. it is easily possible to conclude that although the sizes and engines might be different, one rocket body is pretty much like another. To an extent of course they are. Each is charged with containing fuel and oxidizer, and transmitting the force of the engine's thrust up through the vehicle, all the way to the payload. Despite the outward characteristics, the way a tank or stage structure is designed and fabricated, beginning with the specific alloy chosen, can have a substantial impact on overall system cost, and the history of space exploration is partially told in the development of progressively better metals.

Beginning in the 1950's as the demands of the Cold War arms race pushed the U.S. to field ever faster and higher performing jets, the introduction of missiles and reentry vehicles designed to survive and perform at speeds which were barely imaginable twenty years before pushed the boundaries of what was achievable with currently available metals. Consequently, the technology of metallurgy rapidly evolved as the nation pursued a program of developing lightweight and stronger metals, chiefly by refining alloys of aluminum. This trend continued with the advent of the space age, and in more recent years with the introduction of aluminum lithium alloys employed in the third generation, super lightweight Space Shuttle external tank. Compared to the conventional aluminum external tank which first boosted Columbia to orbit in 1981, and the Lightweight Tanks which were employed beginning with STS-6, the aluminum/lithium alloy Super Lightweight Tanks in use at the end of the Shuttle program offered a weight savings of nearly 7,500 lbs. every ounce of which increased the Shuttle's payload capacity to orbit by an equivalent amount and making it possible to service the highly inclined orbit of ISS. [61]

In developing the Falcon 1, SpaceX had employed a conventional aerospace grade aluminum alloy for the first stage tank, but with the unique twist of its flight pressure stabilized design. It had also experimented with aluminum lithium alloy for the second stage fuel tank, using it as a test bed prior to proceeding with the much larger tank which would be required originally for the Falcon 5, now turned Falcon 9. For a time, it appeared the alloy would not be available, as the Shuttle External Tank was the metal's single largest user, and with the last set of tanks built years before the Shuttle's final flight, manufacturer Alcoa indicated it might halt production, or at the very least, was going to substantially raise its pricing. Fortunately however Alcoa had an opportunity to rethink its strategy when it began to face an overall loss in the sale of aircraft grade aluminum as both Boeing and Airbus started incorporating progressively more composite construction into new aircraft projects. Changing course, Alcoa countered by investing heavily in further refinements and production of aluminum lithium alloys as an alternative.

As the lightest of the elemental metals, the introduction of lithium into aluminum provides the benefits of reduced weight, increased stiffness and lowered density. The drawback, which is frequently the case with introduction of new alloys is that they also require the development of new welding techniques, as well as new procedures to verify the weld's integrity. One of the difficulties encountered in conventional welding is the fact that it changes the nature of the material, and can often weaken it, requiring heat treating to restore lost strength. Certain alloys however, do not respond well to the process, requiring a different solution. For aluminum lithium alloys, the answer was friction stir welding, a relatively new technique invented at The Welding Institute in Britain in 1991. FSW is a steady state welding process in which a high speed rotating bit made of conventional hardened steel, backed by considerable pressure, allows the joint to be welded not to a melting point, but to approximately 80% of that level causing the metal to plasticize and to flow together without creating any voids or introducing any impurities. Given that it does not require flame or gases, it is a welding technique which could very conceivably find its way to in-space and planetary applications.

One of the drawbacks of FSW is that whenever the bit is withdrawn, it leaves a potential weak spot in the form of an exit dimple. This problem was overcome by engineers at NASA MSFC who designed a shoulder-less conical tool which allows the bit to be withdrawn without leaving the weld defect. The development of the special bit is a small, but useful example of one side of NASA that the public generally does not see. As opposed to the high profile agency which is frequently publicly skewered over budget busting flagship programs such as the James Webb Space Telescope, or Project Constellation, there is another NASA, one which regularly performs valuable research and technology development which benefits industry as a whole. Despite the rhetoric of Old Space vs. NewSpace which can easily underline a story like that of SpaceX, this was only one of many circumstances in which SpaceX, along with other companies, has benefitted from technical progress made by NASA in solving many of the underlying challenges of the space era. It is a reminder that sometimes the agency is at its best when functioning more like its aviation predecessor the National Advisory Committee on Aeronautics, than the heavy handed bureaucracy which attempted to function as sole source space delivery service in the early days of the Shuttle program.

Besides the decision to use the aluminum-lithium alloy, and friction stir welding for both the first and second stage tanks, SpaceX elected to keep the overall design and construction of the tanks simple, building on the experience gained in the Falcon 1 with the partially pressure stabilized design. The Falcon 9 tank structure consists of three dome sections per tank, upper, lower, and an internal dome acting as a bulkhead to separate the liquid oxygen and kerosene. The tank body is made up of a series of barrel sections which are welded together in a building block fashion. This was an important feature, one which opened the door to producing longer tanks and a more powerful rocket by simply adding in more barrel sections as required without making tooling changes. The original first stage, produced for the first five Falcon 9 flights is approximately 87 ft. long, whereas the second stage tank is made of the same three dome structures, but with fewer barrel sections required to comprise a tank structure, measures a mere 17 ft. in length. Although seemingly obvious, the decision to make the second stage almost identical to the first, with the only real difference being its length, marked a major distinction for the Falcon 9 which sets it apart from almost every other launch vehicle, offering an enormous savings in production and tooling costs. When asked about the benefits of his approach, as often as not, Elon Musk responds with honest credulity that he is at a loss to explain why the practice is not the industry standard.

The stage sections start out as flat rectangular sheets of rolled aluminum stock to which stiffening longitudinal furring strips are added, after which the section is rolled into a barrel shape, with the upper and lower seams welded together to create a single structure. Next, supported by a rounding hoops and a stabilizing frame to keep it from collapsing, each barrel section is carefully maneuvered into place by overhead crane to a rail mounted dolly which conveys it to the heart of the operation, the specially designed circumferential friction stir welding machine. There two sections are welded together, with the process repeated to reach the required length for either the first or second stages, with dome sections added at each end, and one common fuel bulkhead separating the fuel and oxidizer tanks.

Regardless of specific application or industry, one of the principle requirements in any welding operation is that critical welds be inspected for integrity. Ordinarily this is a time consuming process in which other welding jobs are put on hold, with the area cleared out so that the weld in question can be checked via x-ray without unnecessarily irradiating everyone in the surrounding area. With the use of two friction stir welding machines, the first built by SpaceX, the weld is checked via ultrasound transducer which follows the welding head. Afterwards, and once the assembly has moved on to the next piece, the weld is also verified by a conventional die penetrant. The net result is a main tank fabrication process which moves at a steady pace with only modest man power. When operating at full capacity, two people could complete a tank assembly in 17 days.

Between the two stages is the all-important interstage, assigned the challenging task of keeping the separate segments of the booster connected through the highly dynamic flight environment.

The primary challenge is first bearing the forward thrust and vibrations during first stage ascent, and then maintaining integrity for what had to be a perfectly reliable stage separation, one of the most hazard-fraught elements of the launch process as SpaceX had witnessed on Falcon 1 Flight 2. In order to provide the greatest strength at the least overall weight, engineers again built on the heritage of the Falcon 1, designing the Falcon 9 interstage connector as a light-weight carbon fiber outer layer over a series of aluminum honeycomb stiffeners. The inter-stage, like both first and second stages, is a constant 12 feet in diameter, and almost 15 feet in length, with much of the interior volume occupied by the Merlin 1C's oversized niobium vacuum expansion skirt.

The original F9's stage separation system uses a series of 9 collets equally spaced around the circumference of the inter-stage to hold the pieces together, and a pneumatic pusher system consisting of three arms, also equally spaced to provide a firm but gentle push to drive the stages apart. Although simple, the collet and pusher system, held together by pneumatically released nuts, weighs more than the other primary approach to stage separation, a series of explosive bolts which mechanically separates the two halves. The advantage is the fact that it is straightforward, presents a safe working environment and can be tested repeatedly, which was accomplished easily enough by suspending the interstage from the factory ceiling, using counterweights to simulate the flight environment. Explosive bolts on the other hand, like solid rocket motors, can be fired only once, making it impossible to test a flight model. The interstage, like most of the rest of the vehicle is entirely fabricated in house, and each separation is tested repeatedly prior to being used in flight.

The initial design and fabrication of the first tank assembly, serial number 001, took place in 2007 at the original facility in El Segundo. After craning it out of the factory, the next stop on the long road to space was the new structural test stand, recently constructed at the McGregor, Texas proving grounds. If it met its objectives, the tank structure would return home to California, but this time to the new factory in Hawthorne, for further testing. Like other new launch vehicles, the Falcon 9 would have been subjected to extensive testing under any circumstances, but in this case, because it was designed from the ground up to eventually be certified for human flight, SpaceX engineered the structure to meet NASA's published human rating requirements which require a launch vehicle to be able to withstand stresses 40% greater than would ordinarily be encountered in an off nominal launch. Existing EELV boosters by comparison, were designed to the lower threshold of 25% in excess. Designing the Falcon 9 to withstand such loads necessarily meant testing it to the same degree, a regime which began at McGregor with basic leak checks, and proceeded through a series of experiments using hydraulic actuators to test the loads which might be imparted from any given direction.

The purpose of the testing was to answer two vital questions. First, to determine whether or not the first stage met the basic design objectives for structural soundness; i.e. could it withstand the various challenges of the launch environment? Second, in addition to being sound, was the design sufficiently mature that it met both performance and manufacturing objectives? If the answer was yes, then the design could be "frozen," and the factory could be authorized to begin production. Before it could get to that point however, it had to pass through a rigorous and creative series of tests to see if it can withstand individually, the cumulative factors which would be happening simultaneously during launch.

Taken together with the even more exhaustive regimes of engine testing, by the time the first stage section was ready to leave McGregor, SpaceX could be reasonably assured that its new vehicle was almost ready. Once reaching Florida there would be a series of tanking and detanking tests, countdown and static fire tests before the match was ever lit for real. This path, initially established with the first Falcon 9 would then become the procedure for future flights. Tanks and engines, tested together and separately were combined into a system which was fairly straining at the bit by launch day. Before it finally climbed away from the pad in Florida, the Falcon 9 would have "flown" several times, leaving engineers to wonder not about those conditions for which they had tested and were confident, but instead those which simply could not be tested anywhere but in an actual flight.

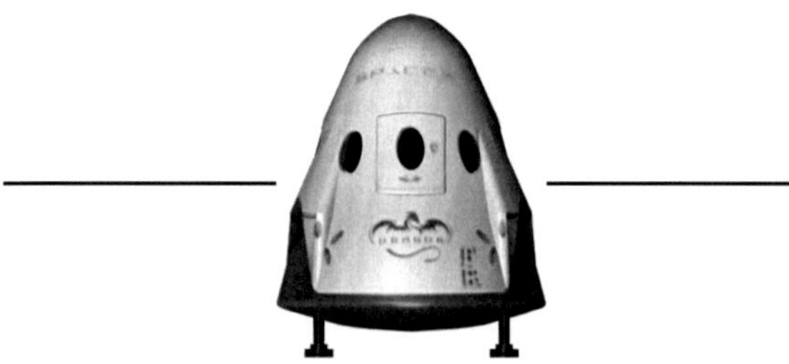

Chapter XII: Orbit at Last

Despite the building sense of anticipation as the Falcon 9 booster began coming together, as well as progress being made on the Dragon spacecraft and its subsystems, coming into 2008, SpaceX faced the more immediate challenge of completing development of the Merlin 1C engine and employing it in a successful launch for the Falcon 1. With the prospects for the Falcon 9 absolutely dependent on the new engine, the next Falcon I flight would serve a dual purpose and hopefully reduce much of the risk for the larger booster.

The Merlin engine however, hadn't been the source of Demo Flight 2's difficulties in reaching orbit, leaving a number of corrective actions to be taken on the Falcon 1 booster itself before the next effort. Shortly after the flight, DARPA once again convened a review board, and as before, returned its findings in very short order. In the first place, the review board concluded just as Elon Musk had asserted, that based on its objectives as a test flight, Demo Flight 2 had been a success, and the fact that it had not achieved orbit could be assessed as failure to achieve a "secondary mission objective." Furthermore, the board concluded that had it not been for the upper stage control issue, the booster would have achieved orbit.

On the other hand, the review also found that there were a total of 8 "anomalies" associated with the flight, only some of which were related to the main issue of interstage contact which led to a loss of second stage propellant reaching the engine. [62] Two of the eight issues were with the quick disconnect (QD) hoses which separate from the booster as it is leaving the pad. On the first stage, both the LOX and fuel QD hoses travelled further than they should have with the rocket before they finally separated at the quick disconnect panel. A similar problem occurred with the second stage LOX QD, but it was potentially more serious. In that case, the QD failed to disconnect at all. Instead the ascending booster tore off the QD panel and several inches of hose. Fortunately a check valve functioned as designed, preventing a potential launch pad disaster.

Another issue took place later in the flight, after stage separation, when one of the two bands holding the two piece payload fairing failed to separate despite telemetry indicating that redundant pyrotechnic bolts had fired. As the other half did operate as intended, the fairing successfully separated from the vehicle, but any anomaly related to fairing separation was a source of concern. Also, for reasons which were somewhat unclear, the second stage Kestrel engine propellant utilization slightly underperformed. Had the stage control issue not arisen, the booster may have reached a slightly lower, but still acceptable orbit as a result.

Regarding the main problem of an accelerated first stage pitch over which set in motion the chain of events leading to second stage fuel starvation, the review board found a secondary cause to be that fact that SpaceX had loaded the wrong flight information file into the computer, causing the engine to underperform ever so slightly, running "lean on lift off and rich at altitude" resulting in slightly diminished thrust. Although minimal, when magnified by gravity loss, it was still enough to cause main engine shut down "lower and slower" than expected. The net result was that the atmosphere was marginally denser, enough so as to impart additional drag which in turn led to the enhanced pitch of the stage. Also, in an unrelated issue, the

1st stage ullage tank pressure dropped near the end of the burn, causing the LOX turbopump to cavitate.

One other item which was of considerable interest but not critical to the flight itself, concerned the effort to recover the first stage, which had failed when the recovery ship proved to be out of position. According to the report, the failure was due to the Army's launch range control failing to take aerodynamic drag into account when it provided an estimated location for the waiting recovery ship. Adding to the problem was a failure of the GPS locator beacon in the stage itself, a circumstance which SpaceX attempted to correct by making the system triple redundant. Each of these issues required corrective action before the Falcon 1 could launch again, but for once there was ample time as the Merlin 1C development program stretched through the summer and fall of 2007. With the three versions of the Merlin 1C destined for use on two boosters, the development plan called for qualifying the lower thrust version for the Falcon 1 first, then progressing to the higher thrust Falcon 9 first stage version, and finally completing the program with the second stage Merlin 1C Vacuum engine. All would have to pass through the three tiered gauntlet of development, qualification and production.

As the testing program unfolded in McGregor, the combination of a more powerful turbo pump and an actively cooled thrust chamber led to impressive results. The new engine generated a sea level thrust of 95,000 lbs., with a corresponding vacuum thrust of 108,000 lb., and a thrust to weight ratio of 92 to 1. Specific impulse was increased to a respectable 304 seconds. The regenerative cooling system cycled 100 lb. of kerosene per second, while absorbing nearly 10 MW of heat.

Over the course of its development, the McGregor team conducted 125 hot fire tests with a cumulative run time of 3,000 seconds. The testing schedule averaged out to nearly 1 test per working day with 21 test firings in the month of August alone. The first engine up however, was the flight engine for the upcoming Falcon 1 flight 3, and hopefully the company's first success. Compared to the Falcon 9 version, the Falcon 1 engine was derated so as not to outstrip the smaller booster's tolerances. Lifting off with sea level thrust of 78,000 lb. (90,000 vac), the Falcon 1 offered the advantage of a very conservative flight test for the new engine.

In February 2008, during the final day of qualification for the Merlin 1C engine designated to debut on the Falcon 1, engineers conducted four full mission length test firings, with a cumulative firing of 27 minutes. With this version of the Merlin 1C now certified to proceed into production in support of future Falcon 1 flights, SpaceX turned to a new regime of tests to progressively increase the engine's capability to achieve the Falcon 9's higher threshold and then begin the challenging task of precisely firing multiple engines at the same time.

In expanding its testing infrastructure to encompass the new nine engine configuration, the decision to acquire its own test facility paid huge dividends, allowing SpaceX to conduct an aggressive, sustained test program in a controlled and private environment which simply would not have been possible had it chosen another route. For the residents of McGregor, the regular pace of firings became a window rattling reminder that somewhere nearby progress was underway in a sense which could be literally felt. Even as the U.S. economy was inexorably slowing down in the beginning of the Great Recession, residents could take some measure of satisfaction in the work being undertaken locally. It was just a taste of what was to come, but it was also a validation of the decision to eschew the NASA Stennis testing center, at least for now. Justifiably proud of its role numerous propulsion development efforts, NASA likes to assert that "all roads to space lead through Stennis." Sitting as it does right off I-10, and just off the route from McGregor to the Cape, where the Merlin engine family was concerned, the agency might be reduced to claiming that all roads to space at least pass by Stennis.

Six thousand miles away from McGregor Texas, in the middle of the Pacific Ocean, it was finally time to puts its efforts to the ultimate test, and the one thing which could not be duplicated at McGregor, an orbital launch attempt. Almost 18 months after Demo Flight 2, and following seemingly endless engine testing in Texas, a third Falcon 1, safely stored aboard its transport cradle and this time shrink wrapped against the elements, was loaded aboard a ship at Long Beach for the long journey to Kwajalein by way of Guam.

Much like the multi-faceted correction taken following the Flight 1 failure, engineers addressed the problem which had doomed Flight 2 from multiple angles as well. In the first place, powered by the Merlin 1C, stage separation would occur higher and faster than on the previous flight, well beyond the point of atmospheric interference. Nevertheless, in order to make certain the problem did not re-occur, programmers adjusted the main engine shutdown to take place over a longer time period. This correction it was hoped, would significantly delay the inevitable rotation and pitch over of the first stage until well after the second stage was safely clear. Just to be sure however, baffles were added to future second stage tanks to contain any extraneous fluid movement regardless of the source. In addition to the new engine and software changes, several other issues identified by the review board resulted in minor changes to the second stage, including the substitution of better aluminum alloys and a slightly reworked and more efficient Kestrel engine. Finally, the lanyard system and first stage insulating blanket was eliminated with a change to foam insulation bonded directly on to the tank.

After characterizing Demoflight 2 as 90% successful, and with the new engine thoroughly tested beyond any reasonable doubt, Flight 3 was designated as the first operational mission, carrying a U.S. Air Force satellite as well as two smaller payloads. During the interval between flights 2 and 3, the military had decided to postpone the launch of what had been intended to be the next payload, TacSat. Now, the sponsor for the primary payload was the Operationally Responsive Space Office (ORS), based out of Kirkland AFB, New Mexico. ORS was a new program designed to test industries' ability to conduct what amounted to a la carte launches of a payload chosen and integrated in minimal time. The intention was to develop the capacity for battlefield commanders to launch virtually on demand a small satellite selected from a number of possible ready-made selections. The program was somewhat similar in nature to the Office of Force Transformation program which has sponsored the first Falcon flight, after that office had been closed down following the departure of Secretary of Defense Don Rumsfeld. The first satellite picked by ORS, one of three possible payloads to be chosen, was the Trailblazer 1, an 83.5 kg spacecraft built out of off the shelf components by SpaceDev, and selected by the ORS office only two weeks before launch. Other payloads included two NASA cubesats as well as a small space burial payload for the commercial company Celestis. One of the NASA cubesats "Nanosail D," was a groundbreaking spacecraft in its own right, designed by NASA Ames to test the deployment of a solar sail measuring 110 square feet.

Following the previously established script, the launch team conducted a wet dress rehearsal, and flight readiness review, after which one small component in the second stage was swapped out "in an excess of caution." Following a successful static fire on June 25th, the first available launch window, lasting four days, was not until August 1st. With no issues to resolve, launch day was set for Saturday, August 2nd during a 5 hour window from 4:00 – 9:00 p.m. PT. Based on the changes made, and coming off the first 9 engine test fire of the Falcon 9 just two days before, spirits were high, and there was every reason to expect a much different outcome than on the previous flight.

Liftoff was initially scheduled for 4:30, and then pushed back to 5:00. Shortly after 4:00, the pad crew departed Omelek for a quick boat ride across the lagoon to the exclusion zone on Meck Island as liquid oxygen loading began. After a hold for collision avoidance with ISS, planned liftoff moved to 5:55. Following another hold at T-16 minutes to allow completion of the helium chilling sequence, managers decided to off-load part of the booster's fuel supply to prevent it from cooling below flight parameters, adding another 45 minutes to the process.

With live launch coverage and commentary by SpaceX coming from the cavernous new headquarters in Hawthorne California and a rowdy crowd of employees, now numbering nearly 700, and their families calling out the final countdown, the launch attempt went all the way to T-zero, only to end in a last second abort triggered at the moment of engine ignition. After troubleshooting the problem, which turned out to be a minor (0.5 psi) variance in the turbopump purge pressure, the launch team reset the sensor and recycled the rocket for another attempt in a window which was now down to about 40 minutes.

This time the countdown and liftoff occurred without a problem, and compared to the two prior flights, the third Falcon seemed to leap off the pad, climbing into a bright blue sky punctuated only by isolated small clouds. After going supersonic and passing though Max Q effortlessly, the booster continued to accelerate smoothly, with the roll control actuator making regular adjustments in a field of view which still centered on the rapidly shrinking tropical island below.

As it passed T plus 1 minute 40 seconds, with the vehicle now heading down range at an altitude of 35 kilometers and velocity of 1,050 meters per second, the new and improved Falcon approached its own moment of truth, a main engine shutdown and stage separation taking place higher and faster than it ever had before. For a moment, everything seemed to go as planned. The first stage shut down right on schedule, followed immediately by clean, decisive stage separation. Suddenly however, the now released and supposedly inert first stage *accelerated,* striking the second stage which had successfully ignited and reached chamber pressure, but before the thrust could drive it out of harm's way. The blow fatally damaged the Kestrel engine, separating the thrust chamber from body and sending the stage spinning out of control and into a steep plunge into the Pacific. With the video feed dark and the crowd fallen silent, it was once again time to pick up the pieces and find out what went wrong.

Unlike Demoflight 2, where it took some time to determine that the booster had not made it into orbit after all, a window which allowed a bit of hope to remain, this time there could be no doubt. In the blink of an eye, all of the elation and sense of progress in winning the COTS award and testing the Falcon 9 only days before simply vanished. Instead of the triumph which seemed so close at hand on the day the first privately developed liquid fueled rocket reached orbit, events had taken a much more ominous turn, and with it, the narrative accompanying SpaceX, three flights and three failures. As Musk had presciently asserted prior to the decision to move to Kwajalein for the first launch, it suddenly looked like the critics were right, and his company did not know what they were doing after all.

Fortunately for everyone involved, particularly for Elon Musk, who was now out of money and literally in debt to friends, after a long night poring through data, by the time the sun rose over Kwajalein the next morning, the cause of the failure was clear, and a plan was already coming together. The reason for the failure was almost immediately obvious, and as Elon Musk described in a posting only days later "four methods of analysis – vehicle inertial measurement, chamber pressure, onboard video and a simple physics free body calculation – all give the same answer." The loss could have only been caused by one thing. It was the main engine shut down transient, the same flight elements which had afflicted the preceding flight. This time however, the problem was different, and was caused by the fact that compared to the Merlin 1A ablatively cooled engine, the new actively cooled 1C engine had a small amount of fuel remaining in the cooling jacket at the time of shutdown. Combined with residual oxygen, it was enough to overcome the separation motion imparted by the pneumatic thrusters, allowing contact between the stages in the near frictionless realm of the upper atmosphere.

The really frustrating fact was that SpaceX engineers were aware of this possibility and had even taken it into account during engine testing in Texas, which had seemingly confirmed a safe shutdown. If the devil is in the details, then the relevant detail here was that the test stand in Texas operated at near sea level atmospheric pressure of 14.5 lb. per square inch, whereas actual stage separation took place in a near vacuum. After shutdown, the remaining fuel vapor was still at 10.5 lbs. too low to register on the stand, but still several times the local atmospheric conditions. The result was continued thrust which sent the first stage crashing into the second stage and spinning out of control. Bump drafting works great at 180 MPH in Daytona and Talladega, but not so much at 4,000 mph and 60,000 ft.

Ironically, if SpaceX had conducted the third flight with another Merlin 1A engine shortly after the second failure investigation concluded, rather than delaying it to incorporate the Merlin 1C, the result would have almost certainly been a success. On the other hand, if it had followed that course, the accident, which Elon Musk's personal assistant bluntly summarized as "You mean we rear ended ourselves?" may have only delayed the accident to the next flight, resulting in the loss of a scheduled commercial payload, or worse yet, occurred on the first launch attempt of the Falcon 9.

In the moments following the loss of the video feed at SpaceX headquarters in Hawthorne, a pall descended over the assembled workforce. For the very first time, employees who had worked tirelessly over the previous months and years, began to sense that perhaps their moment had passed, and the wall really was too high. As they waited for Elon Musk to emerge from the Falcon 1 control trailer parked at the back of the cavernous building, everything hung in the balance, and on the next words to come from the customarily relentless leader.

Addressing his employees, a visibly shaken and bone tired Musk sought to gather himself and gain control of the moment. His response, smoothed over and echoed in an official statement subsequently released was straight forward.

"The most important message I'd like to send right now is that SpaceX will not skip a beat in execution going forward. We have flight four of Falcon 1 almost ready for flight and flight five right behind that. I have also given the go ahead to begin fabrication of flight six. Falcon 9 development will also continue unabated, taking into account the lessons learned with Falcon 1. We have made great progress this past week with the successful nine engine firing.

As a precautionary measure to guard against the possibility of flight 3 not reaching orbit, SpaceX recently accepted a significant investment. Combined with our existing cash reserves, that ensures we will have more than sufficient funding on hand to continue launching Falcon 1 and develop Falcon 9 and Dragon. There should be absolutely zero question that SpaceX will prevail in reaching orbit and demonstrating reliable space transport. For my part, I will never give up and I mean never."

It was Musk's equivalent of Winston Churchill's "We shall fight on the beaches" speech following Britain's military disasters in 1940, and for a beleaguered SpaceX workforce, it had a similar effect. Energy, optimism, and perhaps most important of all, confidence, flowed through the crowd.

The immediate disappointment swept aside, a chagrined SpaceX team of analysts quickly recognized the problem and sought to resolve it by increasing the stage separation time from 1.5 to 5 seconds, allowing ample margin for the additional energy to be expended. Confident that the rest of the vehicle was sound, four days after the Flight Three failure, Musk made good on his word and announced that his company, which had averaged more than a year between flight attempts, would not wait for one of two formal reviews which would soon be getting underway, but instead would try again quickly, possibly within a matter of weeks, this time completely at its own expense, loaded with only a dummy payload.

Barely a month later, another Falcon 1 was on its way across the Pacific, shipped in its entirety directly to Kwajalein Atoll from LAX, along with the launch team, in the belly of a specially chartered USAF C-17 cargo plane departing on September 8th. After a brief transfer across the lagoon from the Kwajalein landing strip to Omelek, the latest Falcon was ready for a test firing on September 20. Following a successful test which resulted in a precautionary replacement of a single second stage LOX component, SpaceX set the next orbital attempt for Sunday, September 28th at 4:15 Pacific Time.

Leading up to its fourth launch attempt, SpaceX had suffered three consecutive launch failures, providing ample ammunition to critics who ridiculed the company, the founder and the rocket. Nevertheless, only one failure, the first, had been due to an outright failure of the hardware itself, the second and particularly the third losses were procedural. Perhaps more importantly, the Merlin engine had proven itself in each circumstance. Even on the initial launch, failure occurred due to a corroded nut, the engine itself had functioned flawlessly until it was shut down by the loss of pneumatic pressure, and the flight control system had gone on working even as it was engulfed in flames. Furthermore on each of the next two launches, the first stage had done its job of delivering the entire vehicle to the point of stage separation. Consequently, despite the 0-3 record, or perhaps because of the data gained in the course of it, SpaceX had reason to be optimistic that the fourth attempt would see its goal attained at last. What it did not have was a commercial payload of any type, this time the launch expense would be covered entirely by SpaceX. With a number of potential contracts waiting on a successful attempt, and all eyes focused on the COTS contract, SpaceX had little choice but to proceed.

Following each of the first three launch attempts, Elon Musk had confidently assured SpaceX team members that in the event of a bad day, he would proceed undeterred. And just to be sure, he had sought outside investment from the Founders Fund, a venture capital investment fund run by Elon Musk's PayPal co-founder and close friend Peter Thiel. Although Musk was personally out of money, SpaceX still maintained cash reserves, but the infusion of $20 million provided a much needed safety net. Despite this, the fourth launch clearly marked what he had originally outlined as the outmost limits of his commitment. Years later he would confide in an interview on 60 Minutes that despite outward assurances, this fourth launch attempt would be the last. Elon Musk was betting his company, his reputation and what little remained of his fortune on one last gamble. Luck had so far been an elusive commodity at the beautiful Pacific island setting, but perhaps having paid his dues, things would turn out differently this time.

On September 28th, 2008 the launch control team was ready to find out for sure. With a sharp breeze snapping the small American flag at the top of the launch tower and venting gases strung out like a rope from the Falcon 1, the crisp white rocket in contrast to the bright blue tropical sky, it hardly seemed the setting for one of the most significant moments of the space age, but back at the new headquarters in Hawthorne, the tension was obvious. As the countdown proceeded, this time without interruption, Elon Musk, flanked by propulsion chief Tom Mueller, gazed intently into the monitor relaying the images from Omelek.

The launch countdown proceeded smoothly, and at 4:15 pm PST, the Merlin roared to life, swiftly carrying the Falcon from the pad. Once again Merlin and Falcon 1 went supersonic and then passed safely through Max-Q. Up next was the Achilles heel so far, a flawless stage separation and ignition occurring at T plus 2 minutes 37 seconds with main engine cutoff. Five seconds later, the stages separated and the pressure fed Kestrel engine flared to life, followed almost immediately by the separation of the two part payload fairing, which this time only protected a 360 lb. aluminum payload simulator. Free of unnecessary mass and steadily accelerating through the near vacuum of the upper atmosphere, the Kestrel engine, exhaust nozzle glowing cherry red, passed through its previous boundary and kept right on firing until the moment of second stage cut-off at T plus 9 minutes 30 seconds. SpaceX had just made history with the flight of the first privately developed liquid fueled rocket to achieve orbit.

As cheers of joy and no doubt considerable relief erupted from employees gathered to watch the launch from a large screen inside the factory, and well as from space enthusiasts watching via internet feed, the Falcon soon propelled itself out of range of tracking at the Reagan Test site, on a nearly equatorial 9 degree trajectory which would bring it back over the launch site 90 joyous minutes later. There was still work to be done however, and the little pressure fed Kestrel engine had one more critical milestone to achieve.

Reaching any orbit was a remarkable accomplishment, particularly considering the trouble it had taken to get to that point, but it was equally important to demonstrate the Falcon's commercial viability by reaching the precise orbit intended. For once the good news kept on coming. The Falcon 1 had hit the bull's-eye with an initial orbit of 330.5 kilometers and an inclination of 8.99 degrees. It was remarkable targeting, but it was far from the only flight objective. The Falcon also needed to demonstrate the ability to conduct an in-space ignition and secondary burn to achieve a payload specific orbit. As the second stage completed its initial trip around the world and came back within range of the command center on Kwajalein, the order was initiated to begin the second stage burn. With the stage already in a safe circular orbit however, SpaceX needed to make certain that the second firing did not inadvertently create a large elliptical path which would carry it within collision distance of other space assets. It was a problem easily avoided by rotating the upper stage perpendicular to its orbital path via cold gas thrusters and essentially "wasting" some of the thrust on the most energy intensive aspect of orbital maneuvering, the plane change. The final result was an orbit 621 x 643 km at a new inclination of 9.3 degrees.

It had been a great day for SpaceX, one which earned it a permanent place in history regardless of anything that followed, but what came next was already on everyone's mind. As spokesperson Diane Murphy succinctly put it at the end of the webcast, "Now our attention turns to Cape Canaveral."

Epilogue: The Falcon 1

Though the rocket was small and the payload inert, it is difficult to overstate the significance of what SpaceX had achieved with its first successful flight. The entire aerospace establishment of the United States, the country which had put a man on the Moon, as well as the establishments of other space faring nations, were heavily invested in space launch systems developed in the traditional agency/contractor relationships which had served the very unique needs of the Cold War and the space race. Yet, this method, highlighted by cost plus contracts which featured little real incentives or initiatives to seek the lowest possible cost at any stage in the process had produced an environment in which it was impossible to substantially expand the basic level of space commerce and infrastructure. In going to the Moon and launching probes to Mars, NASA had shown the world a glimpse of what the future could hold, but then agonizingly kept that same future at bay through its heavy handed and failed effort to commercialize the Shuttle launch system.

For that portion of the population which held on to an interest in space after the heyday of Apollo, as well as scientists, researchers and people pursuing potential avenues of space commerce, the issue of launch costs and lack of progress had been infuriating, and to some a matter of betrayal. In the course of one small launch, SpaceX undercut many of the key tenets of the aerospace establishment. The first was that of complexity. From the bottom of the Merlin bell housing to the top of the nosecone, the Falcon 1 rocket was both an elegant solution and a simple machine, intelligently designed. And yet, simplicity did not necessarily exclude technological advance. To take just one example, by using an on board Ethernet, the Falcon 1 substituted a more modern, and more advanced solution in the place of a traditional wiring harness.

Flying in the face of the rest of the defense establishment, manufacturing and assembly was accomplished primarily in house with the absolute minimum number of outside vendors. Whereas both NASA and legacy contractors such as ULA routinely issue PowerPoint images which actually brag about the far flung nature of its subcontractor base and the number of different states producing components of its launch vehicles and spacecraft, SpaceX was conceived as vertically integrated from the outset for the simple reason that it had to be in order to control costs.

In terms of its immediate impact on the commercial launch market, the first launch was but a pinprick, but in simply achieving orbit as a privately developed liquid fueled launch vehicle, designed, built, tested and flown with new components, the Falcon 1 achieved a stunning accomplishment. It was in effect, a well-aimed shot that proved once and for all that the monster was not invincible, that it could be wounded, and it showed the way.

With an indisputably successful launch finally under its belt, the next challenge for the Falcon 1 was to repeat the process, but this time with both an actual satellite and a paying customer. SpaceX's originally scheduled payload, the Naval Research Laboratory's TacSat-1 had been cancelled following three years of delays, but its first commercial customer, the Malaysian government had been patiently waiting through the Falcon's growing pains, and was still looking for its ride to space. Consequently the next and what turned out to be the final launch of the Falcon 1 took place the following year. The customer was Malaysia, and its payload, a small remote imaging satellite called RazakSat 1.

Malaysia lies in an equatorial zone which is blanketed by heavy cloud cover up to 50% of the time. As a result, remote sensing via typically polar orbiting satellites which pass over the small country is frequently impeded. RazakSat, being launched in an equatorial orbit would pass over its home nation on a much more regular schedule offering an otherwise unavailable opportunity for its Medium Aperture Camera to take high quality images for both governmental and private customers.

This time, with a success to its credit and no immediate rush as it waited completion of the payload, SpaceX was able to save the outrageous expense of a chartered C-17, and ship the first stage to the island complex by water, with the second stage arriving by Air Transport International's "Combi," a Douglas DC-8 jet equipped to haul freight in the rear with 32 passengers in a conventional cabin section to the front, the regularly scheduled means of supply between Hickam Field in Hawaii and the remote Pacific range.

Unlike the drama of its first four launch campaigns, the only noteworthy incident in the pre-flight scheduled was the removal of two small cubesats from the payload ring adapter due to a potential communications conflict. Reflecting experience earned in the first four launch campaigns, Falcon 1, Flight 5 lifted off from Kwajalein (Omelek) without incident on July 14, 2009 placing the 180 kg RazakSat into a nearly equatorial orbit similar to that achieved by the test payload on flight four.

As it turned out however, a combination of bad planning, bad design and bad luck resulted in a short, unhappy life for the satellite. In the first place, although SpaceX placed the spacecraft precisely in the intended orbit, there were two problems with the orbit itself. In the first place, designers had not taken into account the simple fact that even though the spacecraft would cross over the nation several times a day as intended, the crossing would frequently occur either at night, or when the sun was at a bad angle, rendering many photographs useless. Second, the orbit carried the satellite squarely through the South Atlantic anomaly, a region of high radiation in space, located as the name implies over the southern Atlantic Ocean. Unless heavily hardened against radiation, which this satellite was not, the radiation wreaks havoc on multiple components severely shortening a satellite's lifetime. As if that were not enough, there was trouble with the spacecraft itself. A problem with the pointing system resulted in an error margin of nearly thirty miles, meaning the satellite consistently missed the target of whatever it was supposed to be imaging. Though the launch was a proud moment for Malaysia, the satellite was essentially useless, and was turned off a year later, never having produced a usable image. For SpaceX however, the RazakSat launch was a culmination of six years hard work, and brought an end to the frustrating drama which had played out on the tiny Pacific Atoll so far from the cool hills of coastal California which were to have been the setting for the inaugural launch campaign. The next SpaceX attempt would take place from neither, but instead from the epicenter of American space history, Cape Canaveral. And following the trials on Kwajalein, the striking success would take everyone by surprise.

As for the Falcon 1 itself, the diminutive booster whose effect is still being felt today, Flight 5 was regrettably the end of the line. Despite the almost herculean effort required to gain the first successful launch of the Falcon 1, and more than a few statements from Elon Musk that booster would remain a vital part of the SpaceX family of launchers, the history making rocket would in fact never fly again, and along with its intended second generation successor, the Falcon 1E, an extended version meant to take full advantage of the Merlin 1C's greater performance, the little rocket soon disappeared from the SpaceX manifest altogether. Over the course of the long development campaign, several customers found other solutions. The first intended payload, the Naval Research Laboratory TacSat 1, delayed when the SpaceX moved from Vandenberg to Kwajalein, was ultimately canceled when its successor, TacSat 2 flew first. Furthermore, the Operationally Responsive Space program, which offered the potential for multiple launches, was never popular within certain Air Force circles, and had been put on programmatic life support as SpaceX struggled to qualify its booster. Significantly, several early customers, including Canada's MDA Corporation as well as small sat fleet operator Orbcomm, indicated they were amenable to having their satellites launched on the new Falcon 9 instead, effectively sealing the fate of the historic booster. [63]

For anyone who was listening, Elon Musk had made it clear from the outset his purpose in founding SpaceX was transformative, extending far beyond the small launch vehicle which marked the company's initial entry into the field. It was hardly surprising then that the company had already turned its attention to a more capable rocket before it had even achieved its first success. Contrary to earlier analysis which indicated a small but reliable market for small launch vehicles such as the Falcon 1 and its main competitor, the OSC Pegasus, launch opportunities began to evaporate, in part because of cutbacks of small NASA small science missions which had previously been a staple of the industry. Before long, the increasing costs of maintaining such a limited production line drove the price of a single Pegasus mission to nearly $40 million, creating the distinct likelihood that where two younger space companies might have once engaged in an epic battle for bragging rights in the small launch vehicle business, both would be permanently retired.

The experience in designing and building the Falcon 1, and particularly the trials on Kwajalein as the company prepared to fly the vehicle, marked a sort of childhood for SpaceX, in which the first steps towards reaching space occurred in the relative protection of a government sponsored program in which failure, if not acceptable, was at least not fatal either. To the credit of Elon Musk and the development team, which took an enormous gamble, the company boldly responded to the Falcon 1 flight three failure by scheduling another flight as soon as possible, and as a result passed its first major flight threshold, a completely successful orbital launch. It was a major step on the path to maturity as a launch vehicle provider, but it was only a single step, and the next one, already well underway, would place the young company squarely in the middle of one of the most confused, turbulent and dysfunctional eras in the 50 year history of the American space program.

Even though the Falcon 1 would not go on to ignite a revolution, or even make a ripple in the small satellite market, its considerable legacy lies in the context of the learning curve it offered SpaceX as well the actual hardware and software contributions which were directly applicable for the Falcon 9 already underway. As Elon Musk would come to point out repeatedly, it was a hell of a lot smarter to make mistakes and go through a learning curve on a $6 million dollar rocket than one costing ten times as much. Arguably, the greatest contribution, and the real genius behind the SpaceX approach, was the choice to base its future plans on the upgraded version of the Merlin engine. In that regard, the patience exhibited in the long delay between flights two and three while the 1C was undergoing testing would yield valuable dividends in the future.

Legacy of the Falcon 1

SpaceX achieved orbit with the Falcon 1 on the strength of its vision, the practical engineering of its rocket and the money of its founder. And though the armed forces of the United States, including the Army, Navy and Air Force and the Office of the Secretary of Defense all provided valuable assistance in different ways, it was an accomplishment primarily won on the undeniable strength of its own effort. The path beyond September 28, 2008 was made possible by a unique confluence of events and decisions in which SpaceX was as much the beneficiary as the cause of action. Unlike the failed efforts of the previous decade, when the commercial satellite market collapsed and killed several new launch vehicle efforts almost overnight, for once a company was quite simply, in the right place at the right time.

We cannot see past the doorways not entered, but for all its promise, it is difficult to see how SpaceX might have achieved its ultimate success if not for the fact that it was already well on the way to developing the Falcon 1 when the bizarre and highly improbable chain of events which led NASA, nearing its 50th anniversary, to simultaneously have no independent access to its orbiting space complex which had been the focus of its efforts for the past 15 years, and yet a nearly perfect script for integrating a private solution which it otherwise would have already abandoned.

To that extent, the diminutive rocket played a critical role in history, putting SpaceX in a position to answer the call when the COTS competition finally emerged after years of speculation and aborted attempts. Although early progress in the Falcon 1 development cannot take credit for opening NASA to the possibility of COTS, it certainly made it possible for the agency to get the program off to a solid start, in spite of damage inflicted by the financial demise of fellow competitor RpK. History might have taken a different and all too familiar course were it not for the presence of SpaceX and the Falcon 1 as a counterbalance.

COTS 2/3 Falcon 9 first stage in the hangar awaiting integration with the second stage which has just arrived in its shipping container. Credit: NASA

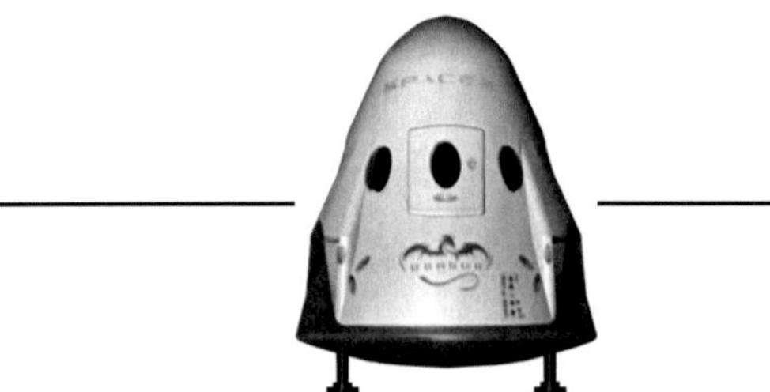

Chapter XIII: Establishing Operations at Cape Canaveral

As it turned its attention to Cape Canaveral following the first successful flight of the Falcon 1 on September 28, 2008, SpaceX was not only shifting venues for the introduction of its flagship Falcon 9 booster, it was also preparing the stage for what would be a major component in its effort to reduce launch costs. Cape Canaveral would be the setting for both the initial NASA missions as part of the COTS program, and also for an unprecedented effort to simplify and streamline the launch process for what soon became a steadily growing manifest of commercial launch agreements signed by companies who had been waiting in the wings to see evidence of a successful flight.

In much the same way that designing the Falcon 9 for low cost manufacturing was a critical part of the challenge in shattering prevailing launch costs, it was equally important to develop an infrastructure and methodology for what happened after the booster's components rolled out of the Hawthorne factory on their way to the launch facility. Despite the general perception of the unique complexity and inevitable high cost of performing real life "rocket science," many of the costs in conducting an orbital launch are no different from that incurred in any other business; in personnel and man-hours used in producing the product or service. Historically, the time spent on and around the launch pad in integrating and testing a rocket has been a significant source of the industry's expense, and in the effort to drive as much of the cost out of the overall system as possible, it was an area in which some of the largest gains could be made at the lowest expense and risk. For SpaceX, with both governmental and commercial ambitions, the primary considerations were devising a process which was both practical and affordable, yet was still flexible enough to accommodate an accelerating schedule as well as allowing for potential upgrades to the vehicle itself. The goals were achievable, but accomplishing them on what was a historically meager budget, meant standing tradition on its end, literally.

Building on its experience with horizontal integration for the Falcon 1, SpaceX applied the same method for the much larger Falcon 9. In doing so, it broke tradition with almost all of the previous large launch vehicles which had thundered off the Cape over the years, a change both deceptively simple and yet profound. The most obvious example of established technique was the Saturn family, and in particular the Saturn V, which necessitated the construction of the gargantuan Vertical Assembly Building at Cape Canaveral in order to provide a space in which the rocket could be pieced together. Designed to simultaneously accommodate up to four Saturn V's in various stages of construction, the result was a truly huge structure which dominates the landscape for miles, and is even instantly recognizable from the window seat of commercial airliners plying the busy East Coast flight corridor.

The Saturn V was a three stage vehicle intended for a unique mission, one that was certainly never designed nor promoted as offering rapid assembly and low launch costs. That was precisely the claim however, for the vehicle which would inherit much of the Saturn's ground infrastructure, the Space Shuttle. The new combination of the Shuttle Orbiter, External tank and Solid Rocket Boosters would be integrated within the same VAB, but in a very different technique which NASA at one time claimed could ultimately achieve a flight rate of once every two weeks. Instead the Space Shuttle became a sort of poster child for just the opposite, a prohibitively expensive launch operations procedure requiring massive infrastructure, a standing army of over 17,000 contractors and civil servants and a history of expensive launch delays. [64] Much of the expense was driven by the sheer complexity of the system itself and what was still a pioneering effort at re-using both the spacecraft and first stage main engines. Launch processing, with the necessity of carefully mating the orbiter to the tank, the inherent danger in handling solid rocket boosters, a long, slow trip to the pad aboard the massive Apollo era crawler transporter, and finally the frequently troublesome task of loading liquid hydrogen and managing weather delays resulted in a system which was anything but rapid.

Although the Falcon 9 was in fact the next American vehicle after the Saturn and Shuttle to be both designed from the outset to carry astronauts, and to actually see service, it is not possible to draw a fair comparison between either in terms of the concept of launch operations. A useful contrast can be drawn however, between the Falcon 9 and the Atlas-V booster which began its operational life with an August 2002 mission out of Cape Canaveral, followed by a 2005 West Coast debut out of Vandenberg. Although they share in common an expendable, unmanned two stage architecture, including kerosene/oxygen first stages of roughly similar dimensions, the different approach taken by veteran Lockheed Martin only a few years earlier is instructive.

At the time it began developing the Atlas V launch pad at the Cape Canaveral as a result of winning a share of the EELV program's initial launch orders, Lockheed Martin was already launching the balloon skinned Atlas IIAS out of Cape Canaveral and Vandenberg, using a conventional vertical assembly technique relatively unchanged since the earliest days of the Atlas booster's original flights. Even though the new Atlas V employed an entirely different first stage tank assembly which made it considerably more rugged for ground handling operations, Lockheed Martin, working with the Air Force, followed precedent and built a new vertical assembly building and mobile launch tower which it financed with Space Florida for $294 million dollars, an amount which exceeded NASA's entire original COTS award to SpaceX. [65]

Although Lockheed Martin's original EELV award also included two West coast launches, with the presumption of more orders to come in the future, the defense and aerospace giant initially declined the award because it did not want to bear the cost of building a new pad at Vandenberg, California. Rebuffed, the Air Force re-assigned the launches to then bitter rival Boeing and its equally new Delta IV. Lockheed Martin only relented several years later, when it received a number of launch orders, a portion of which were on the West coast, rescinded from Boeing as part of that company's penance for the alleged possession of nearly 25,000 pages of illegally obtained proprietary data regarding the Atlas V. Even at that point, Lockheed Martin resisted until the Air Force agreed to provide a low interest loan of $225 million dollars to underwrite the expense of converting the Vandenberg site which was adjacent to that occupied by none other than SpaceX with its Falcon 1, a change which led to the move to Kwajalein. Having successfully evicted the smaller company from the area, Lockheed Martin substantially modified the pre-existing vertical assembly building, but anticipating a lower flight rate, elected to construct a retractable building which moves away from the rocket once it is prepared for flight.

By comparison, SpaceX, which also intended on returning to Vandenberg as soon as possible after first proving the Falcon 9 at Cape Canaveral, received an initial infusion of only $278 out of an ultimate total of $396 million from NASA's COTS program, was somehow supposed to not only pioneer the development of its first ever EELV class launch vehicle, as well as the recoverable and reusable Dragon spacecraft, but also establish from essentially nothing the requisite launch infrastructure which had cost Lockheed Martin an equivalent amount for just the launch facilities alone. There was no way SpaceX could afford to duplicate the expensive infrastructure installed by Lockheed Martin for a similarly sized vehicle, nor did it have any desire to.

SpaceX had already taken a new direction in designing the Falcon 1, originally envisioned as a "ship and shoot" system, in which the rocket left the El Segundo, Ca. headquarters for a drive by tractor-trailer to the launch site at Vandenberg. This process never actually took place due to the need to divert the initial launches to Kwajalein. Having established its spartan facilities at the remote island location, SpaceX originally contemplated using it to conduct the original Falcon 9 flights as well. [66] With clear launch access to all orbits and a nearly equatorial location, Kwajalein offered unrivalled potential for an American launch vehicle in terms of orbital mechanics for spacecraft headed towards equatorial geostationary orbits, which was a market SpaceX was targeting heavily with the Falcon 9. It was not necessarily a superior location than Florida's Kennedy Space Center for launching to the highly inclined orbit of the International Space Station however, and as the custodian of the COTS and CRS programs, NASA clearly preferred that the Dragon, laden with supplies for ISS, and possibly one day astronauts, launch from the vicinity of one of its own facilities. With less efficient but still reasonable trajectories to GSO, world class payload handling facilities and a local workforce facing steep cutbacks as the Shuttle program was coming to an end, the area around the Kennedy Space Center was an obvious choice due to its dual use potential for commercial launches. Fellow COTS competitor Orbital Sciences Corporation, on the other hand, lacking a clear commercial market for its smaller Antares booster, and headquartered in Virginia, eventually elected to stay closer to home in preparing to launch its Cygnus spacecraft to ISS, and instead base its operations out of the Mid-America Regional Spaceport at NASA's Wallops Flight Facility on the Virginia/Maryland coast.

Stretching back to a frenetic program of missile testing which began shortly after the end of World War II, the U.S. government made extensive use of the unique geographical feature which was Cape Canaveral, beginning with the first missile launches in 1950. Over the next 15 years, as the nation plunged headlong into the Cold War, nearly every missile under development for the vast U.S. arsenal underwent testing from one of more than 20 launch pads strung out along the barrier island occupied by Cape Canaveral Air Force Station. Later, as the space race emerged as the high frontier of the Cold War, a number of the Air Force pads were converted to serve NASA's needs for launches both manned and unmanned, including all of the Mercury and Gemini flights, and the first manned flight of the Apollo program, Apollo 7, which lifted off aboard a Saturn 1B from Pad 34 on October 11, 1968. This would prove to be the last manned launch from the Air Force station for more than 40 years. Today, a number of those same pads which once made history are overgrown with vegetation, with long obsolete infrastructure steadily rusting away in the heat and humidity of Central Florida's Atlantic coast.

With the unique needs of the Apollo program and the Saturn V booster, NASA at last began launching from its own facility at the Kennedy Space Center on Merritt Island, lying to the immediate west of Cape Canaveral across the Banana River and now connected physically to the Cape itself by the massive crawler road which runs from the VAB to the two Saturn V launch pads. These were in turn subsequently converted into Shuttle launch pads, 39A and B, which supported the next 30 years of U.S. manned space flight. As SpaceX began to make plans for a launch facility on the East Coast, the Space Shuttle was still flying out of Pad 39A, and the less frequently used pad 39B was in the process of being converted to support the Ares-1 rocket. Consequently, if it wanted to make use of any existing infrastructure rather than starting from scratch with all the environmental permitting and earth moving that would entail, SpaceX would need to look just to the south, and to one of the vacant pads on Cape Canaveral Air Force Station, managed by nearby Patrick Air Force Base.

Among the available sites, the most desirable was without a doubt Space Launch Complex 40 (SLC-40), which had previously served as the launch site for the Titan III and Titan IV rockets, America's first "heavy lift" boosters, hosting in that role numerous Air Force launches as well as several unforgettable interplanetary missions, including most recently the Saturn Cassini probe. As the launch pad for a comparatively large and powerful rocket, with recent history, SLC-40 was built to sufficient specifications to not only easily handle the Falcon 9, but also to accommodate larger vehicles as well.

SLC-40 and the Titan IV booster had seen its last use in 2005, where it was the only remaining component of what had been a truly immense military space program which predated NASA's Saturn facilities and had in fact served as precursor to that model. The entire system was improbably based on a heavy lift version of the same workhorse Titan II rocket which had begun life as an ICBM, and was then adapted to become the launch vehicle for the manned Gemini program, where it served flawlessly. In addition to these two roles, the Titan II then became the core component of the military's chief means of lofting an ambitious series of large spy satellites, whose weight and dimensions far outstripped existing launch vehicles. Flanked by two large solid rocket boosters, the new "heavy lift" Titan debuted in 1965, and began a long history of flights on both the East and West Coasts in a configuration which ultimately influenced both the design of the Space Shuttle System, as well as the French Ariane V. In the heady atmosphere of a surging military industrial complex and a burgeoning space race, the early 1960's was no time for thinking small, as the infrastructure planned for the new launch system bore witness. Starting in 1964, the Air Force began construction of a launch assembly complex which included a massive Vertical Integration Facility, designed to accommodate up to six Titan III cores under various stages of assembly. Once assembled a core would be wheeled along a railroad bed by twin diesel locomotives to another building, the SMARF, Solid Motor Assembly and Readiness Facility, where the solid rocket boosters would be attached. Next the steadily growing rocket would ride along the rails to a final assembly building at one of two possible launch pads, SLC-40 or SLC-41. Here, the upper stages of the rocket would be attached and payload integrated in a procedure which could last for several months. The total distance from the original assembly building to the pad at SLC-40 stretched almost four miles, and its construction required extensive dredging of the tidal basin to construct a series of man-made islands to link the sites. The last piece of this massive infrastructure, where final assembly was performed, was the most unique.

Whereas the Saturn V and Space Shuttle were assembled inside the VAB and then transported to the launch pad on the gargantuan Crawler/Transporter; for the Titan III and IV, the entire assembly building itself, complete with clean room, was designed to be mobile, and was rolled away from the static launch vehicle which remained stationary and connected to the umbilical tower. The ungainly result, multiple stories high and weighing more than 13 million pounds and frequently described as the largest mobile structure in the world, was also among the most expensive. And yet, the military consistently and proudly referred to the system as "cost effective." Designed to launch up to 40 missions per year, it never exceeded 8, and was periodically plagued by reliability problems. Once confronted with advancements in Europe and Russia which threatened to render the U.S. launch industry uncompetitive, the soaring costs and spotty reliability of the Titan heavy system were major factors in the Air Force's decision to pursue the new Evolved Expendable Launch Vehicle Program. When Lockheed Martin selected the old Titan complex at LC-41 for its new Atlas V, SLC-40 still remained available as the most desirable pad at CCAFS.

After winning Air Force approval to lease the facility on April 25 2007, the groundbreaking ceremony was held on the first day of November. Only six days later, SpaceX achieved its first major ignition at the site when a fire broke out during preliminary renovation work. It would not be until the following spring however, that a contractor working for the Air Force explosively collapsed the hulking Mobile Service Tower which dominated the site. After bringing the structure to the ground, the next step was tediously cutting more than 12 million pounds of steel into 4 foot segments so it could be sold at maximum dollar to recyclers, allowing the Air Force to effectively fund the entire demolition without dinging the tax payer.

With the debris finally cleared away, the SpaceX launch pad team, now led by Delta program veteran and newly hired Director of Florida Launch Operations Brian Mosdell, could begin to develop in earnest a launch site infrastructure which was as fundamentally different from that which previously occupied the historic pad as could possibly be imagined. Even though the Falcon 9 was a much larger vehicle than the Falcon 1, with its overall 12' diameter, it could still be shipped cross country in segments by common tractor-trailer trucks which are the backbone of America's ground transportation system. By contrast, the larger diameter Boeing Delta IV cores, although built much closer to Cape Canaveral at its Decatur, Al. facility on the banks of the Tennessee River, require the expense of a specially configured transport ship, the Delta Mariner to reach its Florida launch site. West Coast launches are even more time consuming, necessitating a trip across the Gulf of Mexico and passage through the Panama Canal to reach a small port terminal at Vandenberg which was once designed to receive Shuttle External Tanks. After relocating Atlas V production to Decatur following the creation of United Launch Alliance, that booster, though smaller in diameter than the Delta IV, ships via barge as well.

Following its far more mundane route, arriving at the launch site, the SpaceX Falcon 9 would require only modestly more on-site assembly than the smaller Falcon 1, but in a similar process conducted on a larger scale. The company would still follow the goals of the ship and shoot system, except in this case after being assembled inside an on-site hanger, "ship" meant placing it on a short railway from the hanger to the launch pad, and towing it several hundred feet to the launch pad less than 12 hours before launch, with the ultimate goal being a one hour process.

Although SpaceX had no use for either the Mobile Service Tower or the smaller launch gantry, both of which were now removed, the requirements of the Titan IV had resulted in a number of other, more basic items which remained and were quite useful. First, was a large, well-constructed launch pad, one of only a handful of heavy lift pads at the Cape. It included a usable flame trench, embedded rail line, underground instrumentation room, electrical, lightning towers and water suppression system, as well as a small ready building and surrounding ring road. Each would go to good use, and each saved valuable resources.

Unfortunately, the hypergolic fueled Titan core, and its attached solid rocket boosters left little material in terms of the complex systems required to store, distribute and recover kerosene fuel and liquid oxygen which would fuel the new rocket, much less the nitrogen and helium which were also necessary to clear lines and pressurize systems, all of which would need to be constructed and thoroughly tested in preparation for a first launch. Relentlessly practical, the team assigned to prepare the launch pad scoured the expansive confines of the Kennedy Space Center for whatever might be put to use. The most impressive find was a massive 125,000 gallon liquid oxygen tank which had not seen use since the era of the Saturn V. Striking a deal with NASA to acquire the tank for $86,000, one dollar more than the value of the scrap metal contained, SpaceX had the spherical tank loaded on a tractor trailer and slowly motored down the wide open roadways marking KSC, to its new home at SLC-40 where after a thorough cleaning and repainting a gleaming white, it joined the new tanks which housed the RP-1 to fuel the vehicle. Other notable items were refurbished rail cars which NASA had previously used for transporting helium. The team was also able to make use of several rusty rail carriages left over from the Titan IV, acquiring two more used bogeys on E-bay, each of which would serve as carriers for the launch erector assembly.

Compared to lofty and expensive vertical structures at nearby pads, including the Atlas V vertical integration building and the Boeing Delta IV's retractable launch servicing structure, as well as the Kennedy Space Center with its plethora of buildings dedicated to servicing and supporting the Shuttle, SpaceX ordered the construction of a very modest 225 x 75 x 50 foot tall steel framed, metal sided building, little different from those which serve the everyday needs of American commerce, from plumbing supply warehouses to ever present feed and seed stores dotting the rural landscape. Instead, this building would house a rocket, and the surprisingly bare bones equipment used to assemble it for flight. Even before the pre-fab building was installed however, SpaceX, rushing to meet a self-imposed deadline of having the first Falcon 9 assembled and vertical at the Cape 9 by the beginning of 2009, began shipping both the launch vehicle itself, as well as components of the transporter/ erector to the bare concrete pad.

When the Falcon 1 finally made it to orbit, there was little time for celebration. Due in part to the lengthy time span required to certify the Merlin 1C, SpaceX was already more than a year behind schedule on the Falcon 9 for the COTS program, and coming under increasing pressure to perform. Establishing operations for a first of its kind Falcon 9 launch pad at Cape Canaveral would be a very time consuming process, involving a long series of fit checks and practice runs to bring together the elaborate production of launch day. It was a process which could not even begin until there was something very much resembling a completed booster and launch assembly on site. Consequently, after the Falcon 1 flight, SpaceX began a nearly frantic campaign to make up for time lost time, and hopefully meet the mid-2009 first launch date which it had agreed upon with NASA. The program proceeded on two fronts; the first, in Florida, focused on preparing pad operations for flight. The second mostly took place in Texas, where both the first and second stages still needed to be qualified, and then acceptance tested for the inaugural flight.

Even though the initial rocket being sent to the Cape was not an actual flight vehicle, it did consist mostly of flight components, including an engine assembly and first stage. Before it departed for Florida however, SpaceX needed to verify that the basic design was sound, as any structural changes to the vehicle could require a change in the pad configuration. The process began in November 2007 with a single Merlin 1C engine firing at the full thrust level demanded by the Falcon 9 as opposed to the de-rated engine which was simultaneously being acceptance tested for the upcoming Falcon 1, flight III. The first ever multiple (two) engine firing came on January 18, and was followed by a three engine, full duration mission length burn of three minutes, taking place on March 8. This was followed by a five engine firing on May 29, which marked the first time in the regime that engineers added more than one engine at a time, and in the process conducted what would have been a full engine test for the now cancelled Falcon 5. Buoyed by the progress, and encouraged by successful static firing of the Falcon 1 flight engine thousands of miles away on Kwajalein, as it prepared for an imminent launch, engineers made preparations to add in the remaining four engines all at once for the next scheduled firing, and attempt to pull off what Elon Musk had characterized as “the most difficult milestone in the development of the Falcon 9.”

The nine engine firing took place without incident on July 30, generating 832,000 lbs. of total thrust. Two days later, a successful second full power, 9 engine firing pushed the company past its critical COTS 12th milestone with two months to spare. With the Merlin 1C performing perfectly, the August 1 test was supposed to be the opening act for the Falcon 1 flight scheduled the next day, and a one-two punch of good publicity. Instead, it became the silver lining in what appeared to be a very dark cloud when the Falcon 1 failed on its third attempt. As it turned out, the Merlin’s flawless performance and successful test firing of all nine engines only days before played an important factor in Musk’s make or break gamble to fly another Falcon 1 to the remote Pacific atoll for another attempt.

What had been years of frustration with the Falcon 1 and slower than anticipated development of the Falcon 9, quickly turned into a string of much needed good news which began when SpaceX received its operational license for Cape Canaveral and SLC-40 from the USAF on September 9. Less than three weeks later, the fourth Falcon 1 and the second to be powered by the Merlin 1C finally roared into space on September 28, clearing the way for the company’s first ever full length burn of all nine engines on the first stage. The test took place on November 22nd, generating a maximum of 855,000 lbs. of force as all nine engines burned for 160 seconds. At this mark, two outboard engines successful shut down, just as they would in flight to prevent over-acceleration of the now very lightweight rocket. The remaining 7 engines burned for another 18 seconds, and then they too shut down right on cue. Significantly, in demonstrating the ability to safely shut down engines in flight, the company was building the case for reliability based on engine-out capability. Having now passed the challenge Elon Musk described as the “highest hurdle” in Falcon 9 development, the first stage could briefly return to Hawthorne for cleaning and analysis before joining all the other components for a journey across the country and a first ever visit to the Cape.

Back in Hawthorne, engineers rushed to complete fabrication of the launch structure which would provide the base and strongback for the new booster, working around the clock over the Thanksgiving holiday. Several days later, the inter-stage and aft engine skirt were test fitted to their respective ends of the first stage tank which, freshly painted in gloss white was loaded on a truck and sent to Florida. Following the first stage, strung out along the nation's interstates was a fleet of 12 tractor trailers, hauling other components of the Falcon 9, including the two halves of the 17' diameter payload shroud, which although unused on the initial flights, would eventually see service in large satellite launches.

With the major pieces of hardware arriving at Cape Canaveral just before Christmas, the integration team set to work, foregoing any semblance of a normal Christmas break, but celebrating nonetheless a very unique holiday season. For once, work proceeded ahead of schedule, and on December 30, a complete Falcon 9, mostly comprised of actual flight components took its place among the launch vehicles dotting this most unusual strip of the lengthy American East Coast. Next up was completing assembly of the transporter/erector, the final step in hoisting the rocket to vertical for the first time ever.

The launch erector assembly consists of three separate sections. The massive square base which bears the weight of the fully loaded launch vehicle is assembled from four welded steel box beam sections weighing together more than 97,000 lbs. Connected to the base, the launch strong back which serves as the "tower" is built as a truss structure and is comprised of two sections. The lower section, also made of steel, is hinged to the launch base, while the upper section of the structure is built from lighter weight aluminum. With the erector assembly bolted together during the first week of the New Year, and hydraulic systems attached, two large cranes lifted the waiting Falcon 9 off the rail assembly where it had been assembled, and placed it on the launch erector for the first time. At 12:45 EST on January 8, the erector steadily lifted the Falcon 9 to a vertical launch position.

For the most part, the particular version of the rocket which greeted the second week of 2009 was quite a bit more photogenic than the plain white rocket which would conduct the initial launch campaign. Rather than the stubby Dragon spacecraft which would mark the first 4 launches, this Falcon 9 was topped with the larger 5.2 meter fairing, with the rocket itself highlighted by contrasting white fuel tanks and black inter-stage and thrust skirts. Together, it made a far more striking presentation than the small Falcon 1 which had so recently put the company in space, and SpaceX wasted no time in taking a number of glamour shots under different lighting conditions to show the world that a major change in the launch industry was finally underway.

Once construction of the vehicle integration hangar was completed, work commenced to install the pressurized kerosene systems which would be used to test the Falcon's hydraulic actuators as it was being prepped for flight. Other projects included installing the environmentally controlled clean room for payload integration, and the mobile system which maintains temperature and humidity for payloads once they are integrated into the vehicle right up until the moment of launch.

The interval between erecting a partially completed Falcon 9 in the first week of 2009 and actually conducting the first flight was one which would be dominated by a continuing series of tests following the time worn path of development, qualification, and acceptance testing conducted in their respective order. While the Merlin 1C main engine was already qualified at a lower thrust level, and performed well in the first full length mission firing, it was still awaiting a turbo-pump upgrade to bring it up to the target thrust level. The second stage Merlin 1C vacuum engine, absent from the fit check vehicle at the Cape, was still in development and would clearly be the tailing item.

The SpaceX business plan was built on meeting three critical criteria; simplicity, safety and affordability. With 80% of the Falcon 9 designed and built in house, the rocket was simple, and the revolutionary pricing assured it was affordable, but with no heritage, a new stage structure, a controversial (to some) 9 engine first stage architecture, and the increasingly important COTS/CRS station re-supply program on the line, establishing the third criteria, safety, could only come through exhaustive testing in the Lone Star State.

What had begun in McGregor, TX in 2003 with a single horizontal test stand for the original Merlin 1A engine was now a sprawling facility consisting of the Big Falcon Test Stand, a new upper stage test stand for the Merlin 1C Vacuum engine, a vacuum test chamber for the Draco thruster being developed for the Dragon capsule, as well as a structural test stand used to test the upper and lower stage structures to their physical limits. Altogether it comprised a complete facility capable of subjecting each component of the Falcon 9 to both structural and performance testing in order to prove its worth long before it ever got to the launch pad. The launch environment is a famously unforgiving business, and everybody tests, but compared to other development efforts, what stands out about the SpaceX approach is the sheer relentless nature of the testing, and the confidence in the final product it elicited.

The plan for the first flight called for each of the 9 first stage engines to undergo acceptance testing in Texas, after which they would be shipped back to California for integration with the thrust structure and skirt assembly. Then, they would be shipped back to Texas for integration with first stage tank structure for yet another test firing before finally heading to the Cape where it would await arrival of the second stage, which was undergoing the same process. Acceptance testing on the first stage engines proceeded smoothly, with six engines completed by the end of June 2009, and the remaining three units formally accepted in July.

At the same time the first stage tank and connected interstage assembly were undergoing their own grueling series of structural qualification tests on the vertical test stand. Qualification tests for the first stage structure consisted of more than 150 pressure cycles, testing for pressure loading at the point of maximum dynamic pressure, and stiffness tests, with each geared towards meeting or exceeding NASA's standards for man-rating. The structural test stand was also used to test the stage separation system for the carbon fiber interstage. Placing a mass simulating the weight of second stage, technicians demonstrated the pneumatic separation system, a larger version of that used on the Falcon 1, under alternate load conditions. In each case, the collets holding the stages together released as intended, with the three pusher arms confidently projecting the mass simulator safely away from the first stage.

SpaceX announced the structural testing program complete as of July 29, setting the stage for acceptance testing for the maiden flight. After integrating the 9 acceptance proven flight engines into the thrust assembly back in California, the complete 17,000 lb. unit was shrink wrapped and shipped to Texas where it was integrated with the waiting first stage tank structure and hoisted onto the 235 ft. tall Big Falcon Test Stand in Mid-October for firing an as integrated assembly, the last test prior to being certified as ready to ship to the Cape. Testing included two firings, the first, on October 12 and lasting 10 seconds, followed by a 30 second test conducted on October 14. [67]

The remaining item was the second stage Merlin Vacuum engine, which was encountering initial difficulty with excessive nozzle temperatures, resulting in a "de-tuning" to bring the readings within acceptable limits. With the completion of a new Upper Stage test stand specifically designed for the Falcon 9 second stage and the Merlin Vacuum, SpaceX conducted the first 40 second test firing in November, initiating a steady series of testing culminating in a full duration test firing on the second day of 2010, the beginning of what would be a very good year for the company. The firing took place without the nine foot tall expansion nozzle installed, and lasted a full 329 seconds, the length of time required for orbital insertion. In the process of generating 92,500 lbs. of force, the Merlin Vacuum engine demonstrated the highest specific impulse, (336 ISP) ever recorded for a hydrocarbon engine built in the United States. Following the first stage to Florida, the arrival of the flight second stage on February 11 signified that all the pieces were at last in place to begin integrating what all hoped would be a history making vehicle.

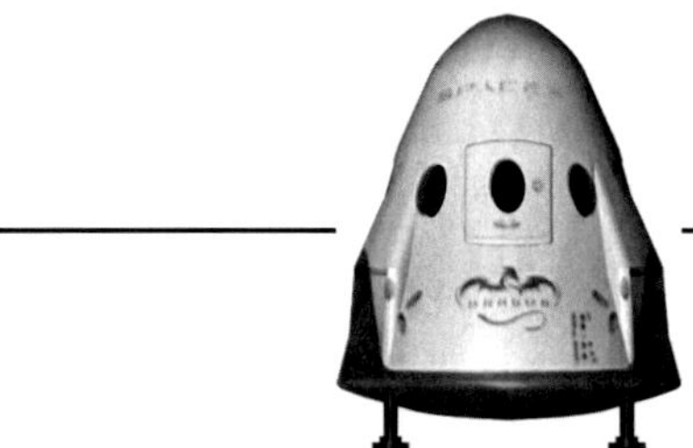

Chapter XIV: Integrating the Falcon 9 for Flight

Following a tumultuous and thrilling 2008, the year 2009 had seen steady progress for SpaceX, beginning with the initial raising of the Falcon 9 at Cape Canaveral in early January, followed by the company's first successful commercial operation on July 14, with the fifth and final launch of the Falcon 1 and its RazakSat spacecraft. Behind the scenes, the various elements of the first Falcon 9 flight vehicle had progressed through a grueling series of certification and acceptance tests in Texas even as all the infrastructure necessary to conduct operations was coming together at the SLC-40 launch site.

With the Falcon and Dragon running 18 months behind the initial COTS timetable though, 2010 was dawning as another make or break year in which so much of the company's efforts would come down to a moment of truth with the first launch for the Falcon 9. In the greater context, the beginning of the second decade of the new millennium was also proving to be a major inflection point in the space program as a whole. Even though it had only been a few years since the COTS program was announced in 2006, the nation's space policy was already in the midst of a major transition.

In the run-up to the 2008 Presidential Election, then candidate Senator Barack Obama initially offered a mixed and disturbing message regarding his views on space policy. Early in the campaign, the candidate indicated that he would prefer to see a major portion of the NASA budget pay for a new program of early childhood education stating "The early education plan will be paid for by delaying the NASA Constellation Program for five years." [68]

Although Obama later changed his position, this brief glimpse into his viewpoint at the time, and the apparent ease with which he was willing to set aside the nation's first serious effort at returning to space exploration since Apollo without offering any plausible alternative, easily rankled many within the agency and aerospace industry at large, and in its own way helped to set the stage for what would become a bitter and unproductive debate in the very near future. Later in the campaign, speaking to a crowd on Florida's Space Coast, Obama offered a somewhat paltry consolation prize, announcing his intention to press Congress for funding for one final Shuttle flight to the ISS, which was then scheduled to end with STS-134, Endeavour. With the program already almost past the point of no return, the addition of a single flight, without being placed in any greater context, seemed little more than a thinly veiled attempt to buy votes among the soon to be displaced Shuttle workforce. As it turned out however, when STS-135 Atlantis lifted off on the final flight of the Shuttle program in July 2011, it was anything but superfluous, carrying the Multi-Purpose Logistics Module Raffaello, packed with a full load of supplies and equipment sorely needed to maintain the station while SpaceX and Orbital Sciences rushed to complete their respective cargo vessels.

Space is rarely high on the agenda for new administrations and it has not been uncommon for incoming Presidents to wait a year or more until finally getting around to appointing a new NASA Administrator. Obama's transition team soon made it clear however, that in this case, current administrator and Bush appointee Mike Griffin would quickly be shown the door. What is usually a polite, civil transition in which all outgoing heads formally tender their resignation with the outgoing Administration, some in hopes of being asked to stay, took an unexpected turn when the head of newly elected President Obama's NASA transition team, and soon to be NASA Deputy Administrator, Lori Garver got into what was described as a "heated discussion" with outgoing Administrator Griffin during a reception for a book signing at NASA Headquarters shortly after the election. Reportedly, the conversation erupted when noted author Dr. John Logsdon observed that as part of his research into a new book on President John F. Kennedy's decision to go to the Moon, that Kennedy had completely disregarded the advice of his own NASA transition team. Griffin responded that he wished the incoming transition team would talk to him, to which the answer from a transition team member came back "well Mike, we're here now," after which the fireworks ensued. [69] At issue was the fact that as a normal part of the questioning process in preparing various policy options for the new president, the transition team asked whether or not it would be feasible to cancel Project Constellation entirely. Given the President Elect's previous campaign statements suggesting a willingness to delay the program for 5 years, this should not have come as a surprise, but Griffin's response was anything but calculated. Instead of pointing out the drawbacks in cancelling the program he had been brought in to champion, Griffin instead suggested that the transition team, which contained no engineers, lacked the technical chops to understand what they were looking at. As job interviews went, this was hardly a hall of fame performance.

As Garver's questions implied, and Obama's previous positions indicated, the new Administration was not overly enthralled with Project Constellation, which by late 2008 was running well behind schedule and way over budget. Most significantly, it was painfully obvious that even if NASA overcame a number of technical difficulties, the resources available did not match even the most basic goals and timelines attached to the plan. The choice was to either ignore the problem, which would have made the contractor community happy as long as the money kept flowing, increase the budget or find a plausible excuse to make a change in priorities. The time honored way to build political cover for undertaking a major policy change certain to cause resentment is not to tackle the problem directly, but instead to form a Blue Ribbon commission to study the issue and make recommendations to be ignored or adopted as needed. This is of course, precisely what happened.

The commission, formally named the Review of Human Spaceflight Plans Committee was announced by the White House Office of Science and Technology Policy on May 7, 2009, with respected aerospace veteran and former Lockheed Martin CEO Norman Augustine as its chair. It was in fact the second such commission he chaired, the first had been in 1991, and both are casually referred to as "The Augustine Committee." The new committee held public hearings throughout the spring and summer of 2009, focusing on the goals and objectives of human spaceflight, with a particular examination of the projected costs for Project Constellation. The summary report was released on September 8, 2009 reaching what had become to many a foregone conclusion. Under predicted budget scenarios, there was simply no way that NASA would have the funds available to develop and enact Constellation on a meaningful timeframe. NASA could possibly complete the Ares-1 Crew Launch Vehicle, but had made negligible progress on the much larger Ares-V which would rendezvous in low

Earth orbit with the Orion capsule, and send its crew to the Moon. Furthermore, work on a new lunar lander, named Altair, had not progressed at all. Looking past Ares-1, NASA could either build the Ares-V, or build the payloads it might carry, but it did not have the money to do both. [70]

Close up shot of the first Falcon 9 V1.1 flight, carrying the CASSIOPE satellite into a clear blue sky above Vandenberg, Ca. Credit: SpaceX

In the absence of an increased budget baseline, which was clearly not forthcoming, the Augustine commission saw little practical choice but to recommend restricting near term planning to missions which did not require the fuel or hardware to land in a deep gravity well. Constellation's plans to return to the Moon and establish a base, followed by an eventual journey to Mars would instead have to give way to the somewhat less inspiring objectives of conducting a series of undefined flights to various locations in Cis-Lunar space or perhaps a near Earth asteroid. In the opinion of the commission's members, Mars would remain the inevitable goal of the U.S. Space Program, but circumstances required that the nation would need to follow a "flexible path," to get there, one which could take a number of decades to do so. To critics, the timeframe was so far out as to be meaningless for policy planning.

Early in 2010, as the final pieces of the first Falcon 9 arrived at Cape Canaveral, Administration insiders, speaking on conditions of anonymity, began indicating that the soon to be released Administration Budget request for FY2011 would strip funds for Project Constellation. By the time President Obama traveled to Cape Canaveral to introduce his new space policy, one loosely based on the conclusions of the Augustine committee, word was already out. Standing before a crowd of NASA personnel at the Kennedy Space Center, the President unveiled his proposal on April 15, 2010. In place of Constellation's concrete plan of a return to the Moon, which the President summarily dismissed with an airy "we've been there before," the U.S. would proceed along the so called "flexible path" outlined by the Augustine Commission and focus its efforts instead on a mission to an asteroid sometime in the next decade. [71]

Rather than continuing to fund Constellation, the new policy called for an end to the troubled Ares-1 and largely undefined Ares-V launch vehicle, as well as the proposed Altair Lunar Lander which was still years from any meaningful development. In its place, the new direction would offer an increased emphasis on commercial space through the COTS program, which was well underway, and most significantly, the full-fledged development of a commercial crew capacity to return U.S. capability for domestic manned flight which would soon be lost with the Shuttle's retirement. The Orion crew capsule, now without a launch vehicle, after originally considered for cancellation, was to remain as a lighter, bare bones version of its former self ostensibly intended to become the long sought emergency crew return vehicle for the station. Though soon lost in the controversy which erupted from the policy, the plan also called for increased research into new propulsion technologies, an item President Bush listed in his original speech introducing the Vision for Space Exploration and subsequently emphasized by the Aldridge Commission. The President's plan also called for a budget increase of 6 billion dollars over the next five years, including 3 billion to spend on studying advanced heavy lift technologies with a development decision on the new booster to be made in 2015, a date which according to the Administration, would actually advance the debut of a new big booster compared to the timeline unfolding under Project Constellation.

To underscore the potential inherent in the new policy, President Obama preceded his speech at the Kennedy Space Center with a visit to the newly activated SLC-40 and the Falcon 9. Initially, the visit was intended to take place at SLC-41, the United Launch Alliance Atlas V pad, but the venue was changed to SpaceX at almost the last minute, when the Secret Service became concerned that a high velocity sniper's bullet could penetrate a liquid oxygen tank at the ULA facility, potentially endangering the President. [72] Instead, President Obama toured SLC-40, where a visit by the President of the United States gave added emphasis and publicity to a launch campaign which was already well underway. It also pinned a bull's eye squarely onto SpaceX, hardening opposition and provoking in some cases barely concealed hostility from the Administration's Republican foes in the Senate. It did not matter that it was actually President Bush who both cancelled the Shuttle program, and provided the impetus for the COTS program, or that SpaceX was emerging as an entrepreneurial success story which should have delighted purportedly anti big government Senators. What was, and should have been nothing but a test launch of single rocket, was turning into something much larger.

The launch campaign had started in earnest two months earlier with the arrival of the second stage on February 11, marking the first time all the components of a flight ready Falcon 9 were at the Cape at the same time. Following the script established with the Falcon 1, the schedule for beginning operations called for vehicle integration, a series of fueling and ground tests, followed by a "hot fire" hold down test of the first stage, and finally launch certification by the Air Force.

President Barack Obama tours SLC-40 with Elon Musk on April 15, 2010. Hours later, he formally announced the cancellation of Project Constellation before an apprehensive crowd at the Kennedy Space Center. Credit: NASA

The key component of the SpaceX integration process is the fact the entire procedure takes place horizontally. Inside the rather plain building at SLC-40, each of the major components; thrust structure, first stage with interstage, second stage and finally payload, are placed horizontally on what amounts to oversized shop dollies, after which they are maneuvered into place for assembly along the remaining Titan IV rail segment built into the shop floor. Secured on the dollies by a circumferential band, the stages can be rotated 360 degrees along its axis for precise alignment with other components. Once the Falcon 9 is integrated, the entire rocket can then be rotated 360 degrees along its longitudinal axis to allow ground level access to any point. Although a simple change, the results are substantial. Compared to traditional vertical integration, with technicians spread out over multiple levels, the result is a significant difference in efficiency and workflow. In the first place, communications between technicians can now take place in plain sight of one another and in a normal speaking voice rather than shouting or through the disjointed and sometimes garbled cadence of radio. No longer separated by elevators and steel floors, each requiring their own set of tools, technicians are also able to use a reduced tool set, and a dropped wrench falls only a few feet, as opposed to plunging more than six stories to the ground, endangering everyone in its path.

Altogether it is a safer and more comfortable working environment, requiring far less manpower, and cost, than conventional vertical assembly. It is also far more flexible. Each time NASA elects to make a section of the VAB available for a different launch vehicle than the one which preceded it, the change requires the removal of carefully engineered floor sections precisely matching the height of a specific rocket, followed by the installation of new sections matching the dimensions of the new rocket. It is an expensive and time consuming process which the agency perversely presents as a sign of progress; out with the old and in with the new. The case is much the same with both EELV launch integration efforts as well. The Delta IV, unlike the Atlas V is integrated horizontally, but at the insistence of the Air Force, all payloads must be mounted and integrated vertically.

There are a number of other benefits to horizontal integration however. One of the most common ways to get more production out of a given booster is to first improve the engine, and then "stretch" the fuel tanks to produce a taller rocket with greater propellant capacity. For a stacked rocket, this means re-arranging the levels on your vertical assembly structure, another in a series of roadblocks which serve as dis-incentives to upgrading vehicle performance. With its eye on possible upgrades to the Falcon 9, SpaceX instead used horizontal rail integration, a procedure indifferent to the length of the rocket. Besides the obvious gain from not having to pay to build and then amortize a huge vertical assembly structure at Cape Canaveral, adopting horizontal integration from the outset insured that the costs of establishing launch operations in other locales such as Vandenberg, Ca. would be similarly contained. Already considering a potential third launch site as well, a fully commercial launch facility unhampered by the conflicting demands of a military test range like Cape Canaveral, the best way to preserve these possibilities was to minimize the investment required to bring them about.

Yet another reason for using horizontal integration and minimizing time on the launch pad was to avoid tying up the pad unnecessarily once the frequency of launches began to increase. With the same launch pad ultimately intended to service a heavy lift version of the Falcon booster as well, maintaining flexibility at the pad was a legitimate concern. Scheduling an orbital launch is often a balancing act between preparing the launch vehicle on one hand, and integrating the payload on the other, with launch delays frequently arising from difficulties encountered by the manufacturer in having the payload ready in time. This is particularly the case for mission specific scientific payloads, but it is not uncommon in the commercial satellite industry either. Establishing flexible launch operations would allow SpaceX to re-arrange schedules as payload issues arose and limit the opportunity for one problem to boomerang through an entire year, as frequently occurred in both the Shuttle and EELV programs. Commercial satellites are very expensive items, which are almost always financed. Once a satellite is commissioned and launch is ordered, the satellite owner begins accruing interest and expense which can only be offset after the payload is launched and service begins. Delays mean money. Designing for rapid and flexible launch operations might have seemed somewhat ironic given the undeniably slow start to its efforts, but as in everything else, Elon Musk was taking the long view of things, and considering future opportunities as well. Launch contracts often contain a reserve clause which allows the satellite owner to switch providers if their launch is unduly delayed, and with two of its three potential competitors, Sea Launch and Proton both suffering major failures in recent years, designing for a flexible launch schedule opened the real potential for picking up supplementary orders in the wake of a future failures or delays.

A typical Falcon 9 integration procedure begins with the installation of the thrust structure to the first stage. Next, technicians working on the second stage attach the fragile niobium skirt extension to Merlin Vacuum engine, and then install the avionics package into its bay on top of the second stage, whereupon it is carefully fit into the awaiting space of the interstage structure which has already been attached to the top of the first stage. Finally, the payload itself is integrated into the now complete launch vehicle, a process which begins with vertical integration of the payload onto the adapter ring, after which it is encapsulated inside its protective fairing and then rotated 90 degrees to be mounted to the second stage.

The key enabling piece of the Falcon 9 launch process is the transporter/erector assembly, which carries the fully assembled rocket from the hangar to the pad, and then functions as the launch gantry itself. Once the integration process is complete, the fully assembled rocket is lifted by two overhead cranes and gently placed on the transporter/erector, where it is secured both at the vehicle's base and around the second stage with a hydraulically operated C shaped collar. The truss structure is connected to the box section base, which provides both stability and connecting points to secure the transporter erector to the reinforced concrete pad. In another one of those curious moments when the everyday world collides with high tech expectations, the erector assembly, rocket and all, is towed along the 600 ft. railway to the launch pad by a very well-used aircraft tug no different from those in use every day at hundreds of airports around the world. It is somehow not difficult to imagine that if the price had been right (and the hay plentiful) relentlessly practical SpaceX personnel would have just as readily employed a mule team to tow the rocket to the pad.

Once the erector assembly is in place, the ground crew installs six locking pins which secure the assembly to corresponding holes bored into the concrete pad itself. After the process is completed, it is safe to begin rotating the structure to the launch position. Going vertical is an equally simple procedure. A pair of hydraulic cylinders are attached to the lower portion of the transporter erector, and they gradually raise the box frame cradling the rocket towards the launch position. Once over center, two opposing cylinders pull it the rest of the way and with little time and less fanfare, the rocket is pointing towards space.

Next begins the process of hooking up the various umbilical cables which bring the rocket to life. Again, following the discipline of simplicity, all of the umbilicals are already routed to the launch vehicle before it leaves the hangar, and terminate in quick connect fittings at the base of the erector. Finally, the pad supply systems are attached to transporter/erector via the quick connect fittings, and Falcon 9 is ready to be fueled and launched, at which point the strongback tilts 15 degrees from vertical, allowing room for the vehicle to lift off from the tower. Just how soon that process took place for a first launch depended completing a series of wet dress rehearsals, a successful test fire, and receiving Air Force approval for the launch to take place.

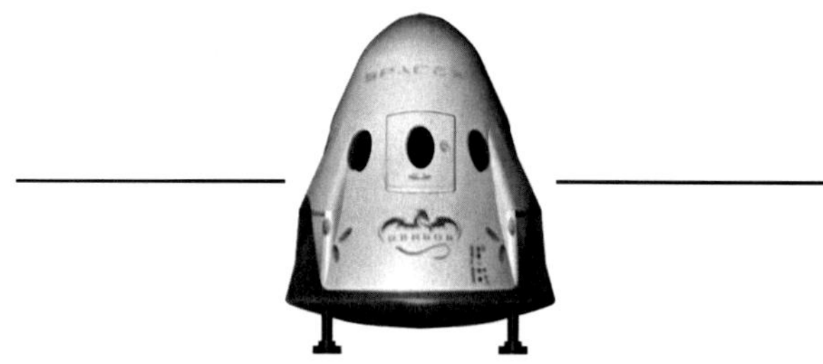

Chapter XV: Counting Down to Launch

With the arrival of the second stage after acceptance testing in Texas, the maiden launch campaign for the Falcon 9, years in the making, began to come together quickly with integration of the upper stage. The payload for the first flight was a simple boiler plate model of the Dragon capsule, its chief purpose to provide an aerodynamically accurate facsimile of the functional models slated for the rocket's next four missions. Following a careful check of second stage hydraulics with the pressurized RP-1 system inside the hangar, and installation of the flight termination system, a flight ready Falcon 9 rolled out to the pad for the very first time on Friday, February 19th. After securing the launch transporter to the pad, the team rotated the rocket to vertical the next day, setting the stage for the first of what would be four wet dress rehearsals prior to launch.

As immense but comparatively fragile containers loaded with thousands of gallons of flammable liquids, launch vehicles are somewhat problematically equipped with rat's nests of tubing, connections, valves, check valves, seals and sensors, all of which present an opportunity for disaster in the form of a leak. Consequently one of the most important steps in preparing for a successful launch is to take every possible precaution to make certain that once the count-down begins in earnest, a leak does not occur. The best way to do that, as SpaceX had done with all five launches of the Falcon 1, is the wet dress rehearsal; put the rocket into flight configuration, simulate a countdown up until the moment of ignition, and verify integrity of the complete system. With its ethereal, super cold hydrogen fuel, this was always a point of frustration for the Space Shuttle system, and a frequent source of delays as technicians searched for the guilty component. Though fueled by RP-1 in part to avoid such difficulties, the new Falcon booster had its own potential liability in the form of all the piping and connections required to feed 9 separate first stage engines. With a new booster, a new pad, a new range control to work with, and coming up soon enough, a new partner in NASA, SpaceX faced a host of challenges in preparing for Falcon 9's first flight.

Launch day for any rocket is a carefully choreographed series of events involving participants in multiple locations, all interacting with the rocket, the ground support systems and each other. Like any production, it requires a number of rehearsals to get the timing or "cadence" right, and for early launches, the initial run-through often bears little resemblance to the final product.

For this first wet dress rehearsal, SpaceX maintained a partial media blackout. By now, Elon Musk was growing annoyed with the criticism which ensued each time the company missed one of its own well publicized deadlines, and as the first launch date approached, the SpaceX founder repeatedly cautioned that any scheduled launch date was merely an assertion of the first day that operations could theoretically begin and not a statement of intent. This policy extended to the testing sequence, and it was not until later in the evening after the initial test was complete that the company made a public comment.

The launch operations team conducted the first wet dress rehearsal on Friday,

February 26, beginning the process at T-minus 2.5 hours in the countdown with an electrical power up of the rocket and verification that all systems were functional. This was followed by an engine purge check to make sure that all engines were in ignition configuration and prepared to receive the onslaught of fuel and oxidizer for an actual launch. A little over an hour later, the fueling process began with liquid oxygen flowing first. Once the upper and lower stage LOX tanks were 98% full, topping off procedures began to achieve the final loading, releasing the tell-tale white vapors, and covering the rocket in a sheet of ice.

Next, 29,400 gallons of rocket grade kerosene flowed to the respective tanks. With the booster fully fueled, the on-board guidance aligned as it would for launch, the engine steering systems proceeded through the pre-launch checkout and the Falcon transferred to internal power. At the T-40 seconds mark, the propellant tanks were pressurized, and the Falcon 9 came fully alive for the first time in its history. With 10 seconds remaining in the countdown, computers conducted a pre-planned pad abort, and moments later, the process to safe the rocket began. [73] All things considered, the first WDR went surprisingly well, with SpaceX personnel in the launch control centers rehearsing the communication links with Eastern Test Range officers at Patrick Air Force Base, as well as with personnel in Texas and back in Hawthorne with Musk summing it simply by saying "It was a good day."

The one exception concerned an issue with cork insulation attached to the LOX tank. Fully fueled, the rocket contained nearly 76,000 gallons of kerosene and liquid oxygen, and the loading begins with the latter. As with any cryogenic fluid, it is added slowly at first to allow the entire structure to chill down as evenly as possible. Once a working temperature is reached, the rapid fill cycle begins. As might be expected, even when added in a controlled manner, the introduction of large quantities of a cryogenic liquid has an effect on the entire structure of the rocket, which has hopefully been engineered to absorb the shrinkage induced with no meaningful distortion. During this initial test, contraction of the first stage LOX tank caused small pieces of space grade cork to peel off the tank. The cork, which was intended to provide protection during atmospheric reentry for first stage recovery efforts, had been applied once the booster arrived in Florida, and was not present for previous stage testing in Texas. Intent on recovery of the first stage if possible, SpaceX consulted with NASA engineers regarding the de-corking issue, and in the weeks after the first test, subsequently applied a new layer of cork, using a different adhesive.

The second major pad test would repeat the steps followed in the first WDR, and extend the final countdown a few more seconds to a brief 3.5 second firing of all 9 first stage engines, to verify performance, including the pad hold down system. The initial attempt came on Tuesday March 9, and proceeded smoothly until a few seconds before ignition, when a high pressure helium valve failed to open, rendering turbo pump spin-up inoperable. After waiting out two days of thunderstorms with the rocket still on the pad, drained of fuel and protected by lightning from four large lightning towers and a spider web of barely visible cables which cover the immediate pad area, the team was ready to try again on Friday. This time, the countdown went as planned, and the engines roared to life in a 3.5 second test, briefly reaching 800,000 lbs. thrust with the rocket safely secured to its erector base by four hold down arms attached firmly to the strongest part of the rocket, the thrust assembly. The fire of the 9 Merlin engines was accompanied by the billow of steam as the super-heated exhaust struck the wall of water provided by the pad's acoustic protection system, dubbed "Niagara" which floods the flame trench to dampen out harmful shockwaves. All 9 engines cut off on cue, ushering in a sudden silence. The Falcon 9 was ready.

With the test successful, the sole remaining impediment was launch approval and installation of the flight termination system. SpaceX however was also eager to demonstrate the feasibility of first stage recovery, and wanted one more chance to verify that the new ablative cork would stay adhered during flight. So, one day after President Obama toured the pad, the engineers took the occasion to fuel the vehicle and practice the countdown procedures one more time before a launch date which was then scheduled for no earlier than May 8. Once again the test went smoothly, and the cork stayed where it was placed.

By the beginning of May, SpaceX had completed every test except those that would be conducted as part of actual launch operations. Two elements stood between the Falcon and a moment of truth. First, the FAA still needed to give its final o.k. and the Air Force had yet to certify the range safety system, a procedure which required that each element had to be tested sequentially, and each step required completion of the one prior. In the case of mishap and flight off the intended trajectory, the range safety officer would transmit a signal to the flight termination system, a shaped charge running the length of the booster, designed to instantly rip the tanks apart and destroy the vehicle. Before turning its attention to SpaceX, the Air Force wanted to complete the first launch of an enhanced Block 2F Global Positioning System Satellite aboard a United Launch Alliance Delta IV, scheduled for late May. In other words, the Falcon wasn't going anywhere until the Air Force processed the final paperwork, which would not take place until after the GPS launch. The Delta IV lifted off at 11:00 PM on Thursday, May 28, placing the Falcon 9 in the on-deck position.

On Tuesday, June 1, SpaceX announced that the initial launch attempt would be scheduled for Friday, June 4, beginning at 11:00 am with a window lasting four hours. In the event of a one day scrub, the range was reserved for the following day as well. Weather forecasting estimated a 40% chance of "no go" conditions on both days. By design, the range certification would not come until two days prior to liftoff, after the vehicle had passed the final flight certification and the Air Force confirmed the ground to vehicle radio communications system in a final pre-flight countdown.

At the time of Tuesday's announcement, the Falcon 9 was still in the hangar, awaiting final checks, which took place later that same evening. The all-white Falcon 9 was rolled out of the hangar on its short journey to the launch pad on Wednesday morning under blue skies, and by later that afternoon as the typical Florida afternoon rain began to develop, the erector lifted the rocket to its launch position.

At 4:40 pm, the 45th Space Wing commander formally approved the launch scheduled for Friday.

With launch day rapidly approaching, Elon Musk once again took pains to play down expectations and to emphasize that this was a test flight only. Although the company was coming off its first, albeit ever so short "string" of back to back successful launches with the Falcon 1, memories of the three previous failures were a constant reminder to everyone involved that there was still much which could go wrong. As with the first two Falcon 1 flights and in keeping within the strict definition of what was actually meant by the term "test launch," success in this case was relative, and could be measured as anything other than a launch pad failure. Specifically however, it would be defined as a percentage of the number of individual flight objectives which were met. Musk himself assessed the probability of success as no more than 70-80%, adding with characteristic humor, that the figure was somewhat less than the odds of a successful outcome in a single turn of Russian roulette. In the final analysis, a successful first stage flight would qualify as a good day.

As Friday, June 4th 2010 dawned, SpaceX had a lot on the line. Compared with the company's previous experience with the Falcon 1 deep in the Pacific open and far from prying eyes, this event would take place in one of the more populated areas of the United States. Any mishap early in the flight would be plainly obvious for all the world to see, juxtaposed against the Shuttle and all its sweeping infrastructure, and captured in imagery which would doubtless be played over and over again on the 24 hour news networks and as well as on YouTube videos. The launch would also be taking place from in between the United Launch Alliance Atlas and Delta pads, the home field of the United Launch Alliance monopoly whose supporters in Washington had devoted a good amount of energy in trashing SpaceX and blocking its attempts to secure a foothold in the EELV market. A failure would not only prove expensive and embarrassing, it would almost certainly have serious political consequences. With President Obama's newly announced, and none too popular space strategy only six weeks old, and already the subject of intense arguments, in part due to a remarkably inept and poorly managed roll-out on the Administration's part, Elon Musk knew all too well that public comments to the contrary, a bad day could spell more trouble than was rationally warranted. It simply did not matter that the original Atlas had failed 12 times prior to its first successful flight, or much more recently that the soon forgotten Delta III failed on its first two flights. Even the most successful vehicle in the world, the Ariane V had failed on its first flight as well. With the Space Shuttle program irrevocably committed to shutting down, SpaceX and the Falcon 9 were more than a company and a rocket, they were the centerpiece of NASA's post shuttle plan for the International Space Station, and perhaps most importantly, the Dragon capsule which the new rocket would boost on its next flight, was the only foreseeable means of returning cargo and experiments from ISS. On another level, SpaceX and the Falcon 9 were also the most visible symbols of a NewSpace movement struggling to survive in an implacably hostile environment, and the movement as a whole stood to either fall or rise on the outcome of the day's events.

With SpaceX withholding any release of information regarding the flight until the start of a webcast which was supposed to begin 20 minutes before liftoff, launch observers were left to fill in the blanks on the event timeline until the webcast began. The weather proved a challenge, with condition red caused by thick clouds early in the launch window, which cleared up only to make room for high cumulus clouds and a storm front approaching from the west as time ticked down.

Public relations finally released a press kit at 10:00 am, and a few minutes later fueling was underway. Altogether the process would take approximately 90 minutes, and conclude once 39,000 gallons of LOX had been loaded into the first stage and 7,300 into the second, along with 25,000 gallons of kerosene in the first and another 4,600 in the second. Also due to load was helium for tank pressurization and turbine spinup, as well as nitrogen for line purging, cold gas thrusters and finally, TEA-Tab ignition fluid.

The countdown went into a preprogrammed hold at T-15 minutes, staying there as SpaceX and the Air Force attempted to improve weak signal strength between the launch pad at SLC-40 and the range safety officer several miles away. The solution proved to be a matter of switching to a different antenna. At 12:29 p.m., the tower rotated back into the launch position, and liftoff seemed moments away. Unaware of events proceeding on land, a sailboat strayed into the immediate safety zone around Cape Canaveral, and helicopters were dispatched to warn it away. The launch was subsequently reset for 1:30, and this time the countdown went without incident until just a few seconds before ignition when an automatic abort brought events to a sudden halt.

Not surprisingly, one of the many thousands of vehicle checks showed an engine parameter on the number three engine had fallen below limits. During a press conference the day before, SpaceX advised that such an event was likely to occur, as the company had deliberately set a whole host of parameters more narrowly than it ordinarily would in order to gain as much information as possible on the state of the vehicle right up until the moment of launch, knowing that in many cases it could look at what had caused the alarm, and then go back and reset the limits if conditions and prudence allowed. Anticipating this procedure was one of the reasons Musk cited in not releasing a launch countdown, he did not want to give competitors a list to "nit-pick" each point of the launch.

As it turned out, this was exactly the case and in a turn of events which surprised many of those observing the launch, including an Air Force officer who reportedly began promptly shutting down his station upon seeing the abort, SpaceX announced at 2:22 p.m., with less than an hour remaining in the launch window, that it was resetting the launch for 2:45. The sudden reset with only moments left, did little to dispel the perception among some that once again, the company was behaving as rashly as its critics predicted.

What observers did not see was the rapid fire sequence of events which took place between the launch control team at the Cape and engineers back at SpaceX headquarters in Hawthorne, Ca. Suspecting a sensor issue rather than an actual problem, the engineers at the Cape downloaded control software from the vehicle and transmitted it to California, where engineers ran it through their mission simulation program, verified it, and sent it back to Florida. Understanding their vehicle, and what had caused the hold, SpaceX saw no reason to halt operations for the day, as there was still enough time for recovery crews stationed off the Georgia coast aboard the NASA ship Freedom Star to conduct operations in daylight. With the software uploaded into the Falcon 9's flight computers, the countdown started again. [74]

At 2:32 the launch control team polled and the word was go. Three minutes later, the Falcon 9 once again entered computer controlled automatic countdown at T-10 minutes.

At T-60 seconds, Niagara's 30,000 gallons of water per minute began flooding the flame trench to provide acoustic protection. Fuel tank relief valves closed, allowing tank pressure to build to flight levels.

Next all 9 engines went through a steering check, followed by pre-valves opening to allow the first flow of fuel and oxidizer to fill the engine plumbing.

T-5 seconds: Nitrogen purge

T-3.5 seconds: Ignition!

T- 3 seconds: Turbo pumps spun by initial combustion

T-2 seconds: 9 engines begin generating horsepower in earnest

T-1 second: Engines reach full power

T- .5 seconds: Computer queries all criteria and command hold down release

Liftoff!

Generating more than 800,000 pounds of sea level thrust, the white, almost non-descript Falcon 9 clawed its way off the pad, illuminated by the tell-tale ruddy orange glow of kerosene oxygen engines, gradually building speed until it merged into the hazy white background of a partly cloudy Florida summer afternoon, heading east/northeast over the Atlantic and straight into the pages of history.

At T plus 60 seconds, accelerating rapidly, the Falcon 9 went supersonic, and then approached the Max Q threshold. Seconds later two outboard engines shut shown as planned to prevent the vehicle from over accelerating. Twenty seconds later, and fifty miles above sea level traveling at 4,500 miles per hour, the remaining 7 Merlin engines cut off right on cue. Regardless of whatever happened next, the day was already a success by the standards SpaceX laid down. What did happen next was that after an appropriate delay of 5 seconds, hard won on the Falcon 1 Flight Three failure, the separation bolts fired and the pneumatic arms built into the inter-stage stage disengaged the collet clamps and pushed the second stage clear, sending the expended first stage arcing ballistically towards a hopeful ocean landing somewhere off the Georgia coast.

The next critical moment was second stage ignition, and the proof test of the new Merlin 1C Vacuum engine and its huge expansion nozzle, which was already safely clear of the detached inter-stage. An event which was not precisely re-producible on the ground, it would be happening live for the first time. The extra months of testing in Texas finally paid off, and stage ignition occurred right on time, with the vehicle navigation system successfully acquiring its new vector.

At T plus six minutes, and 2:51 p.m. local launch time, the second stage was halfway through its planned burn and well on the way to placing itself and the Dragon test unit in orbit.

Two minutes later, with just three minutes remaining, one hiccup in the launch occurred when on board cameras indicated that the stage was developing a slight rolling motion which appeared to be accelerating. Though noticeable, the motion was nothing like the corkscrew motion on Demoflight 2, and gratefully did not materially affect the second stage vacuum engine which continued to burn right up until its planned cutoff, moments before the vehicle passed out of communications range.

While it would take some time to verify the exact orbital parameters, it was clear that the Falcon 9 had achieved what many had decried as the impossible or at least the highly unlikely for such a young company on the maiden flight of its flagship rocket. The Falcon 9 had performed almost flawlessly, achieving a clean lift off. To the joy of space enthusiasts around the globe, and the possible chagrin of observers immediately to the north and south of the launch pad, a new era in space access, which had begun 22 months ago, far away on the Kwajalein atoll, had now entered the main stage and could not be so readily dismissed.

Later that day, an ebullient and visibly relieved Elon Musk announced that the Falcon 9 had hit an orbital "bull's-eye" at 155 nautical miles and inclination of 24.5 degrees in a highly circular orbit with the apogee only 1 degree higher than intended and a perigee a remarkable 2/10 of one degree lower than planned.

Once it passed beyond the range of U.S. tracking stations, SpaceX would have to wait along with everybody else until the vehicle's orbit brought the second stage and attached test article over Australia, where communications could be re-established. Having done so, there was still one more test to perform. The Merlin 1C vacuum engine was designed to be restartable, a necessary capability for it to place commercial satellites in specific orbits. Consequently, SpaceX performed a "burp test" briefly re-firing the second stage engine. As a result the orbit was shifted from the original 155 mile circular orbit to a slightly more elliptical orbit with a maximum altitude of 174 miles.

As the spacecraft's orbit brought it over Australia, where it was briefly illuminated by the rising sun just over the local horizon, the craft sparked several reports of a strangely luminous UFO crossing the pre-dawn sky. As it turned out, the Falcon second stage was venting tank gas at the time in order to reduce internal pressure and eliminate the potential for a debris creating explosion, and caught the sun's rays at just the right angle to illuminate the gas in a well-defined halo as the craft rotated in orbit.

The only real disappointment of the day came when the crew of NASA's Space Shuttle Solid Rocket Booster Recovery ship Freedom Star, on station to retrieve the first stage, found only floating debris. Originally considered a consequence of a parachute not deploying, subsequent analysis revealed it was actually the result of the first stage re-entering the atmosphere and tumbling violently, a clear indication that stage recovery would be a considerably more complicated task than the company had hoped. Back in orbit, and running on battery power, the Dragon payload simulator relayed a wealth of data on its condition as it settled into an orbit which would keep circling the planet for the next several years.

One interesting unresolved issue surrounding the first flight was a persistent suggestion that perhaps a party other than SpaceX had partially funded the initial flight, and when asked about it at a June 16 press conference, Elon Musk laughed, stating he could neither confirm nor deny the rumor. Considering that the Falcon 5 was originally upgraded to the Falcon 9 due to the request of an "unnamed government customer" it seems entirely plausible that the customer was either DARPA, which had sponsored the first two flights of the Falcon 1, or perhaps the U.S. National Reconnaissance Office which operating with a clandestine budget might have had an interest in both gauging the launch capabilities of the Falcon 9 as well as potential applications for the Dragon capsule.

Reaction to the Flight

There can be little doubt that the initial flight was wildly successful, surprising everyone including SpaceX. In a post-mission teleconference with reporters, Elon Musk stressed again that he would have been happy with just a successful first stage flight. Lost in the elation was the fact that the purpose of the flight was to gather data, specifically atmospheric flight performance data regarding the Falcon and the Dragon configuration aerodynamics. Musk at one point even mused that in some ways the flight was a little too perfect in that by performing as advertised it did not return a lot of data which demanded correction, although between two roll control issues, one in space and another slight rotation in the vehicle's first few feet of ascent, there was at least something to work on. Even so, Musk confidently predicted that the next flight could take place as early as the end of summer. The second Falcon 9 rocket, intended for that flight, which would be the first under the COTS program, was already complete, and the Dragon capsule it was slated to launch was almost to the same point, 99% complete.

While NASA, space leaders, organizations and enthusiasts were quick to heap praise on the company, with Florida Senator Bill Nelson calling to offer personal congratulations, some of his "distinguished colleagues" in the Senate were not quite as forthcoming, damning it with faint praise, and even former NASA Administrator Michael Griffin, who had overseen the initiation of the COTS program, but perhaps was bitter over the recent cancellation of Project Constellation, chose to downplay the significance of the flight, electing instead to stress the fact that it would take many more launches before it could be considered a reliable system for commercial crew transport. It must have been a difficult experience for Griffin, watching SpaceX complete an orbital launch the entire development cost of which was considerably less than the expense of the nearly $500 million dollars his agency had spent on just the launch tower alone for the recently cancelled Ares-1.

Of particular note was the petty response from Texas Senator Kay Bailey Hutchison,

"This first successful test flight of SpaceX's Falcon 9 rocket is a belated sign that efforts to develop modest commercial space cargo capabilities are showing some promising signs. While this test flight was important, the program to demonstrate commercial cargo and crew transport capabilities, which I support, was intended to enhance not replace NASA's own proven abilities to deliver critical cargo and humans to low Earth orbit. Make no mistake, even this modest success is more than a year behind schedule, and the project deadlines of other private space companies continue to slip as well. This test does not change the fact that commercial space programs are not ready to close the gap in human spaceflight if the space shuttle is retired this year with no proven replacement capability and the Constellation program is simultaneously cancelled as the President proposes. "[75]

While generally avoiding the temptation to respond to critics directly, when asked about the Senator's response during the post launch teleconference, Musk seemed genuinely troubled and perplexed, asking why she wanted to hurt a Texas company, particularly one which was growing at a steady rate.

The Senator's statement neglected to point out the fact that the "gap" which seemed to trouble her so much, was substantially less than that which had developed under NASA for Project Constellation. Furthermore, it ignored her own culpability in the Senate's inability to pass a budget and allocate the necessary funds to move beyond the COTS program into Commercial Crew. Ironically, the SLS program for which she was working so feverishly to make the centerpiece of the NASA FY 2010 Authorization bill, would not have any crew capability until after 2020. More than one blogger wondered if Ms. Hutchinson, never regarded as one of the Senate's leading intellectuals, was even aware that SpaceX operated its testing facility in her own state.

Alabama Senator Richard Shelby's response was even more acidic:

"Belated progress for one so-called commercial provider must not be confused with progress for our nation's human space flight program," Shelby said. "As a nation, we cannot place our future space flight on one fledgling company's definition of success." [76]

The COTS program to which the Senator was immediately referring was of course not limited to "just one" fledgling company, but also included veteran launch provider and spacecraft manufacturer Orbital Sciences Corporation. And even more bizarrely, the other launch provider which most clearly stood to benefit from NASA's Commercial Crew program as it began to emerge was United Launch Alliance, which builds its Atlas V and Delta IV in Decatur, Al, the Senator's home state. It remains something of a mystery how the Senator was able with a straight face to attack the Commercial Crew program on the basis of SpaceX's inexperience and unproven launch record, while studiously ignoring the fact that EELV program's Delta IV and Atlas V had now racked up a string of 34 successful launches, against one partial failure which occurred on a test flight.

For two Senators who seemed so concerned with "belated" progress, it must be pointed out that by the time of SpaceX's June 4, 2010 launch, the recently cancelled Project Constellation had been underway for a year longer than COTS, yet with a cost of $9.4 billion dollars, neither the Ares-1 nor the Orion Crew Vehicle was anywhere close to flight. At the time President Obama effectively cancelled Ares-1, the rocket still did not have even a completed design, lacking among other things a second stage engine. It was also battling both a vibration problem which threatened to shake the crew like rag dolls and a launch escape simulation which strongly suggested that in the event of an explosion the crew would plummet to their deaths as shrapnel from the steel casing and burning chunks of solid rocket fuel severed parachute cords. Even after it was cancelled, Alabama Senator Richard Shelby succeeded in forcing NASA to spend nearly an additional $500 million on the project by inserting language into the NASA FY2010 authorization bill keeping the project on life support until Congress passed a FY 2011 budget. When it failed to do so, the language remained, driving another year of fruitless spending. By contrast, NASA had invested a sum total of $396 million dollars in SpaceX's allotment of the COTS program, yet had just witnessed the first flight of a booster which had been designed from outset to carry crew. Furthermore, NASA had not paid for this flight, with the entire cost being borne by SpaceX.

For these two Senators, repeatedly elected running against big spending liberals and in favor of the American free enterprise system, and otherwise implacably opposed to government involvement in the private market, SpaceX somehow clearly represented a free enterprise which they could not abide. Apparently they preferred the old Soviet system, as subsequent votes hamstringing funding for Commercial Crew inevitably led to the United States paying inflated prices to Russia to continue launching American crews on Russian Soyuz, even as a series of launch failures indicated that the Russian aerospace base was in danger of breaking down.

Assessing the Flight

The immediate result of the Falcon 9's maiden flight was that pending a thorough review of the flight telemetry, the countdown to the first flight under the COTS program could begin.

This was an important step both for SpaceX and for NASA. As of June 4, 2010 there were two officially scheduled Shuttle flights left in the program, with a strong possibility that a third and final flight would be added. Beyond that however, and irrespective of what became of the Commercial Crew program, with Constellation cancelled, and any follow on large launch vehicle many years in the future, NASA needed commercial re-supply services to begin as soon as possible in order to avoid a genuine logistical nightmare, not to mention a public humiliation.

The Orbital Sciences entry to the program was also important, but with OSC getting off to a late start following its substitution in place of Rocketplane Kistler, and offering limited performance in terms of a lack of return capability for its Cygnus transfer vehicle, SpaceX, having won the lion's share of launches, was NASA's best hope. Furthermore, with the battle heating up over Commercial Crew and the impending lack of an American ability to conduct a manned space launch, the debut performance for the Falcon 9 went a long way towards suggesting that at least from a distance, the concept of "commercial" astronaut transportation was sound.

For the country at large, public perception was beginning to follow a building narrative that the nation was turning its back on its own history, and that NASA was abandoning the Shuttle system way too early, with no replacement in sight. The President's Kennedy Space Center address had not been received warmly, and was accompanied by repeated protests, even coming from the normally friendly AFL-CIO President Richard Trumpka. Accompanied by the obligatory newscasts deploring the loss of thousands of high paying jobs as a result of the Shuttle closeout and the cancellation of Constellation, the successful launch of a new rocket whose principal claim to fame was that it didn't require 10,000 or even 2,000 people to get it off the ground was hardly welcome news.

In contrast to the Falcon 1 launch campaign, in which the outcome of the effort principally concerned SpaceX only, this launch had broad implications with definitive external consequences. For the Administration, which had taken a significant risk in proposing to turn transport duties to low Earth orbit over to commercial providers at the same time it attempted to kill NASA's own, separate program in Project Constellation, the SpaceX success provided much needed momentum to pass the core of the President's funding request. On July 15, the Senate Commerce, Science and Transportation Committee unanimously approved the NASA Authorization Act of 2010, officially initiating the Commercial Crew Program. In a harbinger of things to come however, the Senate cut the funding request from $500 million to $312 million, effectively beginning a foot dragging process led by adherents to the traditional NASA spaceflight model. Although less than requested, $312 million was more than a token, and signified that at least one aspect of an American space policy so frequently in limbo was finally proceeding in a definite direction. Had the outcome of the SpaceX launch the previous month been radically different, it is entirely possible, even likely, that the committee would have taken a different course as well.

From the point of view of the overall launch vehicle industry, the maiden launch of the Falcon 9 indisputably validated the 9 engine first stage approach to an intermediate class vehicle, and along with it, the technical capability SpaceX was bringing to the field. Symbolically, the launch also marked three successful launches in a row, and a break-even point in the SpaceX launch record. For the company's opponents, the narrative began to change from SpaceX could not perform, to SpaceX still had a long way to go to prove itself, (which was true) and that in the long run it would have to raise prices. It was a line of attack that only time and success could effectively combat.

NASA was not the only one paying attention however. Before the initial Falcon 9 launch, SpaceX was enjoying some success in signing up commercial customers for its new rocket, but it was primarily with new operators who were unwilling or unable to pay the premium for reliability demanded by the world's premier providers. And while the first flight led to a series of new orders, including the announcement on June 15 of the largest single satellite order in history, a $492 million dollar deal with Iridium, other orders were out there; waiting on the company to prove the first flight was not a fluke.

With the debut flight of the Falcon 9, SpaceX entered a very exclusive club, and in one critical way, it was already the most unique member. Every other American launch vehicle was the product of a contracting system in which the prime contractor assembled components from different sources to arrive at a single system, and regardless of their final form, most launch vehicles were variations on components built from the same limited group of suppliers. Throughout the space era, a steady series of mergers and consolidations ensured that the name emblazoning the side of any given vehicle, was rarely the name of the company that originally designed it. For instance, what is now the United Launch Alliance Delta II, began life in 1957 as the Douglas Aircraft Company's Thor Intermediate-Range Ballistic Missile, later evolving into Thor-Delta, and then the McDonnell Douglas Delta II, which was then purchased by Boeing, which then consolidated its operations with those of Lockheed Martin to become ULA. The Atlas followed an even more circuitous path, originally built by Convair Corporation, which then was sold to General Dynamics which then was sold to Lockheed Martin, which was itself the result of a merger between Lockheed and Martin Marietta. The progression of corporations was driven by the increasing costs and gradual scale back of the rate of new weapons systems after the massive buildup of the 1950's and 1960's, as well as an overall business environment which values mergers and acquisitions as a way to increase market opportunity and reach economies of scale.

SpaceX, in designing and building a system almost completely in house, marked a radical departure from the status quo which separated it from contemporary and historical American launch vehicles. Although at one point Boeing had acquired Rocketdyne, which produces the RS-68 main engine powering the Delta IV launch vehicle, the engine manufacturer was never integrated into the Boeing culture and the division was sold to Pratt and Whitney with little fanfare in 2005, citing a decision to pursue a more horizontally integrated business model. Standing in stark contrast, on its next flight SpaceX would seek to extend its experiment in vertical integration even further, launching and recovering a fully autonomous and equally home built Dragon spacecraft, and in doing so, form a complete end to end space transportation system.

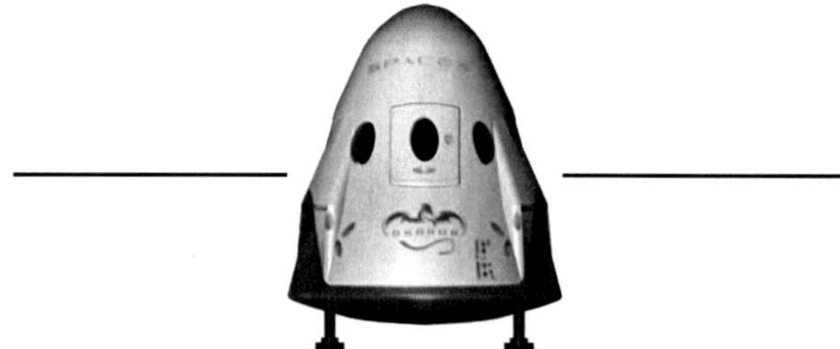

Stewart Money

Chapter XVI: Supplying the International Space Station

Perhaps one of the least appreciated aspects separating the current age from the rest of human history is the speed and ease with which materials and goods are shipped around the planet on a regular basis. It is through a network of interlinked transportation modes that casinos and restaurants along the Strip in Las Vegas, or throughout the various resorts in Disney World can offer a stunning variety of fresh foods from around the planet all with use of a logistics system which is essentially invisible to the thousands of people visiting every day. It is a capability influencing almost every aspect of daily life, in which comparatively small markets in remote locations routinely offer a variety of goods and services which were unthinkable even as recently as the beginning of the space age.

In conducting the human exploration of space however, the ultimate expression of technological savvy, we are ironically enough transported back in time to an earlier age in which the well-being and even the survival of expeditions, outposts and settlements depended on the timely arrival of supply ships which could mark the difference between life and death. And although never faced with such dire circumstances, in supporting the first semi-permanent human outpost in space, mission managers on more than one occasion have had to make plans to abandon the International Space Station when the supply line was in question, as happened in the aftermath of both the Columbia accident, and the failure of a Progress re-supply vessel in late 2011. For all the expense and effort put in to the Station however, it was not the supply line to the Station, but rather the supply from the Station back to Earth which stood to suffer the most with the retirement of the Shuttle.

In many ways the architecture pursued with the International Space Station is the polar opposite of the Apollo program. Whereas the approach to Apollo was clearly a destination driven, almost exclusively American enterprise, conducted with a specific date in mind (the end of the decade) and carried out with an all up design in the massive Saturn V booster, the International Space Station is something altogether different, and more difficult to define. Part of the reason is the justifications for the program offered by NASA seemed to shift continuously over the first decade of its existence, frequently leading to the charge that it is a program without a purpose.

Originally proposed as the U.S. Space Station Freedom and accepted by the Reagan Administration in 1984 on the somewhat questionable grounds that if the Soviets had a station "shouldn't we have one too?" the program migrated through a series of design changes even as the political circumstances which prompted its creation, competition with the Soviet space program changed completely with the fall of the Berlin Wall and the dissolution of the Soviet Empire. The dividends of peace proved to be a double edged sword for NASA. With the original rationale now changing and the costs escalating, the Space Station project became a frequent target for budget cutting, with multiple attempts to kill the program off entirely taking place in Congress, including one effort in 1993 which failed by only a single vote. At the same time, NASA did little to help its own case. Like many other large projects, the combination of actual and projected expenditure of funds so outstripped the original estimate, ($8 billion in 1984) that it was hard to determine if the agency had no idea what it was doing, or was deliberately underestimating costs just to keep the effort going.

Following its near death experience, the Space Station program was radically overhauled in late 1993 to become the International Space Station, with the costs and capabilities to be shared between the U.S., Russia, Canada, Japan and Europe. Making the program international in scope provided much needed political cover, but at the cost of increased complexity in terms of design, construction and management. From a logistical standpoint however, the metamorphosis into ISS also removed dependence from the Space Shuttle as the sole source of support and opened the door to what ultimately evolved into a mixed fleet of resupply vessels.

The launch of the Station's first segment, the Russian built Zarya module in 1998, marked the beginning of a 13 year period of construction which saw the arrival of the first two person crew in 2000. While the Shuttle, along with the Russian Proton, was tasked with transporting most of the larger components comprising the Station itself, the equally important responsibility of re-supply was assigned to the venerable Russian Progress "freighter." Shortly after the first crew arrived, the long string of resupply missions necessary to keep the station functional began with the flight of Russia's Progress M1-3 spacecraft in August of that year.

To the casual observer it is very difficult to distinguish the difference between the Progress ship and the Soyuz crew transfer vessel, and for a very good reason. When the Soviet Union lost the race to the Moon in the 1960's, it turned its space ambitions to developing a series of small manned space stations which served both military and civilian purposes. With cosmonauts launched to the Salyut stations via the Soyuz system, Soviet designers took the same basic components and produced an unmanned version which became history's first "space freighter." To make the system work Soviet engineers designed a reliable automated craft which could perform autonomous rendezvous and docking in an era in which the U.S. still depended on manual control to achieve the same result. Even though the Salyut stations and the Soyuz and Progress resupply ships which served them were quite small compared to America's cavernous Skylab which launched to orbit fully stocked in 1973, the Russians were quietly on their way to gaining invaluable long term experience with the logistics of space supply which would ultimately benefit all parties in ISS, including SpaceX.

The Soyuz spacecraft consists of three linked compartments, beginning with forward crew compartment and docking hatch, a middle crew return vessel, followed by a service module. The Progress uses the same architecture but exchanges the crew return capabilities for fuel storage tanks. Once docked to the Station, the Progress freighter is able to transfer fuel from its tanks to the Station's own fuel reserves via fluid connectors in the docking ring. Periodically throughout its stay, the ISS crew raises or adjusts the station's orbit by firing the Progress freighter's own engines. By doing so, rather than using the fuel which has transferred to the on board tanks, the working lifetime of the station engines is extended, and fuel reserves are maintained in the event of an extended period without re-supply. On average, ISS has required 4 Progress re-supply flights a year, with the pace increasing to five flights between in 2009 and 2011.

Each Progress capsule weighs approximately 7,000 kg empty, and can transport approximately 2,300 kg of supplies. For missions where more fuel is required, it is also manufactured in a M1 version which exchanges water tanks in the middle section with two more fuel tanks.

During the course of its stay, the empty spacecraft is gradually filled with trash, and once full, disembarks from the station and maneuvers itself into a steep death plunge over the vast Pacific. For important items which need to find their way back to Earth intact rather than in cinders, the Progress can be equipped with a small container named Raduga, which ejects from the craft as it re-enters the atmosphere, whereupon the container parachutes back to Earth. Thus far, it is a capacity which has not been employed on ISS, only seeing service during the Mir program. All in all, the Progress has proven to be a resilient and reliable system, which though upgraded on several occasions, remains fundamentally unchanged for nearly 30 years.

When it wasn't hauling major components for the Station's assembly, the Space Shuttle also contributed to the resupply effort, but in a completely different manner. In addition to its construction duties, the Space Shuttle could also carry a large, reusable cargo container called the Multi-Purpose Logistics Module (MPLM). The modules, built in Italy by Thales Alenia Space, trace their lineage back to the early days of the Shuttle and the flights of the Europe's Spacelab in a partnership which paved the way for ISS. Carried to space in the Shuttle cargo bay and occupied once in orbit, Spacelab functioned as a re-configurable mini space station which could be arranged for astronaut tended flights of the pressurized modules, as well as for externally mounted pallets which exposed both experiments and hardware to the space environment, and owing to its flexibility, some missions were a combination of both. Furthermore, in addition to the European Space Agency's contribution to the effort over a 29 flight history, there were also two separate and independent German and one Japanese sponsored Spacelab flight as well. Though perhaps overlooked in a historical context, the SpaceLab program laid much of the groundwork for international cooperation in ISS, while also serving as forerunner for the mix and match logistical supply which ultimately emerged.

Once the U.S. embarked on the Space Station Freedom project, Italy, building on the framework of Spacelab, offered to participate by providing a series of small, pressurized resupply containers called the Mini-Pressurized Logistics Module. When Space Station Freedom was re-molded into the International Space Station, the Italian offering changed as well, becoming the much larger and more capable Multi-Purpose Logistics Module. Capable of carrying up nearly 20,000 lbs. of supplies, and nested in the Shuttle's cargo bay, the cylinder shaped MPLMs, 6.6 meters long and 4.57 meters wide, offering a habitable volume of 31 cubic meters, were for all practical purposes temporary Station modules packed full of supplies. The MPLMs flew on 12 occasions beginning in 2001, during which it was attached to either the Unity or Harmony modules via robotic arm. Three units were constructed, each named after one of the Italian masters, Raffaello, Eduardo and Donatello. Unavoidably sharing the monikers with the Teenage Mutant Ninja Turtles, NASA capitalized on the crossover and youth appeal with a series of turtle emblazoned mission logos. The Donatello module never flew, and was instead partially cannibalized when NASA elected to convert Leonardo into a permanent spare room for the Station, named the Permanent Multipurpose Module, which was flown to its orbital home on STS-133, the final flight of Space Discovery, lifting off in February 2011. The remarkable utility of the MPLM's is one of the many unheralded achievements of the ISS program, and if anything points to one of the most frustrating aspects of the Shuttle. If not for the staggering costs of each Shuttle flight which carried it to orbit, the MPLM offered the beginning of a highly effective and mostly reusable logistics system which marked a significant advance over the reliable, but fully expendable Progress. Moreover, as demonstrated by the Permanent MPLM, each segment also offered a potential habitat module for applications in Earth orbit and beyond.

With the impending retirement of the Shuttle, and the subsequent loss of any ability to recover and re-use the MPLM's, ISS resupply needed something to take up the slack. The answer proved to be a pair of cargo vessels built by the European and Japanese space agencies respectively, and based on the MPLM design, but equipped with independent power, propulsion and communications systems, and launched into orbit by expendable rockets. The two cargo ships are the European ATV, and the Japanese HTV. The ATV is a 45,000 lb. loaded capacity spacecraft which offers the potential to haul more pressurized cargo than anything except the Shuttle. Launched by an Ariane V rocket out of French Guiana and capable of transporting more than 16,000 lb. of supplies to the station, the ATV is almost a small space station in its own right, and when docked and then emptied of its burden, offers a temporary add-on room to the orbiting lab. Starting with the pressurized Integrated Cargo Carrier, which is also built by Thales Alenia, the ATV adds solar panels, along with rendezvous acquisition sensors and a propulsion and maneuvering module all derived from Progress and Soyuz spacecraft components to make a complete vessel. Like the Progress, it contains both fuel and supplies, and once emptied is gradually refilled with trash and items to be discarded in an untimely burn over the Pacific as well. To offset the cost of the ATV, NASA barters with the European space agency, trading the delivery of supplies in lieu of cash payments for Europe's share of the station's operating expenses.

As of early 2014, four ATV's have been launched, with one more scheduled, after which ESA will instead shift its contribution to providing a service module for the first flight of the Space Launch System. Each ATV is estimated to cost approximately $300 million, a figure which does not include the operating budget, or the substantial launch costs of the Ariane V. Given its size and relative complexity, the ATV has been the subject of several follow on studies for both a crewed version with an attached capsule, as well as a mini space station in which two ATV's would be linked together to form a single core.

The Japanese HTV freighter is named Kounotori, or White Stork, and is also derived from the MPLM. It is similar to the European ATV, but offers slightly less payload capacity. The HTV is launched on the Japanese H-IIB rocket, a two stage liquid hydrogen/liquid oxygen rocket equipped with 4 solid boosters. Unlike the ATV, the Japanese vessel is not equipped to transport fuel, but instead hauls a larger quantity of dry goods as well as bulkier scientific experiments, and includes both pressurized and unpressurized cargo. One critical difference is that whereas the ATV performs an automatic docking with the Russian segment of ISS using derivatives of the same Russian software and hardware employed by Progress, the Japanese vessel, employs a more "hands on" method. After an approach to within 10 meters of the station, the HTV is then captured by the Canadarm2 remote manipulator arm, and is maneuvered into its berthing location by the station's crew. Once berthed, astronauts enter and unload the HTV, which can transport physically larger items than ATV due to its wider Common Berthing Mechanism docking adapter. At the conclusion of its mission, the HTV is also filled with trash and sent to plunge into the upper atmosphere. Both the ATV and HTV are seen by their respective space agencies as a possible precursor vessel to an independent manned spaceflight capability in coming years, with Japan appearing more likely to pursue a next stage in development with an unmanned recoverable capsule, the HTV-R. HTV missions began in 2009, and the program is proceeding at a pace of one per year through 2016. The financial arrangement with Japan is slightly different than that NASA shares with the European Space Agency, with the U.S. reserving cargo space on the Japanese freighter in exchange for launching that nation's Kibo laboratory to ISS aboard the Shuttle.

With the introduction of the ATV and HTV re-supply ships, the International Space Station partnership was able to spread the challenge of resupply over two new launch vehicles, and to assume much of the capability which would be lost with the retirement of the Shuttle. Nevertheless, a significant supply gap still remained, meaning that at least for a while, NASA would have to purchase cargo services from Russia in the form of an additional Progress flight each year. Even with the mixed fleet strategy, NASA estimated in 2006 that it would encounter a shortfall in the minimum amount of supplies required to keep the Station fully functioning of 48.8 metric tons over 5 years, or roughly 10 tons per year. [77] It was this gap need which formed the basis of the COTS requirements.

In outlining the terms for COTS, NASA had in effect adopted a supply strategy which more closely resembled that of the Russian Progress, with smaller amounts delivered on a regular basis, than the Shuttle derived architecture of large capacity modules delivered only once per year with the ATV and HTV. Adopting the Russian model allowed for a much stronger business case for the COTS winners, one which was built around multiple flights of smaller launch vehicles, conducted on new boosters whose size class maximized the opportunity for finding additional customers to spread the costs and expand the range of operations.

So what were the specific goals being sought?

The cargo requirement was divided between internal and external payloads. This division reflected the fact that a good portion of the research going on at ISS actually occurs outside of the station, with components and experiments deliberately exposed to the space environment. Also, much of the Station's infrastructure is externally mounted as well, and it was for this reason that designers incorporated a number of attachment points connected by a sort of rail system along the length of the facility. Combined with both the Canadian and Japanese robotic arms, as well as occasional spacewalks, externally stowed cargo and experiment can be relocated as needed.

Cargo is also classified by its intended purpose, which can be to either to support ongoing operation and maintenance of the station and labeled as "utilization," or it can be for the "customer;" the ongoing experiments and their associated hardware which comprise the purpose of the station in the first place.

Annual Requirements:

Internal:	8750 kg/yr. customer
	2500 kg/yr. utilization
	1100 kg/yr. water
	300 kg/yr. gases
External:	4500 kg/yr. customer
	2250 kg/yr. utilization
Return:	4000 kg/yr. utilization

Among the consumables utilized by ISS, several points are worthy of consideration. One of the most significant items, fuel was no longer on the list. Although the Space Shuttle had used its own considerable on-board reserves to boost the station's altitude on a regular basis, unlike both ATV and Progress, the Shuttle was never equipped to actively transfer fuel to the station's tanks. Instead, it provided a gentle nudge via its Orbital Maneuvering Thrusters to counteract the inevitable effects of atmospheric drag which grew as the station did, aggravated by its relatively low orbit.

With its operations now transitioning towards an increased focus on research, Congress designated the American segment of ISS as a U.S. National Laboratory in 2005, although with final construction delayed by the stand down and return to flight efforts after the Columbia accident, it did not begin functioning in that capacity until 2012. The National Lab aboard ISS is now managed by CASIS, the Center for Advancement of Science in Space, and provides a platform for organizations and institutions other than NASA to conduct research in the unique environment of microgravity. Whereas NASA sponsored research is generally, but not exclusively concerned with the microgravity environment as it relates to the agency's own goals, the research performed under the auspices of CASIS is broader in scope, but with a strong emphasis on biomedical research with the potential to positively impact life back on Earth. In support of both missions, as a part of the 8,750 kg customer cargo requirements per year, COTS called for the annual delivery of 12 powered and environmentally controlled International Standard Payload Racks (ISPR).

Measuring 79" high x 41" wide, and with a curved rear face corresponding to the station's cylindrical hull mold line, each ISPR is roughly the size of a refrigerator. Overall the station can accommodate 37 such racks which are arranged around the perimeter of each of the cylindrical modules, Kibo, Destiny, and Columbus with either overhead, deck, port or starboard designations. They are the larger counterparts to a concept originally developed during the Shuttle era, in which experiments were generally housed in standardized, containers arranged in the Shuttle's middeck lockers, and called Middeck Locker Experiment, or MLE.

The basic dimensions of the ISPR's form the chief means by which many of the experiments come and go from the station, with one rack able to contain anywhere from one to five or more experiments depending on size. Each comes equipped with standard power, water and gas hookups, and each also features pass-through sections where cables can be routed to desired locations.

Once the Station was finally completed, the quality and quantity of the scientific output would depend in part on the ability to both loft experiments to the station, and then return them in a timely fashion. With much of the NASA sponsored research focusing on life sciences, one of the driving design factors in COTS requirements would be the need to safely accommodate the standardized containers which can contain live plants or animals. As a result of this requirement, even if the COTS D, crew transport provision was never exercised, the resulting craft would necessarily still be capable of providing a life sustaining environment even if it never hosted crew.

At the time the initial winners of the COTS competition were formally announced in August of 2006, the Shuttle has just completed its second "return to flight" test mission, and with the foam problem which had doomed Columbia under control, if not totally eliminated, NASA appeared to be getting back to business as usual. With two new and distinctly different launch vehicles under development through the COTS program, and SpaceX promising a maiden flight in 2009, and Orbital in 2010, the odds seemed good that NASA would keep the supply train rolling. Four years later however, with both COTS competitors behind schedule and each of the Shuttles preparing to make their final flights, things didn't look nearly so certain. The answer would depend on the state of progress with perhaps the most challenging aspect of the SpaceX system, the development of the Dragon spacecraft.

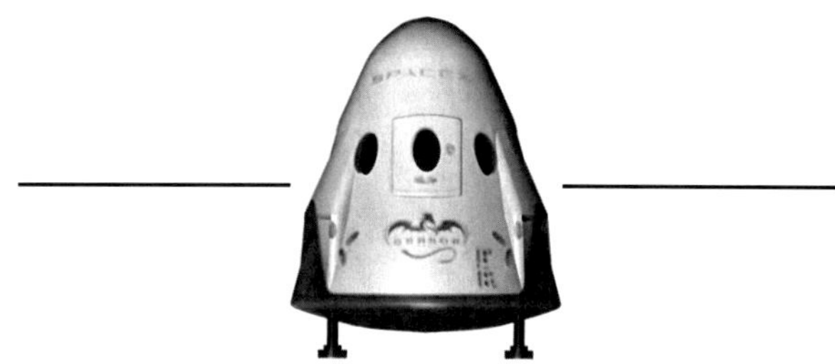

Chapter XVII: The Dragon is Born

In achieving a successful maiden flight for the Falcon 9 on June 4, 2010, SpaceX had already answered one of the major challenges of the program, and demonstrated that a commercial company could design, build and operate a new launch system at a fraction of the cost of any comparable development project. What remained to be answered was whether or not SpaceX, or any private company for that matter, was also capable of designing, building, operating and successfully recovering a fully independent spacecraft capable of supporting crew. If it could, then SpaceX would be credited with a truly remarkable achievement, the world's first, fully commercial end to end space transportation system. For the COTS program though, achieving this goal was still only one half of the challenge. Dating back to the Commercial Space Act of 1998 and the earliest days of the Alternate Access to Space program, the ultimate question was, and still is, whether or not the market for commercial supply services to low Earth orbit could expand beyond ISS and the government as sole tenant to that of anchor tenant for an emerging economy in low Earth orbit. Not surprisingly, the answer would depend on the cost.

For over a decade, the most likely tenant has been North Las Vegas Nevada based Bigelow Aerospace. Founded by hotel entrepreneur Robert Bigelow in 1998 with the hopes of establishing a series of large volume, inflatable space stations based on a NASA concept called Transhab, the original plan was to deploy an inflatable, or more properly, an expandable space station in the first decade of the new millennium. Though a believer in the long term prospects for space tourism, this was not Bigelow's intended market. Instead, he was focused on providing access to space for corporations and smaller sovereign nations which could not afford an independent space program. Bigelow however, had no desire to enter the space transportation business, preferring to partner with established providers to secure passenger service to orbit and back. Initially this meant the Russian Soyuz spacecraft, which had already broken new ground in carrying the first privately funded space tourist, Dennis Tito in 2001. With the destruction of the Space Shuttle Columbia in early 2003, Bigelow found that the additional Soyuz and Progress production capacity he was counting on was being absorbed by another customer he had no hope of outbidding, NASA.

Facing an impasse which could take years to resolve, Bigelow quickly settled on a new plan. Hoping to spur a privately developed space transportation system, in October of that same year, he announced the creation of America's Space Prize, a competition similar to that of the X-Prize model, but with much more demanding goals and a correspondingly higher purse. According to the terms of America's Space Prize, the winning contestant would have to develop an orbital transportation system capable of carrying 5 people to LEO, docking with a Bigelow station, returning them safely to Earth and then doing it all over again within 120 days. The deadline for winning the competition was June 1st 2011, and if a station was not in orbit by that point, a contestant could still win by demonstrating at least capacity to perform a rendezvous and docking. The prize was a very respectable $50 million, but there was one highly problematical catch. Competitors could not utilize government developed systems or receive more than 10% of their overall funding from government agencies. As for the prize itself, Bigelow initially put up $25 million of his own fortune, hoping to entice NASA to put up the remaining $25 million. [78] When NASA, which was already in the beginning stages of formulating the COTS program, ultimately declined, Bigelow stayed the course and personally committed the remainder of the sum as well.

With SpaceX already having signed Bigelow as an early customer for the launch of a subscale demonstrator model of the station aboard the Falcon 5 booster, Elon Musk observed in early 2005 that he believed SpaceX was capable of winning America's Space Prize with a capsule the company had quietly under development. Within months however, NASA signaled that it was finally serious about initiating its own publicly funded competition for developing commercial resupply services, funded at $500 million, and at ten times the amount of Bigelow's prize, an altogether more realistic figure. With the firewall between public and private money remaining in the terms of America's Space Prize, SpaceX, as well as the rest of the aerospace industry, quickly dropped any further pursuit of the smaller purse, which ultimately expired without any meaningful development taking place. For Bigelow Aerospace, this might have been the best outcome that was realistically possible. NASA and the American taxpayer would support the program in a way which offered a much better chance of success, and if it worked, Bigelow Aerospace would be a major beneficiary. It just required a little more waiting.

As it was, the Bigelow competition provided an initial focus for SpaceX's early spacecraft development efforts, which got off to a rather unusual start. Perhaps contrary to some early expectations, SpaceX has never explicitly expressed an intention to pursue space tourism for its own sake, at least on any meaningful time frame. Instead the company has remained relentlessly focused on the fundamentals of a definable business model and building a reliable space transportation system through pursuing established markets, whether the commercial satellite launch business, or new opportunities such as COTS. As a result, SpaceX went about its early efforts at developing a crew capsule in a tentative manner, very different from the public, high profile way it pursued launch vehicle development.

From almost the moment it announced the planned Falcon 5 booster, SpaceX began being approached by outside parties interested in using its launch vehicles for their own proposals, presenting the company with an early decision point as to whether it wanted to collaborate with anyone other than NASA for what would be a very visible and risky program, or whether it wanted to keep developments in house and under wraps. Exploring the issue, Elon Musk reached out to Andy Elson, a high altitude balloon capsule engineer whose products have been used in several record setting balloon flights. With such a close correlation between the requirements of pressurized high altitude balloon capsules and that of a rudimentary space capsule, it seemed an ideal starting point. The initial specification of three seats was soon changed to five with the introduction of America's Space Prize. Another particularly interesting requirement, the capsule should be capable of returning to Earth for a propulsive landing without the aid of parachutes by an aft mounted thruster package which would also serve as the emergency escape engines in the event of an in-flight abort. Elson came back with the Magic Dragon, a cylindrical capsule with five forward facing seats, two pair side by side and one center mounted seat, each with their display screen, and a cabin wall featuring several small, square windows. An ISS standard Common Berthing Mechanism (CBM) was located in the forward end, protected during launch by a disposable fairing. Immediately below the pressure vessel would be the wider, SpaceX supplied propulsion module and heat shield. Interestingly, and reflecting the Dragon design which later emerged, the landing/escape engines would be side mounted, immediately ahead of the heat shield. Launched unmanned, it could deliver three ISS pallet racks. Unfortunately for Elson, who was working out of Great Britain, U.S. ITAR provisions prevented SpaceX from supplying either detailed drawings of the CBM, or much in the way of useful information about the Falcon 5 booster. That seething hotbed of terrorists and longtime sworn blood enemy of the U.S., otherwise known as England, could not be trusted to work on a simple aluminum pressure vessel with all the weapons potential of a 200 year-old steam boiler. Working with what information it did have, Elson's company fabricated a mockup of the Magic Dragon and shipped to SpaceX's El Segundo headquarters for examination in early 2005. [79]

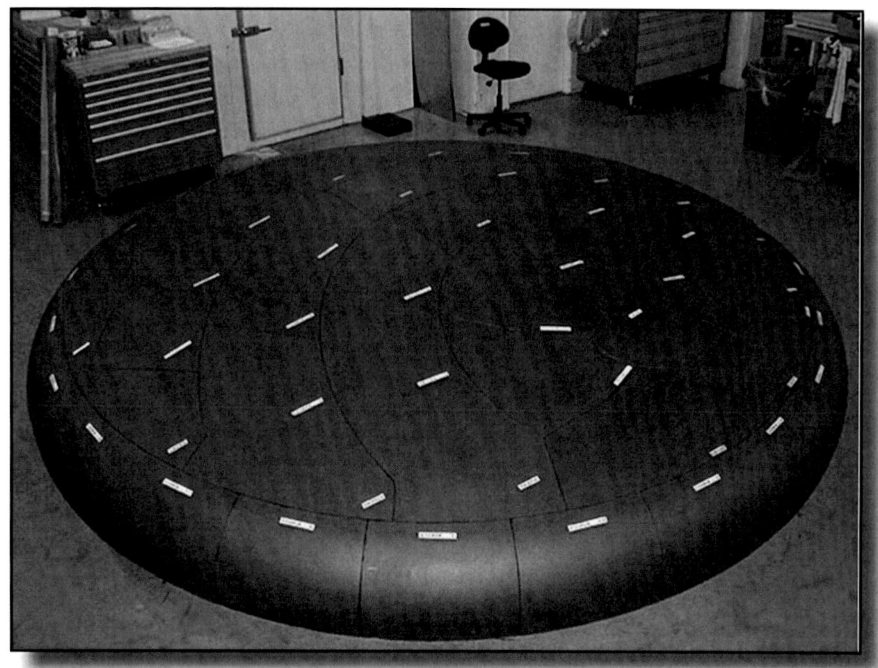

Dragon heat shield. Credit: SpaceX

With the introduction of the COTS program, SpaceX's plans for an eventual crew capable spacecraft along the lines of Magic Dragon underwent a major revision, resulting in the larger and more capable spacecraft which formed the basis of the proposal the company submitted to NASA in early 2006. What had previously been something of a "dark" project for SpaceX was now permanently out in the open. One thing which would remain the same however, was the name Dragon, a moniker chosen from the famous Peter, Paul and Mary song from the very outset as a defiant response to early critics who suggested the company's goals were sheer fantasy.

From its earliest concepts to its latest version, there was never any doubt that driven by the vision of its founder and Chief Technology Officer Elon Musk, SpaceX would pursue a capsule rather than a winged spacecraft or lifting body for its entry into the COTS competition. Still, with other and equally serious contenders for COTS and Commercial Crew just as committed to winged solutions to spacecraft recovery, as well as a newly emerging suborbital space tourism industry developing rocket powered space planes, it is useful to examine why SpaceX elected to pursue a capsule architecture. After all, with the future of space transportation systems still very much unsettled, it just might prove to be one of the few areas where the outspoken founder got it wrong.

Even though NASA had spent the last 30 years designing, building and flying its fleet of winged Space Shuttles, by the time the Orbital Space Plane project got underway in 2002, the experience with the Shuttle's troublesome thermal protection system and its brittle ceramic tiles had already convinced a number of aerospace veterans that for most applications winged spacecraft were simply more trouble than they were worth. It was an opinion tragically re-enforced on February 1st 2003, when the Shuttle Columbia broke up over the Southwestern U.S. on re-entry. While the Columbia disaster shared in common with Challenger the undeniable role of human neglect in addressing a known problem, this time there were no solid rocket boosters to blame, and although the External Tank had been the source of the foam which doomed the ship, the problem was in the nature of the orbiter itself and the horizontal configuration of the stack.

Compared to the blunt capsules which marked the first three US manned space transportation systems, and the design evolving under the OSP program, the Shuttle's swept delta wings allowed for precise and graceful landings with substantial payloads aboard, but they also added considerable weight to the vehicle and significantly multiplied the problems of decelerating through the upper atmosphere, requiring elaborate, computer controlled re-entry maneuvers in addition to the notoriously fragile ceramic tiles. Much of the specific difficulties with the Space Shuttle fleet stemmed from the initial design requirements, originating from the Air Force, that the Orbiters must reenter the atmosphere with a cross range capability of not less than 1200 miles in order to make it possible for the Shuttle to launch and land in a single orbit, or to maximize the chance that in either an emergency launch abort or unplanned landing event, the technologically sensitive orbiter could put down on a friendly military airfield, and not in the middle of a hostile country. Coupled with the enormous payload bay and corresponding airframe, the only way to meet the requirement without adding air breathing engines was a comparatively large Delta wing design which added weight and further magnified the problem of managing the heat and stress of re-entry. [80]

Part of the problem, as the CAIB pointed out, was that no matter how badly NASA wanted to bill the Shuttle as operational, it was still a developmental system, and its design reflected a first effort at solving an incredibly difficult challenge. Subsequent developments, specifically the much smaller, unmanned Boeing X-37B spaceplane have substantially rehabilitated the reputation of winged re-entry vehicles as a whole, confirming the fact that like so much else with the Space Shuttle system, design decisions made in the early phases of an entirely new regime of human flight resulted in a final product which was substantially more expensive, more dangerous and less flexible than it had to be.

Although early visionaries certainly worked on, and dreamed of, winged rocket planes such as Eugene Sanger's antipodal bomber, or Wernher von Braun's own Delta winged spaceplane rendered gloriously on the pages of Colliers in 1952's iconic "Man will Conquer Space Soon series," the pressure of competition and the reality of an imperfectly understood environment, as well as the limitations of launch vehicles themselves led planners on both sides of the Cold War battlefront to select comparatively uncomplicated capsules as the means to reach the high frontier. Though lacking the allure of high performance jet aircraft which the first astronauts were used to flying, mankind's first forays into suborbital and orbital space were conducted in simple capsules because they were a natural fit for the ballistic missiles which would launch them, and most importantly because they were the most expedient solution available.

Forty years later, SpaceX would reach the same conclusion for many of the same reasons, but with an emphasis on one in particular. A winged spacecraft was never in the picture because it was simply "dead weight" for the destination Elon Musk ultimately meant to go, Mars. While the Red Planet's tenuous atmosphere is actually capable of supporting winged flight, and there have been a number of serious proposals to demonstrate the ability to do just that, it requires a design solution of large lightweight wings and very small airframes, more on the order of a powered hang glider, solutions which are completely incompatible with the swept delta of the Shuttle or the stubby, minimalistic wings and lifting body shapes designed to crash back through Earth's atmosphere from low Earth orbit at 18,000 mph. While wings of any sort would be useless in arriving at Mars, they would be an absolute liability on the voyage home, where a spacecraft returning from the Red Planet would slam into Earth's upper atmosphere at the blistering speed of approximately 28,000 mph, a task barely manageable by a blunt capsule with the best heat shield money can buy.

With its goal set from the beginning upon winning all phases of the COTS competition including the crew transfer capabilities under the optional COTS-D provision, certain basic design criteria, such as incorporating both a return capability via parachute and ultimately a launch escape system had been a foregone assumption for SpaceX in crafting its COTS proposal. Still, selecting the actual shape and aerodynamic characteristics of the capsule, its power source and the configuration of the service module which would take the capsule from an initial orbit to Station rendezvous and then deliver both pressurized and unpressurized cargo, was a challenge which offered a variety of potential solutions, beginning with the shape of the capsule itself.

In the early days of the space race, the Soviet Union and the United States chose very different architectures for the first manned spacecraft they sought to develop, and it set both nations on a diverging path which still define them today. They were decisions driven as much by geography as anything else. When the Soviet Union embarked on its space program, it did so from the secrecy of a landlocked launch pad and recovery zone deep behind closed borders on the vast steppes of Kazakhstan where it was already testing ballistic missiles. Facing unpredictable topography and a wide landing ellipse, designers for the original Vostok spacecraft chose an almost spherical re-entry pressure vessel which ejected its occupant for a final, manually controlled parachute descent, while the capsule itself came to a rough and tumble, sometimes bouncing landing on solid ground.

The United States on the other hand, blessed with long Atlantic and Pacific coasts and a navy which spanned the globe, intended from the outset of Project Mercury to launch out of Florida's Cape Canaveral, and to conduct recovery operations over water. For its first two suborbital Mercury launches aboard underpowered Redstone rockets, that meant splashing down in the Atlantic Ocean. Consequently Maxime Faget's Mercury capsule was designed from the outset to provide inherently safe flotation for the astronauts as they awaited recovery. When NASA decided that it would need an interim program, Gemini, between Mercury and Apollo in order to learn the complex skill of rendezvous and docking which the latter would require, designers initially attempted to incorporate a ground landing capacity by including a Rogallo wing, an inflatable, flexible parachute resembling a hang glider. Upon encountering a series of difficulties in deploying the wing, program managers pressed for time by the requirements of Apollo, returned to the simpler expediency of parachutes and water recovery.

As the United States progressed through Mercury and Gemini with its sights set on the Moon, it was clear that with much higher descent speeds, Apollo would need a very broad and blunt heat shield to lessen the G force of deceleration and shed velocity and heat safely on the way to what would necessarily be a water landing. Some forty years after the last Apollo capsule splashed down into the Pacific, SpaceX, racing to produce the nation's next crew capable capsule to actually launch into space and return from orbit, followed both expediency and historical precedent in slating water recovery for its first generation Dragon spacecraft as well.

Another key decision regarded the Dragon's overall shape and the placement of its heat shield, and once again it borrowed some elements from the past. The now familiar truncated cone shape of the Apollo command module, with its broad base and narrow nose, is one which has been closely copied in recent years by both Lockheed Martin for the Orion crew capsule, as well as the Boeing Corporation for its own CST-100 spacecraft. Although at a glance the Apollo capsule does not appear to be an object capable of generating lift in the manner of an airplane wing, its center of gravity is slightly offset to provide a nominal degree of tilt, and thus the capability to generate a modest lift to drag ratio (.037) both assisting with deceleration as well as the ability when combined with thrusters to target a landing spot within a couple of miles. SpaceX wanted this capacity for the Dragon as well, but augmented much further to facilitate splashdown within an even narrower ellipse, a necessary requirement since it would not have the resource of the United States Navy at its disposal for conducting recovery operations.

Like the Apollo designers, SpaceX was also contemplating eventual high velocity returns, both from Mars as well as the Moon and other locations. This also suggested a comparatively broad heat shield, designed to increase atmospheric drag and distribute the heat of re-entry over as large a surface area as possible. As a result of these two converging requirements, SpaceX selected a similar shape to that of Apollo, incorporating a broad heat shield at the base, but with a more vertical cabin mold line which kept the base constrained to the width of the booster at 12'. Such an approach may seem obvious, but it was by no means the only choice.

Until the advent of high resolution digital imagery, the nation's fleet of spy satellites, operated by the National Reconnaissance Organization, like the rest of us, had to develop images on film. For the NRO however, operations were impeded by the fact that the film was anywhere from 200- 600 miles above the Earth, contained on long reels on a massive satellite which was never coming down. The necessary solution was to develop a means of periodically sending capsules of film back down to Earth for analysis. With the film obviously being highly sensitive, and the satellites by design following orbits which maximized the time over potential enemies, the capsules need to return in a tightly defined de-orbit to a very precise location. As if this was not difficult enough, in order to prevent possible interception and damage from a hard landing, the film canisters would be snatched out of the air by military aircraft as they were parachuting through the final layers of the lower atmosphere.

The solution was a small capsule called Corona, which entered the atmosphere nose first, and was shaped more like a lower velocity bullet than the more familiar blunt, rear end first capsule shape. Although perfected for an unmanned capsule weighing only a few pounds, the design is easily scalable to much larger spacecraft as well, and several serious COTS contenders included variations of this approach in their own proposals.

There are a number of advantages in this design. First, it does not require a disposable nose cap for launch operations. Depending on the degree of the angle, and distribution of heat shielding, it is a design which can lead to a much higher level of lift and maneuverability for re-entering vehicles, sitting on the evolutionary path towards true lifting bodies, and ultimately space planes. It also leaves the broad area where the heat shield would normally be placed as the preferred location for a docking hatch, allowing for easier loading of oversized cargo within the limits of the hatch itself. On the other hand, Corona style capsules present an interesting challenge for crew accommodations as the vessel experiences re-entry deceleration in the opposite direction as the launch configuration. With Dragon's basic orientation determined by the long range goal of conducting high energy returns from deep space, the easier solution was the one selected. The next major decision regarded the provision of electrical power, and here SpaceX readily followed the path well-traveled by Russia.

The three previous U.S. capsules as well as the Space Shuttle, were each powered by chemical batteries or hydrogen/oxygen fuel cells, a reasonable solution for the comparatively short duration of U.S. manned space missions. The Soviet Union by contrast, incorporated solar power for early versions of its Soyuz spacecraft, and then temporarily abandoned them when making short duration transfer flights to its Salyut series of space stations in the 1970's. By the advent of the more advanced Mir space station however, both the Soyuz transport craft and the Progress resupply freighters were equipped with deployable solar power panels, a configuration which transferred to the ISS program as well. As with both of the other ISS Visiting Vehicles, ATV and HTV, the cargo version of the Dragon would need a steady supply of electrical power while in transit to the station. With a commercial, free flying version of the Dragon already under consideration and future versions intended for deep space, solar panels were a necessity.

If there was one feature which distinguished the Dragon spacecraft from every other capsule which has previously flown, American, Russian and Chinese, it is without a doubt the unique solution SpaceX devised for delivering both pressurized and unpressurized cargo at the same time, all while maintaining the objective of returning pressurized cargo safely back to Earth while preserving as much of the craft as possible for eventual re-use.

The Gemini and Apollo capsules both employed large service modules containing the fuel, oxygen and other systems necessary to allow them to conduct extended operations in orbit and on the way to the Moon as required by their respective assignments, but in each case the service module was discarded prior to re-entry, with only the crew capsule returning to Earth. Russia and China both took this approach a step further, returning only the diminutive middle segment of a three part system.

While the Dragon would not be able to return everything, particularly the solar panels, SpaceX designers came up with an alternative to the typical service module architecture. Instead, the lower third of the pressure vessel was angled inwards to a narrower base, and the critical systems needed to maintain the craft in space were contained within the mold lines, arranged around the perimeter. Instead of a service module, the Dragon would be attached to a "trunk," a variably sized cylinder connecting the spacecraft to the vehicle second stage, to which the solar panels would be attached. Once detached from the second stage, the lower end of the trunk assembly would be open to space, and available for storage of the external cargo, to be carefully removed by the Space Station's remote controlled external arm once the Dragon was berthed to the station. Alternatively, the trunk could also contain smaller satellites for orbital delivery before or after a visit to the station. By virtue of this unique arrangement, the Dragon could return to Earth with nearly all of its vital systems intact and available for careful inspection and analysis to make sequential improvements in subsequent spacecraft, ultimately leading to re-use.

Another major design decision, and one which would become a production issue as well, concerned the orbital maneuvering and reaction control systems contained in the lower skirt assembly. Once again, near term decisions would be made with long term plans clearly in mind. Although the Soyuz capsules can maintain orbit on station for long periods of time, their ultimate limiting factor is the gradual degradation of the hydrogen peroxide propellants used in the reaction control system. For SpaceX, a propellant mixture capable of enduring extended periods in space, at least up to two years, was a major imperative, and one which meant using the hazardous but extremely effective hypergolic mixture of monomethyl-hydrazine and nitrogen tetroxide. Used throughout the aerospace industry in orbital, geostationary and deep space systems, if ever there was an opportunity to buy "off the shelf" systems for their spacecraft, this was it. Not for the first time however, existing vendors were operating on a cost basis which SpaceX found to be incompatible with the company's approach to business. In what was becoming a repeated refrain, SpaceX instead sought to find a better way to do it themselves, a process which led to the decision to produce its own thruster, named appropriately enough, Draco.

The distinguishing feature of the Draco thruster is the fact that it can be fired in a very precisely metered fashion, with bursts ranging from a fraction of a second to sustained burns of several minutes. Because of this flexibility, SpaceX was able to do away entirely with a smaller set of "station keeping" thrusters normally found on other craft and combine both requirements into a single device. Whether it was in pursuit of ISS once placed in orbit, or minute adjustments in the immediate proximity of the station, the clustered and redundant Draco thrusters would carry the entire load.

Far less visible than the Draco thruster nozzles prominently placed around the

Dragon's perimeter, was the equally unique computer system designed to control its every movement in orbit. As with most other aspects of the Falcon/Dragon combination, SpaceX considered the design of the Dragon control system without constraining itself to precedent and the limited suite of very expensive solutions generally utilized by high dollar, high performance spacecraft. Much as it did with the new Merlin 1C engine controllers, which utilized three flight computers per engine, with each "voting" on every outcome and the majority prevailing, the Dragon's designer team outlined an architecture which incorporated three strings of two computers each. Composed of commercial off the shelf components, programmed in C++ and running on the Linux operating system, the Dragon's software selection offered the widest possible field for recruiting top programming talent to produce a highly redundant, yet readily evolvable system. Achieving reliability however, meant devising a defense against the unique challenge of space, and the lurking unseen presence of radiation, which can wreak havoc on even the best computer systems.

Sitting at home on planet Earth, computers, like people, are generally protected from harmful solar radiation by the thick lower atmosphere, a level of protection which diminishes significantly even at the altitudes flown by commercial jets. As one gets progressively higher, finally entering the near vacuum of space, the radiation environment becomes increasingly more hostile, both for people and machines. When high speed particles pass through the human body, they have a nasty tendency to damage DNA, turning cells into rogue agents which can lead to cancer. When those same particles pass through a computer, they can literally flip binary components, bits, from a zero to a one, or vice-versa, and in doing so alter the programming, with potentially disastrous results.

There are two general approaches to protecting components from space radiation. The first is to produce specially protected, radiation "hardened" components which can literally take the hit and keep on working. These act as little shields of armor, and are used on comsats, high performance military satellites, as well as probes sent into deep space. The downside to rad hardening is that the components are very expensive, but more importantly, they offer overall performance several generations behind that which can be found on the commercial market. Furthermore, once a rad hardened component is incorporated into a system, it tends to stay there, severely limiting the potential for upgrades to accommodate changing requirements or new opportunities. As one might imagine, coming from a background in a computer industry where constant upgrades and sequential improvements are a regular fact of life, the fixed and constrained architectures offered by conventional rad hardened systems were an anathema to Musk's concept of development, and hardly seemed like an ideal solution.

The second approach to dealing with radiation's effect on objects in space is to look at the problem from a systemic point of view, with the goal of protecting the entire spacecraft's overall integrity, even as it takes radiation hits. The answer is to develop redundant systems in which the loss, generally temporary, of any one component is accommodated by having another component assume that function. The computers are required to "vote" before performing a function, and if one component has been affected, it merely loses as 2-1 vote and the operation proceeds normally. Though the flight computers were not nearly as advanced, this was the solution utilized for the Space Shuttle, and was one which proved resilient in the face of multiple recorded radiation hits. SpaceX engineers took this a step further however, assigning the Dragon a two-string system for each of the three overall voting computers. In doing so, the overall odds of taking a strike which took out multiple elements of the system in one blow were reduced to extremely low levels.

Even though it was designed to operate autonomously, Dragon would need to stay in constant communication with controllers at SpaceX throughout its mission, during which direct control could be asserted via S band with NASA's orbiting TDRSS satellite fleet at any point, even in the event of a total computer failure. Upon approaching the International Space Station, Dragon would require both an independent means to accurately acquire the station's precise location in space, as well as target its very specific and narrowly defined approach. Finally, the spacecraft required a means of direct radio communications between itself and the Station, completely independent of NASA's TDRSS system, allowing astronauts aboard ISS to issue direct commands to the Dragon if the need required.

As a result of the redundancy being designed into the Dragon system, SpaceX conceived a remarkably versatile spacecraft which in theory at least, should allow the highest margin of safety for any ISS visiting vehicle. From NASA's point of view though, despite the potential, it was an untried system and one which should be viewed with a healthy level of skepticism which only time and testing could assuage.

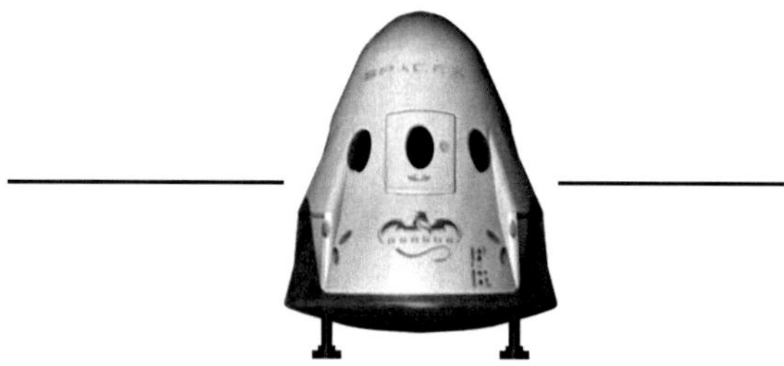

Chapter XVIII: Building the Dragon

From the outside, it is perhaps a bit difficult to comprehend what makes the Dragon such a unique spacecraft. After all, compared to the Falcon booster which launches it, and most assuredly to the Shuttle, some of whose functions it was designed to replace, at 12' width and 14'4" in height without the trunk assembly, the Dragon is a fairly small, unassuming container which quite frankly, one could put on a farm trailer, hook behind the F-150 and tow back to the house, or maybe down to the river for a few days camping. In fact, except for the unusual shape, it might not really attract all that much attention being towed down any interstate, or parked at an RV show. Looks can be deceiving.

From a manufacturing perspective, the first item which stands out about the Dragon was that in winning the COTS award, and then the 12 mission CRS contract, SpaceX was not agreeing to simply build a Dragon, it would have to build quite a lot of them. A minimum of 15 in fact, with the potential for many more if the company eventually triumphed in NASA's entirely separate Commercial Crew competition, and more still if SpaceX received further Station supply awards post 2015. And then there was the price. In signing the firm, fixed price CRS contract for 12 re-supply flights at $133 million per flight, each to be conducted, at NASA's insistence, aboard a new capsule, Elon Musk guaranteed that his company could not only produce and launch 16 EELV class rockets for substantially less than any other equivalent booster in the world, he was also guaranteeing that against equally daunting odds, his company could also design, test, build, fly and recover the 15 separate Dragon capsules as well, and all under a price cap which included the launch vehicle, mission assurance and launch operations as well. Furthermore, by any consideration, NASA's net public investment of taxpayer funds in the development of the cargo version of the Dragon was less than $250 million. By comparison, and over almost the exact same time frame, NASA poured more than 28 times that amount, $7 billion, into the Lockheed Martin CEV/Orion/MPCV project which at best will see one single, unmanned test flight of an incomplete vehicle before the Dragon could be expected to complete its entire initial 15 flight manifest.

And then there was this, in taking on the Falcon 9 booster, SpaceX at least had the benefit of experience, limited as it might be, with its own efforts to launch the Falcon 1, an ordeal which had yet to see success at the time of the COTS award. The Dragon on the other hand, not only represented entirely new and unfamiliar territory, unlike three other major contractors, Boeing, Lockheed Martin and Orbital Sciences, as well as the ISS international partners who were building Visiting Vehicles, SpaceX had no experience in the closely related satellite building industry whatsoever, other than a small financial stake in British Surrey Satellite systems. The latter is a critical and perhaps overlooked point. In terms of much of its functionality, an orbital space capsule is basically a large satellite outfitted with a pressurized storage container rather than instruments, transponders and other mission specific hardware. Recognizing this, in designing its own Cygnus transfer vehicle, Orbital Sciences started with hardware it was already producing for its own very successful series of small and intermediate sized satellites, and added a pressurized module built by Thales Alenia. SpaceX, by contrast, with a complete lack of experience, and no public ambitions for entering the satellite market, elected to design, develop and produce the entire Dragon system, from the beginning, with an absolute minimum of outside hardware. It was no exaggeration then, when Elon Musk observed time and again that the Dragon provided just as much of a challenge to the company as the booster which launched it. Meeting that challenge meant devising an efficient, low cost manufacturing plan for a single model of spacecraft meant to meet multiple mission requirements. Just as with the F1 and F9 boosters; the outcome depended on key early decisions regarding materials and shape.

To begin with, the Dragon system is made of up of two elements, the capsule and the trunk. Structurally, the capsule is comprised of three sections; the protective nosecone which jettisons during launch, the capsule itself, which includes the pressure shell as well as OMS and life support equipment contained in the skirt under the cabin, and finally the heat shield.

With the pressure shell being the key element, engineers took a conservative approach in selecting a bi-conic design which lent itself to ease of construction, a decision further enhanced by the use of conventional aerospace metals rather than potentially weight saving composites. The pressure vessel is fabricated out of flat aluminum/lithium alloy, and in a very different technique than SpaceX applied to the Falcon 9 tank assembly, the alloy is machined into an isogrid structure, exploiting the unique physical properties of the triangle as one of nature's strongest shapes. Working with stock sheets, unnecessary metal is ground out by machine leaving only a network of triangles representing the material's original thickness. Once the isogrid pattern is machined into the surface, the plates are curved into truncated conical sections, and then welded together to form an integral structure.

Arrayed around the base of the structure in separate compartments formed by radial bulkheads are partitions containing the eight spherical hypergolic fuel and oxidizer tanks, with four each, along with two helium pressurization tanks which power the Dragon's Draco orbital maneuvering system. Like the Draco thrusters themselves, the titanium fuel and oxidizer tanks are also built in the Hawthorne facility. Each tank is formed by attaching a round sheet of titanium stock to a rotating steel mandrel, and then gradually heating the sphere which is formed under pressure to a half hemispheric shape. After installing internal tank hardware, the two halves are welded together to form a completed tank assembly. Altogether the Dragon offers 11 cubic meters or 388 cubic feet of pressurized space, with a sidewall angle of a modest 15 degrees, leading to a higher proportion of usable space than offered in the more steeply sloped Apollo, Orion, and Boeing CST-100 capsules.

Following fabrication, each 90 lb. force Draco thruster is tested for consistency, and then as many as 18 separate units are grouped together in 4 "quads" arrayed around the sidewall of the craft just above the heat shield. With the addition of cross-valving from the two independent helium pressurization tanks, the Draco system is dual redundant. In addition to the thrusters and tanks, the Dragon skirt also houses a small, 3 cubic foot sensor bay which opens once the craft is in orbit, exposing its components to the space environment. Mounted on the inside of the sensor bay door is the grapple attachment point which allows astronauts aboard the Station to snag the free flying spacecraft with the remote maneuvering arm, and then guide it to its berthing port on the Harmony module. After release, and prior to de-orbit the bay door is closed and secured for the return trip. The Dragon's descent parachutes are also stored in a separate container within the skirt, with lines routed to re-enforced titanium strong points along the docking collar at the spacecraft's nose.

Like all previous US and Russian capsules, the Dragon uses an ablative heat shield to protect the spacecraft from the extreme heat of re-entry. Earlier heat shields, designed for a single use, were comprised of AVCOAT, an epoxy impregnated fiberglass honeycomb. In the mid 1990's however, NASA found itself in need of a higher performing heat shield than had ever been previously required in order to conduct a successful comet sample return mission. The ambitious mission called for a spacecraft to launch in 1999, and fly into the tail of the comet Wild 2, collect samples from the stream of particles coming from the comet, and then return them to Earth at the scorching velocity of 28,000 mph, the fastest ever re-entry of a man-made object into Earth's atmosphere, and one sure to generate a truly hellish record heating environment. Given the extreme requirements of the mission, NASA Ames elected to develop a superior alternative to the Apollo era heat shielding. The new material, called PICA for phenolic impregnated carbon ablator marked a considerable improvement in the state of the art for heat resistant materials, and the PICA backed heat shield performed flawlessly for the Stardust mission, and again on the Mars Science Laboratory housing the Curiosity rover in 2011.

Rather than proceed with Apollo era AVCOAT and the tedious process of carefully filling each successive void in the honeycomb one after another as Lockheed Martin chose to do with the Orion spacecraft being developed concurrently, SpaceX, working with Ames, elected to update the design properties of the PICA and to manufacture its own variant, at considerably lower cost (10x) which it called simply enough PICA-X. As proven by the NASA comet sampling mission, PICA is more than adequate for return from low Earth orbit, and in fact can handle the considerably higher velocities which would be encountered on a lunar or even Martian return. Manufactured in irregularly shaped segments, the PICA-X heat shield is attached to the Dragon's lens shaped carbon composite carrier structure with thermal expansion joints filled by heat resistant foam.

The second segment of the Dragon assembly is the 9'2" long "trunk" which serves the dual functions of housing non pressurized cargo, as well as connecting the Dragon to the Falcon 9 second stage. The baseline trunk provides 450 cubic feet of cargo storage, with an additional 400 cu feet available in an extended 14' version. Electrical power is supplied by two spring loaded solar panels mounted on either side of the Dragon trunk. The panels, covered by jettisonable fairings during launch, extend the spacecraft's overall width to 54' when fully deployed, and can be rotated 360 degrees to maintain alignment with the sun during the daylight side of the orbit. The panels are capable of generating over 5000 watts to provide power to the spacecraft's functions and charge the on-board, high capacity lithium polymer batteries developed by Tesla, as the craft passes into the Earth's shadow. The trunk section also serves as the mounting platform for the Dragon's radiators, fulfilling the vital role of keeping the spacecraft's temperature within safe limits. The ultimate selection of trunk mounted solar panels probably saved SpaceX the visual embarrassment of its original concept, an alternate arrangement featuring a hinged nose cap which would have opened to reveal both the docking adapter as well as a smaller pair of round, deployable solar panels which likely would have resulted in what was otherwise a rather winsome design bearing an unfortunate resemblance to Mickey Mouse.

One of the deciding factors in favor of the Dragon spacecraft for NASA's COTS program was the fact that it offered a maximum return capacity of 6,614 lbs. when departing from Station, a capability which would be put to use returning scientific equipment, lab samples, and ISS support hardware for refurbishment and re-launch. With ISS intended to transition from a project which was seemingly always under construction to a fully functional laboratory in space, both the volume and pace of research related material to be transported to, and from, Earth orbit was slated to increase significantly once a supplier was in place.

The Mercury, Gemini as well as the current Soyuz spacecraft all returned to Earth under a single parachute with a backup kept in reserve. With a significantly higher dry mass that grows even more ponderous when loaded with return cargo, the Dragon required a more robust recovery system, closely resembling that employed by the Apollo capsule, consisting of a simultaneous three parachute deployment of 116' feet each, preceded by twin drogue chutes. With no U.S. Navy at its disposal, the goal was to target the landing zone as accurately as possible, within a few kilometers of a designated point, and then conduct the splash down at a comparatively gentle rate of descent, with a rapid retrieval from the ocean by a minimal crew onboard a modest vessel. The entire forward face of the capsule is taken up by the NASA Common Berthing Mechanism docking adapter, which is welded into place and protected in flight by a disposable carbon composite fairing.

As an international effort comprised of more than 20 different nations, the ISS is not without its own share of design anachronisms, one of which is that it is equipped with three completely different docking mechanisms. Consequently, incoming spacecraft can only dock with segments which host compatible docking systems. Russian Soyuz crew vessels, Progress supply ships and the European ATV dock automatically (Soyuz controls can be manually overridden), using a probe and drogue system which has been the hallmark of the Russian space program going back to the earliest Salyut stations. It is a robust, mistake tolerant system which actively aligns approaching craft through a male/female cone arrangement. By contrast the Space Shuttle used a derivative of another Russian system which was originally developed as part of the Apollo-Soyuz test project. It is called APAS, or Androgynous Peripheral Attach System, and as the name implies, eschews a male/female arrangement for a system in which either side can play the active or passive role. This was the design the United States originally adapted to allow the Shuttle fleet to dock with the Mir space station. It is a very flexible configuration, but like the Soyuz/Progress docking port, has a comparatively narrow opening which will not permit passage of ISS Standard Payload Racks. The only alternative wide enough for this critical function is the third docking mechanism, and the one used for the Dragon re-supply missions; the NASA Common Berthing Mechanism, which is also used to join all non-Russian segments of the ISS together.

As if three separate docking systems were not enough, after several years of work on a new international Low Impact Docking System, iLIDs, which began under Project Constellation, in late 2012 NASA halted work on the project, and announced its intention to contract with Boeing for a fifth system, SIMAC, which the company already had under development. Like iLIDs, the Boeing design removes the necessity of employing the station's robotic arm to achieve berthing, while also allowing for a comparatively rapid disengagement from the station in the event of a major problem.

Although SpaceX designed Dragon to meet NASA's specific requirements outlined in the COTS program, the space agency was not the only intended customer. Shortly after securing a slot in the program, SpaceX introduced a strictly commercial version of the spacecraft, called DragonLab. While NASA has long touted the potential benefits of micro-gravity research as a justification for the Shuttle program and as a reason for building a permanent Space Station, the agency has incongruously been in the position of both helping and hindering actually performing the research itself. On one hand, the agency enabled the creation of materials research capacity in space by supplying what otherwise did not exist in the form of Space Shuttle Mid-deck Locker Experiments on Spacelab and then aboard the far more expansive ISS. On the other hand, very inconsistent launch schedules coupled with onerous regulations and requirements made it both expensive and difficult to do business with NASA, reducing the incentives to perform institutional and commercial research which offer speculative payoffs at best. Furthermore, though seemingly floating serenely in space, both the Shuttle and Station are rarely quiet due to both onboard machinery as well as human activity. The resulting vibrations can significantly reduce its effectiveness as an environment for some types of experiments. DragonLab is a solution to this problem, capable of functioning as a free flying laboratory and able to remain in orbit for as long as two years before returning to Earth. Offered in this configuration, DragonLab leverages the spacecraft's power and climate controlled capacity to support user payloads. Exploring this market potential, SpaceX hosted a DragonLab user conference at its Hawthorne headquarters in 2006, followed by a second conference in Haarlem, Netherlands, after which the company announced it was booking two DragonLab flights, the first in 2014 followed by a second flight in 2015, (now slipped to 2016 and 2018) with little information released beyond the intended launch dates. It is very possible however, that one of the DragonLab missions will see the first ever re-use of the spacecraft.

Two of the most important considerations in designing the Dragon for Station operations were insuring the ability to maintain reliable communications with both Hawthorne and NASA Mission Control in Houston at any point in flight, and once it reached the vicinity of the Station, to establish direct radio contact with the crew aboard. SpaceX's answer was the Commercial Orbital Transportation Services Ultra High Frequency Communication Unit (CUCU) a two way radio communications system which augments the spacecraft's stand-alone radio communications maintaining almost constant contact between SpaceX Mission Control in Hawthorne and Dragon anywhere in orbit via NASA's fleet of TDRSS satellites, as well as through ground stations. ISS protocols require that the crew aboard the station has direct, line of site radio communications and command authority over any visiting vehicle. Consequently, as Dragon approaches within 23 kilometers of the Station, it switches to direct space to space radio contact, while still maintaining the separate links to ground.

After conducting tests of the CUCU from the ground in Hawthorne, where it demonstrated the ability to stay in touch with Houston Mission Control, the doubly redundant ISS transmitter/receiver and a backup were loaded aboard Space Shuttle Atlantis for STS-129, which launched into orbit on November 16th, 2009, becoming the first piece of SpaceX built hardware to be physically interfaced with the Station. The CUCU unit was installed on ISS in January 2010, beginning a series of on-orbit checkout and verification tests.

As the Dragon comes within the immediate vicinity of the Station for final approach, it employs three different methods to verify its position; differential GPS, LIDAR and thermal imaging. The latter two functions are incorporated into a single unit developed by SpaceX called DragonEye, the Dragon Laser Imaging Detection and Ranging Sensor which provides 3-D imaging for range and bearing. To test the DragonEye in actual flight conditions, NASA mounted a unit aboard the docking adapter assembly of Space Shuttle Endeavour for its July 15th, 2009 launch to ISS. As the Shuttle came in for docking to the Station, the DragonEye performed as designed, paving the way for its first use aboard the Dragon, then scheduled for the COTS 3 mission.

With many of its critical systems developed, the next major step in qualifying the first free flying Dragon for spaceflight was something common to every capsule design, the parachute drop test. The testing sequence began simply enough with a series of 1/3 scale model drops into a swimming pool, designed to grade both the spacecraft's overall stability in the water depending on the initial angle of entry upon splashdown, as well as the overall buoyancy depending on how heavily the small spacecraft was packed for the return trip. Results verified computer simulations showing the capsule maintained stability and buoyancy over the anticipated loading range.

On August 20, 2010, still basking in the success from its first Falcon 9 launch, SpaceX prepared to take the next step, a drop test into the Pacific Ocean. The location was the scenic coastal town of Morrow Bay, California, The test capsule was delivered by truck to the parking lot of the Dynegy corporation, where within a few minutes the tell-tale chop of helicopter blades indicated the vehicle's ride had arrived, an enormous Erikson Air-Crane helicopter. Engineers carefully hooked the Dragon test article to lifting straps, and within minutes, the capsule was on its way to 14,000 feet and nine miles out over a calm Pacific. Upon reaching the center of the circular test zone, located 45 miles north from Vandenberg, the helicopter crew released the cargo.

After an initial free fall, the Dragon deployed two drogue chutes to stabilize and begin to slow the craft. With the drogues released, the orange and white main parachutes deployed from their compartment in the equipment bay on the lower skirt, their harnesses routed in spiral groove leading from the compartment to the connection hard points under the nose cap. Functioning perfectly, the three parachutes deployed as planned, slowing the capsule to a maximum of 18 feet per second, for a controlled splashdown.

The capsule landed without incident, assuming a stable position, and within minutes the recovery teams hooked it to a small support craft to be towed back to port. Once back, Morrow Bay's oddest recreational watercraft was plucked from the water alongside the dock by a waiting crane and deposited back on a truck for the return trip to Hawthorne. With the test proving to be a 100% success, and with nothing on the ticket to clear, further planned drop tests were ruled unnecessary. The next time the Dragon dropped into the Pacific, it would be from quite a bit higher than 14,000 ft.

For veteran observers, the single drop test came as something of a surprise. Other programs in development, such as MPCV went through a number of drop and water tank tests, with more yet to come. The key difference was that free of the higher level of concern which would naturally accompany a crew capsule, one of the advantages built into the COTS approach was the flexibility in affording the competitors the latitude to make many of their own decisions regarding design and testing, and adjust schedules based on the result. After completing the first test mission, there would still be two remaining test flights which presented opportunities for SpaceX to review and potentially improve every aspect of the system, including the recovery capability, prior to beginning the series of 12 flights to the Station under the Commercial Resupply contract. In the end, there was something a little whimsical about the way SpaceX conducted its one and only drop test, ever practical and decidedly frugal. Parsing through the list of credits on its website detailing the event, one sees a picture of a test conducted via short term rentals, a helicopter, a crane, several modest recovery boats, and best of all, a thank you to the Morrow Bay Radio Shack.

Although SpaceX surprised everyone including probably themselves with the unexpected success of the first Falcon 9 flight, the company had also inevitably raised expectations for the next launch, which would take place under the brighter light of public scrutiny which necessarily accompanies taxpayer funded NASA launches. As the first official COTS flight, the objectives were relatively straight forward. SpaceX was to successfully launch and demonstrate control of the Dragon spacecraft with the all new Draco thruster system, which among the other firsts for this mission would require new ground procedures for handling and loading the toxic hypergolic propellants. In addition to demonstrating the capacity to both maneuver and hold station in space, the flight would be a demonstration of the craft's ability to send and receive telemetry via NASA's TDRSS communications network, and finally conduct atmospheric re-entry and ocean recovery.

Two of the key challenges would be new separation events. The first being the separation of the nosecone, which after protecting the Dragon spacecraft's ascent through the atmosphere, detaches from the Dragon during the second stage burn, with several more minutes of flight yet to go. In an operational flight, a failure to successfully eject the nosecone would prevent berthing with the station and result in a loss of mission. After establishing a preliminary orbit, the second and most significant event would be the separation of the Dragon spacecraft from the Falcon 9 second stage, a procedure similar to the first and second stage separation, in this case accomplished by springs mounted on the top of the second stage. Although fully tested on the ground where engineers suspended the Dragon from the Hawthorne factory ceiling with a weighted counterbalance to cancel out the effect of gravity, the first time a new procedure took place under flight conditions was always a cause for concern.

Like the Falcon booster itself, the Dragon spacecraft had also been subjected to a series of pressure, stress, and shock loading tests at the McGregor facility. The results clearly demonstrated that although small, the craft was remarkably resilient and easily capable of withstanding stresses significantly higher than what would be encountered during actual launch and landing conditions, as well as greater than those required by NASA's human rating standards. Selection for crew was far from assured and several years away in any event, but if SpaceX was fortunate enough to win a role, the Dragon capsule would be more than worthy when it came to structural integrity. The real test was the systems which were in it.

COTS-1 Launch. Credit Chris Thompson/SpaceX

Chapter XIX: COTS I

Four years after the COTS program was officially unveiled, and 22 months behind schedule, the successful completion of the Dragon drop test cleared the way for the COTS 1 launch and the beginning of a new era of spaceflight. With much of the attention for the COTS 1 launch focused on the stripped down Dragon capsule and the effort to recover it, the Falcon 9 rocket would still be conducting only its second launch overall, and even though the first flight clearly exceeded expectations, two issues of concern needed to be rectified.

During its maiden flight, the Falcon 9 demonstrated a slight rotation along its vertical axis as it began to push away from the pad. Although the rotation ceased by the time the vehicle cleared the tower, and was in fact anticipated, it was greater than expected. The source of the roll was the torque induced by the turbo-pumps. All liquid fuel rockets experience similar forces, which are easily counteracted by different means depending on the nature of the engine. As an open cycle engine, the Merlin 1C has two exhaust streams; one from the engine itself, and a much smaller exhaust from the combustion which drives the turbo pump. It is this stream which provides the initial source of roll control for the Falcon 9 though a pre-set alignment. Because the Falcon 9 was also using a pad hold down system which kept the rocket firmly in place until it developed full thrust, engineers were concerned prior to the first flight that applying too much initial roll control could damage the hold down system. Applying a fix was a comparatively simple matter of slightly adjusting the roll control program.

The greater area of concern was the completely separate roll observed in the Falcon 9 second stage as it entered orbit with the boilerplate capsule still attached, a phenomenon captured both in the closing seconds of on board video during the launch, as well as by amateur photographers on Australia's Gold Coast who recorded images of a spiral "halo" emitted as it vented liquid oxygen. While it had not substantially affected the outcome of the test launch, a similar occurrence on an operational launch could be a major concern, and video of the roll was uncomfortably reminiscent of the contact induced wobble and slosh effect which had ultimately doomed the second flight of the Falcon 1. Initially thought to be due to radiative heating from the nozzle damaging the roll control actuator, the problem was ultimately traced to the location of a liquid oxygen drain outlet, which had the inadvertent effect of chilling the actuator hydraulic lines, and rendering the stage unable to counteract the natural torque imparted by the flow of fuel to the engine. The solution was to relocate the drain outlet.

Following the procedure established in the buildup to the first Falcon flight, each of the nine first stage Merlin 1C engines intended for the COTS 1 flight were shipped from the Hawthorne headquarters to Texas for full fire testing acceptance. After completing the tests, they were in turn shipped back to California for installation into the thrust structure. Once accomplished, the entire thrust structure was covered in protective shrink wrap and shipped back to Texas for a brief test fire, and then sent on to the Cape. The same process was repeated with the second stage and Merlin D vacuum engine, again without the fragile, 9' high expansion skirt.

Just 6 weeks after the Falcon 9's maiden launch, on July 15, the first stage arrived at the SLC-40 hangar, for a flight tentatively scheduled for September 9. Unsurprisingly, the necessity of thoroughly testing the Dragon's systems led to a series of delays. Pre-launch preparations proceeded along the same path established with test launch 1, with the wet dress rehearsal taking place on September 15th without any issue. It was also without a payload. Back in Hawthorne, engineers continued to conduct a series of hardware in the loop testing, a sophisticated simulation regime in which any component in the spacecraft, (or the launch vehicle) is "tricked" into believing that it is taking place under actual flight conditions, a process as close as one can get to actual flight testing. In this case, further testing on the first Dragon capsule led to a slip from November 10 to the 18th.

Before the launch could legally take place, SpaceX needed to receive a first of its kind commercial spacecraft re-entry license issued by the Federal Aviation Administration's Office of Commercial Space Transportation. The FAA issued the license on November 22nd setting the stage for a pre-launch test fire scheduled for December 3. On the morning of the test fire, the countdown proceeded smoothly all the way to T-1.1 seconds, where it came to a sudden halt with an all too familiar last second abort. Analysis determined the problem to be an incorrect parameter setting leading to a high pressure reading in the number 6 engine. The following morning, SpaceX attempted a second test firing, but this time encountered an abort triggered by a low pressure reading for the gas generator on the same number 6 engine. After troubleshooting the issue, the launch team recycled the countdown, and at 10:50 am, the test firing finally went according to plan. The Falcon 9 was apparently, ready to launch.

As part of the final preparation for launch, SpaceX takes a series of high resolution photographs of its Falcon rocket, both for pre-launch analysis, as well as to provide documentation in the event of a problem. On the evening of December 6, back in California, SpaceX employee Marty Anderson noticed two small cracks in the bell shaped expansion nozzle of the second stage Merlin engine. The cracks were located near a weld joint at the end of the engine, at an area where the niobium alloy shrinks to a thickness only twice that of an aluminum can. Located near the end of the bell skirt, the cracks did not appear to pose a safety risk to either the engine or the mission. The expansion skirt on the second stage engine is designed to allow the rocket to reach peak efficiency in lifting heavier loads or for higher altitude orbits, and would not play a role in the launch to low Earth orbit, but the cracks were still a source of concern.

The solution in this case was to simply trim off the excess material and proceed with the launch. Unfortunately from his point of view, Anderson's reward for discovering the crack was being assigned the task of repairing it. This entailed taking a red-eye from Los Angeles to Florida, an experience heightened by Anderson's somewhat ironic fear of flying. Nevertheless, the next morning, with aviation tin snips in hand, Anderson ascended a platform to work on the now horizontal rocket in near freezing conditions, and deftly proceeded to trim away the excess metal.

Trimming the skirt to eliminate the cracks would effectively reduce the upper stage performance in a planned high orbit ascent test to be conducted after the Dragon capsule was released in parking orbit, but would not have any bearing on the primary objective. The somewhat unconventional repair job completed, SpaceX re-ran mission simulations to account for the slight change and satisfied, turned the results over to NASA for examination. After a series of email exchanges between Elon Musk and NASA head of human spaceflight Bill Gerstenmaier verifying the conclusions, SpaceX received final mission approval.

The flight plan for the first COTS mission called for the Falcon 9 to lift off at 10:43 AM EST following an almost identical trajectory to the first Falcon 9 flight in pursuit of an orbit at 300 km inclined approximately 34.5 degrees to the equator. This trajectory would take the craft over NASA Wallops tracking stations and having passed into the southern hemisphere, over the Australian tracking station at Woomera. After completing one full orbit, the Dragon would proceed to complete almost a second 90 minute orbit before splashing down in the Pacific off the Northern Mexican coast, some 500 miles west of the American Continent. The slip necessitated by the second stage bell housing repair had produced one unanticipated benefit; allowing gale force winds and 15 foot seas in the Atlantic to sufficiently subside to permit another attempt at first stage recovery operations.

Falcon 9 lifts off on the COTS I mission. Credit: NASA

In the interval between liftoff and landing, during the course of two orbits, SpaceX needed to test all of the Dragon's systems, simulating the range of maneuvers which would be required to complete a successful rendezvous and docking with the ISS, including the first ever operational test of the spacecraft's Draco thrusters. Equally important, SpaceX needed to demonstrate constant and reliable communications with the Dragon through both ground based tracking stations as well as with NASA's fleet of TDRSS communications satellites in geosynchronous orbit 22,300 miles above the planet. This was a must to accomplish before the Dragon would be allowed to approach anywhere near the ISS. As part of the communications system test, SpaceX needed to demonstrate the ability to keep the capsule in a very specific orientation to maintain alignment with the narrow as well as omni directional beam antennas, a major test of both the guidance and navigational software.

Accomplishing that, the final and most nerve wracking challenge of all was to wait anxiously as the Dragon performed a de-orbit burn on its way to a controlled re-entry and targeted soft splashdown in the Pacific Ocean. With a spacecraft which might just as easily have been carrying crew it was a unique challenge which had previously only been achieved by the military industrial complex of the three largest superpowers in the world, the United States, the Soviet Union and the People's Republic of China. And all on a rocket which was only lifting off for the second time.

Although not critical for the immediate mission, which was still considered a test of the Falcon 9 vehicle, it was also important that SpaceX, which in its short life had developed something of a history for pad aborts, begin to demonstrate the ability to launch on time to meet the type of narrow, almost instantaneous launch window required to match orbits with the International Space Station.

This time, with no sailboats straying into the flight exclusion zone, and no last second snags, countdown and liftoff occurred precisely on schedule, and as it cleared the tower, the Falcon 9 booster showed no evidence of the slight roll which occurred on the previous launch. The only noticeable event in the first few seconds of flight was a brief flare up as the rocket's exhaust ignited a small cloud of oxygen gas, which was released as the second stage umbilical broke free from the ascending rocket.

Once again the Falcon 9 ascended on a steady plume of orange fire and gradually accelerating, went supersonic and then passed though Max-Q without any difficulty. Moments later the first two Merlin engines cut off on time, and at the 178 second mark, the remaining seven engines terminated as well. During the course of main engine cutoff however, the one meaningful hiccup in the day occurred when one of the Merlin engines triggered a high oxygen, low fuel condition, referred to as "running rich" and automatically accelerated the shutdown which was already only seconds away.

The next step, stage separation, took place on schedule 3 minutes and 2 seconds into the mission and without incident. Pushed away by its pneumatic arms on the interstage adaptor, the first stage began its hypersonic tumble back to Earth and the second attempt to conduct a recovery.

Following the programmed 5 second delay, the second stage Merlin engine safely ignited and began the climb to orbit. Nine minutes into the flight, the second stage engine shut down as the Dragon/Stage 2 vehicle combination entered a parking orbit. Moments later, and for the first time ever, the second stage, flying nominally under precise roll control, released the Dragon capsule into free flight.

Back at mission control in Hawthorne, SpaceX proceeded to put the Dragon through its paces, verifying the communications network, and receiving telemetry. As the team commenced testing the all-important Draco thruster system, it quickly became clear that the Dragon was well on the way to having a spectacular debut performance. Designed to operate on as many as 18 thrusters, but with the capability to lose two quads and remain control authority in all planes, flight controllers were delighted to find a crisp response, with all thrusters, and all systems nominal. Next up was the active demonstration simulating the range of maneuvers required to dock with ISS, and once again the system proved responsive and reliable. Significantly, it demonstrated one of the most demanding procedures, correctly aligning and maintaining a lock with its TDRSS directional antenna.

After completing its battery of tests, and verifying the integrity of each of the systems, the Dragon had one more challenge at hand, hitting a target ellipse in the Pacific Ocean spanning 60 x 20 kilometers. Firing the Draco thrusters again, the spacecraft completed its re-entry burn 2 hours and thirty two minutes into the mission. Twenty six minutes later, the capsule began its plunge into the atmosphere. The flight profile called for the Dragon to re-enter at a 12 degree angle of attack, which would maximize the craft's modest lift ability while controlling the 3,000 degree searing heat. If necessary, attitude control could be maintained automatically by computer controlled thruster firing. As the craft entered the lower, thicker layers of the atmosphere, it encountered what SpaceX considered the two riskiest elements of the entire mission, deployment of first the drogue, and then the main parachutes. After several heart stopping minutes of communications blackout, the little capsule came at last into view, plummeting through the sky in a reassuringly stable attitude. Everything now depended on the parachutes, and a recovery very different from its one previous test from 14,000 ft. With a Navy Cast Glance P-3 Orion aircraft circling the landing zone to take high resolution photographs, both drogues deployed right on time, stabilizing what was already a remarkably smooth descent and slowing the Dragon to within the safe deployment speed range of the mains. Their job done, the drogue parachutes released, floating safely out of the deployment zone as the main parachutes deployed from their compartment near the base of the Dragon, with main straps ripping up through the spiral channel in the outer wall. All three chutes safely deployed, and the Dragon descended at a final, entirely manageable rate of 5.4 meters per second to a picture perfect ocean landing, within 800 meters of the precise target at the center of the landing zone. Completing the flight's events, the Dragon assumed a perfectly even flotation, bobbing placidly in gentle 7-8 foot swells, and waiting retrieval from the recovery vessel which was nearby. Minutes later it was hoisted aboard. Initial analysis of the heat shield revealed that "it was barely scratched."

By any measure conceivable, the day had been an overwhelming success, and following a report by the recovery crew, as well as initial telemetry from the Dragon while the craft was still in orbit, it seemed almost too good to be true. Later that afternoon, a clearly elated Elon Musk, participating in the post launch press conference from mission control in Hawthorne, found himself almost unable to describe the event, observing more than once that "it is difficult to be articulate when your mind is blown. "

Even though it was equipped with multiple backup systems, the Dragon spacecraft functioned so flawlessly that none were required, promoting Musk to observe that "if there had been people sitting in the Dragon spacecraft today, they would have had a very nice ride." Even though obviously delighted with the ocean recovery, Musk followed his observation by pointing out that for the next generation Dragon, the company was committed to a propulsive landing on solid ground.

In the pre-launch interview, Elon Musk had hinted that the capsule would be carrying a surprise payload. When asked about it at the post launch press conference, the SpaceX founder demurred, not wanting to tempt wayward editors with possible distracting headlines, only hinting that if you loved Monty Python, you might like the payload. The next day however, SpaceX finally blew the cover on its "top secret" cargo, a wheel of cheese, now encased in Lucite and on permanent display at the Hawthorne headquarters. A homage to the Monty Python cheese shop sketch about a cheese shop which actually contained no cheese, SpaceX continued to display a refreshing and sometimes irreverent sense of humor unmistakably reminiscent of NASA's early years, but which seemed increasingly scarce of late.

In addition to performing an essentially flawless delivery of the Dragon capsule to the intended orbit, the Falcon 9 vehicle achieved a secondary but essential accomplishment when, after having released the Dragon, mission controllers successfully re-ignited the second stage for a prolonged burn, raising the apogee of the orbit to 11,000 kilometers. Just prior to the re-boost, the Falcon released 8 tiny cubesats, including the first satellite owned and operated by the U.S. Army in 50 years. The secondary achievement of a successful engine restart and subsequent orbit raising maneuver, this time with no unplanned roll, was a necessary milestone on the path to delivering commercial satellites to higher orbits and a critical step in beginning to reduce the steadily mounting backlog of commercial launches.

To the extent that there was any grey lining in this very silver cloud, first stage recovery operations on the Falcon 9 came up short for the second time, but not without gathering some intriguing data. While the stage was again destroyed on re-entry, this time it survived deeper into the atmosphere, allowing a newly installed "black box" to send valuable telemetry regarding the re-entry and descent environment, which was much more extreme than SpaceX had anticipated. Based on the data received, the company would soon be making a major announcement regarding its approach to achieving reusability.

Safe! The first private spacecraft to return from orbit seconds from splashdown off the Mexican Coast. Credit: SpaceX

The first private spacecraft to return from orbit, the COTS I Dragon floats on an azure sea as the recovery team preps it to be hoisted aboard a waiting barge. Credit: SpaceX

As an almost unbelievably successful 2010 drew to a close, it appeared that SpaceX was well on its way to achieving its first mission to the vicinity of ISS on COTS Flight 2 sometime in the middle of the coming year. Whether it could keep that schedule or not depended on a number of factors, including passing the much higher hurdle of NASA oversight where the safety of ISS was concerned. The way forward was also complicated by a request from SpaceX to combine the next two flights; COTS 2 and 3, into a single mission, a request NASA was initially very reluctant to consider. In any event, the first part of 2011 was clearly destined to be dominated by the final flights of all three shuttle orbiters. Even if SpaceX could have been ready in time, which was doubtful, NASA had little incentive to allow the diminutive spacecraft to intrude on the swan song for its proud fleet of Discovery, Endeavor and Atlantis. On the other hand, if SpaceX could mount its next effort in the months immediately following the Shuttle's retirement, both NASA and SpaceX stood to capitalize on the wave of attention which was sure to accompany the nation's goodbye to the program which had dominated the perception of space for a generation, as well as build the case that SpaceX and the Dragon were worthy successors. A successful station run would also be a decisive factor favoring SpaceX in the Commercial Crew competition which was now thrust into the spotlight with the Shuttle's looming retirement, driving an increased sense of anxiety regarding the "gap." Between blasting off from the Cape on the morning of December 8, 2010 and the next COTS launch however, the Dragon spacecraft and its systems needed to pass an extensive and exhaustive series of reviews, as well as a new set of political obstacles, which taken together, would ultimately consume the entire upcoming year and part of the next.

As for the second flight of the Falcon 9 booster, the slightly accelerated shutdown of two engines just prior to main engine cut off led to a brief and altogether nasty controversy which resulted in a lawsuit and unintentionally shed a little light on the political underbelly of the nation's launch industry. The incident arose when according to court documents, Joseph Fragola, a Vice-President for space consulting company Valador Inc. emailed NASA's Chief of Safety and Mission Assurance, in June 2011 saying "I have just heard a rumor, and I am trying now to check its veracity, that the Falcon 9 experienced a double engine failure in the first stage and that the entire stage blew up just after the first stage separated. I also heard that this information was being held from NASA until SpaceX can 'verify' it."

In reality, SpaceX had actually advised NASA of the engine issue only days after December launch, and with the agency having independently performed launch analysis via its East Coast tracking systems, any claim of exploding boosters was demonstrably absurd.

Having raised a safety issue, Valador reportedly then went to SpaceX seeking a "consulting contract" worth up to $1 million to perform outside analysis and assist the company in smoothing over an issue with NASA which it had created in the first place. The SpaceX response came in the form of a lawsuit filed in Fairfax County Va. accusing Valador and Fragola, who had previously performed safety analysis for NASA as part of the Exploration Systems Architecture Study, of making defamatory claims about the reliability of its rocket. The suit was subsequently settled out of court with Valador stating that the company had "confirmed" the rumor was false, and both parties apologizing for any "misunderstanding." [81]

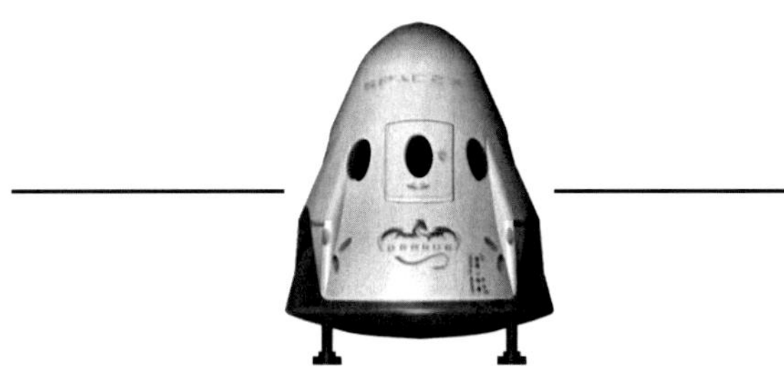

Chapter XX: Interlude in 2011; Introducing the Falcon Heavy and Falcon Reusable

After the excitement of 2010 which saw the first two flights of the Falcon 9, SpaceX entered and exited 2011 without conducting a single orbital launch. Nevertheless, keeping faith with some of his earliest announcements regarding his goals for the company, the year 2011 saw Elon Musk officially introduce two major projects for SpaceX. Separately, each held the ability to significantly alter the basis of current space transportation. Taken together, they offered the potential to change the future of the human race.

On April 5th 2011, speaking to the National Press Club in Washington D.C, Elon Musk formally introduced a heavy lift version of the Falcon 9 booster. It was a project the company openly discussed for several years prior to the announcement, and had even gone so far as to publish performance numbers for when originally introducing the Falcon 9 back in 2005. The rocket, named simply Falcon Heavy, would consist of a single Falcon 9 core, flanked by two first stage Falcon 9 strap-on boosters, with a conventional single Merlin engine powered upper stage topped by an enlarged, all composite payload fairing. To that extent, it closely mirrored the nation's only heavy lift booster besides the Space Shuttle, the United Launch Alliance Boeing Delta IV Heavy.

Unlike the Delta IV, the Falcon Heavy design Elon Musk laid out utilized an ingeniously simple, but mechanically challenging concept, cross feed of propellant. Upon liftoff, all three cores, with 27 combined engines blazing, would produce an impressive 3.4 million lbs. of total thrust. The key distinction is that the central core booster, called the sustainer would initially draw its fuel and oxidizer from the outboard booster's tanks rather than from its own. As a result, at the point in the flight where the strap-on boosters have performed their service and separate from the launch vehicle, the remaining central core is an almost fully fueled rocket lifting off not from zero velocity at sea level, but from a running start of several times the speed of sound and already having passed through the densest part of the atmosphere. It is a screaming hot start which yields a stunning performance in payoff.

Utilizing cross feed of side boosters, and powered by an up-rated version of the Merlin 1C engine now named the Merlin 1D, the Falcon Heavy would be capable of lifting 53 tons, or 117,000 pounds to low Earth orbit, a performance level twice that of both the Delta IV heavy, as well as the Space Shuttle, and fully half that of the mighty Saturn V. Equally jaw dropping was the price, established in a range of $80 million to $125 million, depending on the size of the fairing, the orbit requested and if the cross feed function was required. Government mandated mission assurance fees would be extra, dependent on the contracting agency. Taken together with the performance numbers, the new booster offered a 6 fold advantage over the Delta IV heavy, which costs approximately $500 million per flight, when considering a full accounting of its costs.

In introducing the Falcon Heavy, Musk pointed out that even though the nation would soon be saying goodbye to the Shuttle, scheduled to make its final flight in a matter of months, the United States could look forward to something it could really be proud of, a new booster on a "truly huge scale," one which almost immediately opened up a number of exciting possibilities which otherwise "simply aren't present."

Preparing for Falcon Heavy, transporter/erector at Vandenberg. Credit: SpaceX

He was not exaggerating. Although the price was impressive, it was the combination of price and performance which offered to change the perception of what could be accomplished in space. To get across the idea of just how capable the booster really was, Elon Musk pointed out that at 53,000 kg capacity, the new booster could launch the equivalent of a fully loaded Boeing 737-200 aircraft, complete with fuel, crew, cargo and 136 passengers all the way to Earth orbit. Like the standard Falcon 9 from which it derived, the Falcon Heavy would be designed from the outset to exceed NASA's published crew rating standards. With the entire Falcon 9/Dragon architecture competing in NASA's Commercial Crew competition, a win for SpaceX in delivering crew to low Earth orbit would almost immediately establish a basis for crew launches beyond LEO aboard a Falcon Heavy as well.

To begin with, the Falcon Heavy would come quite close to breaking the seemingly impenetrable $1000 per pound barrier for launch costs to low Earth orbit. Besides offering a far better alternative to existing launch vehicles, Musk pointed out several potential applications for the new booster. For instance, a Mars sample return mission could be performed within the scope of a single launch which would be capable of placing roughly 30,000 lbs. on course to Mars. Often described as the "Holy Grail" of Mars science missions, and a necessary precursor to any human mission, NASA had been discussing a possible sample return mission for decades, but through endless conference presentations and various plans, never came anywhere close to drawing up a serious proposal due to the twin problems of cost and complexity for a minimum two launch scenario. Falcon Heavy solved both.

Furthermore, although President Obama had officially cancelled America's planned return to the Moon through Project Constellation in 2010, the Falcon Heavy would permit an astonishingly affordable lunar mission architecture involving just two flights of the new booster.

It was something of an ironic situation. On one hand, the SpaceX Falcon Heavy architecture seemingly justified the President's decision to kill Constellation by presenting a more viable alternative, but at the same time, the Administration never even gave the appearance of reconsidering its plan based on the new information available. On the other hand, die hard supporters of Project Constellation, though ostensibly motivated by a desire to return America to exploration beyond low Earth orbit, showed absolutely no interest in examining the possibilities enabled by the Falcon Heavy/Dragon combination. Quite the contrary, they instead redoubled efforts to force NASA to build its own heavy lift vehicle out of Shuttle components, even though the most charitable estimates made it clear that it would entail launch costs many times that of the new SpaceX system. Also achievable on a single launch, a lunar circumnavigation by Dragon capsule, replicating or even surpassing the flight of Apollo 8, as well as what would become the first planned crewed flight of NASA's new booster, ten years and more than $20 billion later. SpaceX by contrast, wasn't asking for any taxpayer funds in developing the Falcon Heavy. They didn't need any.

As Elon Musk pointed out, whereas the upgraded Falcon 9 could service approximately half of the range of possible commercial launches, and was quickly building a full manifest of missions to do exactly that, the increased performance offered by the Falcon Heavy would open up the other half of existing heavy and super heavy lift flight opportunities, including every conceivable payload in the DOD's Evolved Expendable Launch Vehicle Program, as well as everything in the commercial satellite industry. In fact, based on that assessment alone, SpaceX envisioned flying just as many Falcon Heavy flights as it did for the Falcon 9.

Given that each Falcon 9 required 10 engines, and a single Falcon Heavy would require a stunning 28, the key to both programs lay in both increasing the production rate and lowering the manufacturing costs of the new Merlin 1D engine, which as of April 2011 was estimated to offer a performance increase from 95,000 to 140,000 lb. of sea level thrust. Anticipating a flight rate approaching 20 per year divided equally between the size classes, SpaceX would need to be able to manufacture 40 cores and nearly 400 rocket engines each year, a rate exceeding that of any nation on Earth. This was a key and somewhat counter intuitive point. By pursuing a heavy lift booster powered by 28 main engines, SpaceX was virtually forcing an automotive style mass production approach, borrowing from the lessons being applied in sister company Tesla, to its rocket engines in an effort designed to drive the per unit costs even lower, allowing the company to maintain the basic price level of $56.5 (2013) million per flight which accompanied the final, production version of the Falcon 9.

The combination of its 21,200 kg capacity to GTO, and a far lower price than any existing competitor, meant that if it worked as advertised, the Falcon Heavy would pose an existential threat to the business case for existing commercial launch vehicles, including even the market leading Ariane V. From a national perspective, it offered the chance to return the United States to a position of global dominance in the commercial satellite launch industry.

It was the potential effect on governmental space policy and existing programs where the introduction of the Falcon Heavy stood to make what could be an even bigger impact. It did not take a rocket scientist to figure out that if you could get fully half the performance of the Saturn V for something on the order of $125 million, then with the notable exception of a smaller payload fairing and volume, the launch performance of the Saturn V could be obtained over two flights for $250 million. By comparison, even if the Saturn V were still in production, the approximate cost would something on the order of $2 billion per flight, plus an untold sum for operating its massive launch infrastructure, including the Vertical Assembly Building, Mobile Crawler Transporters and one of two historic launch pads as well. Put another way, SpaceX was asserting that it could perform the equivalent of 4 Saturn V launches with the Falcon Heavy for what a single Saturn V launch would cost in current terms. While the Saturn V was long gone, and the Ares-V never made it off the drawing boards, NASA, at the insistence of the U.S. Senate was in the process of developing its controversial new mega booster, the Space Launch System. With development cost estimates reaching $38 billion to get past the point of a first crewed flight a decade away, the inevitable comparisons would be even less flattering.

There were a few caveats to consider, mainly related to the total volume of payload available. The Saturn V offered not only amazing launch capacity, but also a large payload volume. The Falcon Heavy by contrast, using the same payload fairing intended for the commercial Falcon 9 was considerably more limited in terms of volume, offering a maximum payload width diameter of 15', with an unbroken length of 37.5' and modest additional volume remaining in the tapering nosecone. The one clear advantage of the gradually emerging Space Launch System, the details of which were still being debated was that although initial versions only called for it to launch the Orion capsule, and a large payload fairing had yet to be either designed or funded, if one eventually emerged, it would be truly huge, possibly on the order of 28' or 33' in diameter and corresponding length of 50 feet. By comparison, the Shuttle payload bay was 15 x 60'.

The question left hanging in the air was just how much could be accomplished within those limitations, and just how important was a large volume anyway. Although NASA had nothing officially to say on the matter, outside of the agency a number of new proposals incorporating the Falcon Heavy began to take shape, and it would not be long until they started to emerge.

Achieving affordable heavy lift is one of the keys to unlocking a new space age, but it is not the only one, and probably not even the most important. The ultimate key is, and always has been, reusability, and six months later, Elon Musk would take to the podium once again to announce yet another new SpaceX project, Falcon Reusable.

Word of the program began to leak out several days prior to the official announcement, when SpaceX filed with the FAA a mandatory public notice of environmental assessment regarding plans to expand the McGregor test facilities operational area and begin conducting low altitude flight tests of a reusable rocket system called Grasshopper. As the company subsequently revealed, Grasshopper is the initial effort in a campaign to develop a fully reusable version of the two stage Falcon 9 booster, called Falcon 9-R, and pronounced "Falcon Niner."

Close-up of carbon fiber landing legs on Falcon 9-R , CRS-3. Credit: Stewart Money

After analyzing data from the first two Falcon 9 launches, particularly flight 2, which was equipped with a "black box" specially outfitted for the purpose, SpaceX concluded that the original plan to recover the first stage booster by parachute was problematic. The basic issue was that once the first stage terminated its boost phase and separated from the second stage, it continued on a ballistic trajectory which causes it to both begin tumbling end over end violently, and to subsequently re-enter the upper atmosphere with all the composure of a fat kid doing a belly flop from a high dive, the results were not pretty, and not easily correctable.

In order to preserve the first stage integrity, it would be necessary to both slow it down, and to regain directional control before it re-entered the atmosphere. This is achievable, but comes at a considerable cost in terms of added margin for fuel which is not going towards launching the craft to orbit, but instead for reducing velocity and landing. Realizing the necessity of re-igniting a main engine, SpaceX dropped the use of parachutes for the stage entirely, substituting an altogether more dramatic powered landing like those which had been successfully demonstrated by the DC-X project, and SpaceX itself was planning for later versions of the Dragon capsule. In fact, even as the Shuttle program was winding down, companies both large and small were already making substantial progress in improving the art of autonomous powered liftoff, transition and touchdown, gradually expanding the flight regime from brief hops of a few meters, sometimes safely tethered to an overhead crane, to progressively longer and higher flights. Whether the purpose is recoverable, reusable suborbital research, or preparation for planetary landings, the ability to safely and routinely conduct propulsive landings with absolute confidence is one of the key enabling technologies in opening up the solar system, and SpaceX was on fundamentally solid ground in it incorporating into flight operations as early as possible.

Grasshopper test vehicle making a 325 meter flight above McGregor. Credit SpaceX

Starting with short hops and extending to a maximum altitude of 11,500 feet, SpaceX would begin to test approach and landing procedures for a Falcon 9 first stage equipped with a restartable and deeply throttleable Merlin 1-D center main engine, as well as fixed landing legs. Short take offs and landings are one thing, but successfully bringing the stage back from the moment it stops boosting to orbit is something else entirely, a process Musk himself described as "super damned hard." Besides adding additional dead weight in the form of landing gear, support structure and supplemental thrusters to guide the booster, as well of course, as the considerable fuel reserve required to direct re-entry and touch down gently, a Falcon reusable would also require a lower staging velocity, something approaching Mach 6 as opposed to the current staging velocity of close to Mach 10. The result, at least for the moment, appeared to be a net reduction of payload to orbit for the second stage, which would now have more ground to make up, of approximately 40%. A significant reduction in performance, it would take a long time to tell if the savings allowed would be offset by the costs of recovery and re-use, and how much maintenance might be required. Nevertheless it marked a significant first step on a road which must be traveled, and even if it initially achieves only a breakeven, the mere fact of establishing first stage reusability (excluding carrier aircraft for air launched boosters) on a functioning orbital system would be one of the most significant developments in the history of space technology, with dramatic implications for even the very near future. Musk however, was not about to stop at partial reusability. The real goal was a fully, "rapidly reusable" system, a challenge which meant making all three stages; first, second and Dragon capsule, recoverable and reusable.

In some ways, after the challenge of developing a recovery technique for the first stage, returning the second stage is a more straight forward proposition which is functionally similar to recovering the Dragon capsule itself. After delivering the payload to orbit, the recoverable second stage, equipped with a forward mounted heat shield, rotates 180 degrees, initiates a de-orbit burn with its single Merlin Vacuum engine, and then plunges back into the atmosphere, headed back towards the designated recovery area, touching down with the aid of extendable landing gear and its own set of SuperDraco thrusters.

The SpaceX concept for full reusability on the Falcon 9 might not have been credible if the company wasn't already achieving some very impressive numbers in terms of improving the mass fractions of major components, beginning with the Merlin 1D engine which comes in at a remarkable vacuum 150:1 thrust to weight ratio. The significance becomes very clear when considering the fact that with 9 first stage engines, as well as tanks, landing gear and the thrust structure to return, an average weight savings per engine makes a material difference. In fact, it is the entire spiral design philosophy taken from the beginning which like everything else along the way, makes the previously impossible, possible.

For instance, by overdesigning the Falcon 9's thrust structure from the very outset, engineers created a strongpoint capable of mounting the first stage landing gear. Likewise, the development of ultra-light weight and highly reusable PICA-X heat shielding initially used on the Dragon capsule is equally applicable to the nosecone of the second stage, and which happens to be the same diameter. The enhanced thrusters SpaceX was developing to power the emergency abort system, became not only the means to save the crew in the event of an emergency, but also the manner to bring the Dragon safely to ground.

The SpaceX decision to introduce the Falcon Heavy and Falcon Reusable program over a six month time frame in 2011 came at what appeared to be a very incongruous time, even as the company was encountering a series of unforeseen delays in getting approval for its next scheduled mission under COTS. To critics it was yet another case of them promising everything and delivering quite a bit less. The case of the Falcon-R was a little more curious. Although other companies, notably Blue Origin were also working on a reusable launch system, there was hardly any rush, and certainly not any of the pressure driving SpaceX to field a heavy lift launcher necessary to fully address the commercial and military launch markets. What was perhaps obscured after parachute based recovery attempts for the first two Falcon 9 flights failed, was that Elon Musk was just as dedicated to developing a reusable launch system as ever. In what was emerging as a spiral development program NASA could only dream about (and sadly no longer was), the needs for the Falcon Heavy program, a stretched Falcon 9 core with upgraded, throttleable engines, also presented the ideal test bed for the initial element of a fully reusable launch system, a recoverable booster which exchanged increased performance in its expendable version for the allowable margin necessary to attempt recovery. To put it in conventional terms, as it appeared at the time of Musk's announcement, by using the same core booster, SpaceX could either match the performance of an expendable Atlas V, or alternatively that of the Delta II, but with a recoverable and reusable first stage.

Though it would be some time until either concept could be fully proved, the twin announcements of both a heavy lift and a recoverable version of the Falcon 9 invariably influenced public perception of the company and what it could mean if SpaceX successfully addressed the more immediate challenges ahead. For anyone interested in the future of human spaceflight, from 2011 on, we were all along for the ride.

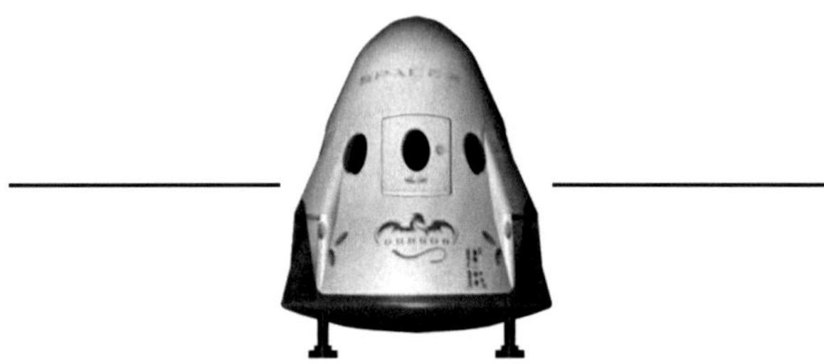

Chapter XXI: Preparing for Station

When President Barack Obama called for the end of Project Constellation in the course of announcing his Administration's space policy at the Kennedy Space Center on April 15, 2010, the most immediate beneficiary may have been the International Space Station program, given a reprieve from the looming specter of termination in 2016. As NASA worked to formalize an extension of the ISS orbital lifetime through at least 2020 in negotiations with its international partners, the role of the COTS/CRS program for resupply took on added importance, and program managers began to become increasingly concerned about the potential of unanticipated problems in qualifying the SpaceX and OSC systems. To alleviate those worries, NASA proposed and received an additional $300 million dollars in "risk reduction" funding as part of the FY 2011 budget. From that amount, $236 million was divided equally between SpaceX and Orbital Sciences, with the agency retaining the balance for administering a new round of supplemental milestones to remove as much risk as possible prior to flights to the station. [82] For SpaceX, that meant 18 new milestones, focusing on its untried LIDAR system, as well as upgrades to the Dragon capsule itself to allow an overall cargo increase, including a more robust powered cargo capability. Although the program's critics pounced, seizing on the risk reduction funding as evidence of cost overruns being picked up by the taxpayer, this was simply not the case. Indisputably, the new funding was for milestones beyond those already agreed to, and in neither case had the companies come asking for money. [83]

Supplemental COTS Milestones

23	Modal Test Plan and Setup	11/2010	$5 million
24	Modal Test	12/2010	$5 million
25	Light Detection and Ranging (LIDAR) Test (open loop)	12/2010	$5 million
26	Solar Array Deployment and Component Thermal Vacuum Tests	12/2010	$5 million
27	Light Detection and Ranging (LIDAR) Test Plan (closed loop)	03/2011	$5 million
28	Thermal Vacuum System Test Plan and Procurement	03/2011	$5 million
29	Overall Infrastructure Plan and Long Lead Procurement	03/2011	$10 million
30	Thermal Vacuum System Tests	07/2011	$20 million
31	Test Site Infrastructure Implementation	07/2011	$5 million
32	Dragon Trunk Acoustic Test	07/2011	$10 million
33	Light Detection and Ranging (LIDAR) Test (closed loop)	08/2011	$5 million
34	Design Review of Enhanced Powered Cargo Accommodations	08/2011	$5 million
35	Design Review of Pressurized Cargo Volume Increase	08/2011	$5 million
36	Full Dragon Electromagnetic Interference/Capability Test and Second Flight-Like Hardware in the Loop Simulator	07/2011	$10 million

For SpaceX, the COTS 1 flight on December 8, 2010 had been performed in partnership with NASA, but to some extent, and after a great deal of preparation, the agency's role was to take a fairly hands off "wait and see what happens," approach to the flight, with 15 of 17 milestones having been met before the rocket ever left the pad. The next mission, COTS 2, was altogether different, complicated by the SpaceX request to actually combine the COTS 2 and COTS 3 objectives into a single mission, which would first bring the Dragon within close proximity to the Station and if all went smoothly, brought in for a historic berthing. The latter aspect triggered a dramatically higher level of scrutiny from NASA, and somewhat exaggerated alarm from Russia, which as a supplier of launch services to the U.S., was not without an interest in seeing the U.S. remain on the sidelines for as long as possible.

It is easy to focus on the development of the Falcon 9 launch vehicle as the most difficult and expensive part of the overall system SpaceX introduced for supplying ISS, but preparing its companion Dragon spacecraft for fully autonomous flight was a comparable challenge, and one which necessitated perhaps the least appreciated part of the development process; the hundreds of thousands of hours and lines of code written in developing and then exhaustively testing the flight control software. It is one thing to develop the software necessary to control a single spacecraft operating alone in a vast empty vacuum. If the launch vehicle and transfer stage has done its job and placed the spacecraft in the intended orbit, then there is very little risk of accidentally striking another object through an equipment malfunction. Even on complicated interplanetary missions requiring very precise mid-course maneuvers to achieve the exact trajectory desired, flight controllers generally have the luxury of both time and space to make incremental changes, and it is not at all unusual for a spacecraft to depart Earth en route to another planet without all of the necessary code having even been written. Excruciatingly long interplanetary flight times offer ample opportunity to finish the work on Earth and upload the new software on the fly.

When on the other hand, the spacecraft in question is going to dock with another spacecraft, or even fly within close proximity, there is essentially no room for error, and even the slightest miscalculation or unforeseen event such as a stuck thruster, can be disastrous. Manned spacecraft, including the International Space Station in all its remarkable bulk, are still fragile, pressurized structures designed primarily to keep atmosphere and people in, and are not so nearly well suited to keeping large foreign objects out, particularly ones with considerable mass and inertia.

On June 25th, 1997, during a docking test aboard Space Station Mir, a Progress supply ship spun out of control and struck the station. The crew on board, which consisted of two Russian cosmonauts and American astronaut Michael Foale, were immediately presented with one of the least welcome sounds that can be heard in space, the telltale hiss of air leaking into the vacuum beyond. If hearing the sound was bad enough, the next challenge was locating it, a very difficult task under any circumstance, but made all that much harder in the cluttered confines of a station module when almost every bit of "wall" space is blanketed by a confusing array of wires, cables, panels, lockers and compartments. Fortunately for the crew, they were soon able to trace the source to the Spectr module, and working quickly, severed cables blocking closure of the open hatch leading into the damaged module and sealed it off. Unfortunately, the cables being cut were power cables coming from Spektr's solar panels which generated approximately 50% of the station's electrical needs. Not only had the tumbling Progress ship punctured the hull, on the way in, it badly damaged one of the four solar arrays, temporarily limiting overall power available even further. Several months later, cosmonauts conducted two "internal" spacewalks in which they suited up, re-entered the damaged module and installed new electrical connections through the now permanently closed hatch.

The collision came near the end of what had been a particularly difficult tour of duty, which was already noteworthy due to an even more frightening event, the eruption of an onboard fire which quickly filled the station, as well as the attached Soyuz escape craft with smoke. Although safety systems came on line and quickly cleared the air, the press surrounding this unfortunate event went a long way towards securing the Russian station's negative reputation and making it the butt of many late night jokes. Even the normally staid PBS joined in producing an on-line documentary labeled "Terror in Space."

The Spectr collision was the most widely reported collision on Mir, but it was not the only one, and in each case, human error played a significant role. With Dragon under automatic control as it approached the much larger ISS, the greatest opportunity for human error to cause a similar problem was in the software code controlling the capsule. Even though it resulted in a lengthy delay following the very successful COTS 1 launch, which was uncomfortable to both sides of the partnership, NASA was determined to ensure that the risk reduction funding be put to good use to ensure that when SpaceX finally launched to station, it would have taken every possible precaution to ensure that any mishap would be limited to Dragon only.

For the first time the Dragon would be directly interfacing with a manned spaceflight system, the International Space Station, and if successfully berthed, would become a temporary extension of that system, with astronauts working inside the attached spacecraft in a shirt sleeved environment. Consequently and for the first meaningful time, SpaceX would have to satisfy not just the NASA COTS program personnel, but also two additional oversight boards charged with assuring the safety of crewed space systems. It proved a difficult challenge.

Over the course of the five years since SpaceX originally won a share of the COTS contract, the two radically different cultures, one staid and bureaucratic, and one entrepreneurial, had through the Technical Interchange Meetings overcome initial obstacles and learned to appreciate each other's respective talents to form a close working relationship. As often happens, members from both sides of the partnership developed what might be characterized as converging viewpoints, and clearly understood each other's communications, even as differences remained. It came as something of a surprise then, when an August 2011 joint meeting with the ISS Advisory Committee and the Astronaut Safety Advisory Panel turned into a near disaster.

The ISS Advisory committee is a NASA committee, chartered with overseeing safe operations of the International Space Station. With the station partnership consisting of U.S., Russian, Japanese and the 17 member European Space Agency as partners, with four different ground stations and five different visiting vehicles, all with different requirements, it is a complex job in which decisions have to be routed around the world and agreed upon by all the respective bodies with a stake in the process. Up until this point however, despite significant cultural differences among member countries, the various space organizations were all governmental in nature. For the ISS advisory committee, SpaceX was something new.

The second advisory board was the Aerospace Safety Advisory Panel, an independent oversight board first chartered after the fatal Apollo 1 fire which took the lives of astronauts Gus Grissom, Roger Chaffee and Ed White in January 1967. ASAP was operating under a new, stricter ultimatum including mandatory Congressional reports, levied in the wake of the Columbia accident. In each of the three fatal accidents which marred America's space program, NASA ignored clear and present warning signs which might have prevented the tragedies from occurring at all. The ASAP was chartered in an attempt to make certain that NASA followed its own rules. Considering the two fatal Shuttle accidents which occurred after its formation, the ASAP's record is open to interpretation, but even the best advice is of little use if ignored. Facing the dual prospect of new hardware systems and an entirely new operating methodology, the ASAP was perhaps even more than usual, acutely aware of its responsibilities where commercial providers were concerned. The purpose of the August 9 meeting, conducted at the Johnson Space Center, which included representatives of both SpaceX and Orbital Sciences, was to allow members from both the ISS Oversight and Aerospace Safety Advisory Panel to ask specific questions regarding the development process for safety measures on the new spacecraft which would be approaching the station.

Throughout the question and answer sessions, committee members were taken aback by SpaceX personnel's overall confidence and attitude regarding the development of flight software which would control the Dragon, Two comments in particular caused considerable concern. One was by the Chief Software engineer who asserted to the committee that "his software didn't contain any errors." The second troubling remark came from another programmer, who when asked about safety requirements observed "We don't set requirements, we just do coding."

Although not meant quite as literally as it sounded, another statement that "NASA was responsible for safety, mission success was the responsibility of the contractor" suggested to some members of the oversight boards that SpaceX's attitude as it approached this next flight was overly glib. [84]

In some ways the issue was more a matter of tone than of substance, with much of it due to the immense cultural divide between SpaceX and the oversight bodies. With its aggressive entrepreneurial approach and a younger workforce which even dressed more casually than its counterparts, the easygoing confidence that came with being among the best software engineers on the planet was almost certain to clash with the staid, conservative, even dour oversight boards which many within NASA itself sometimes view as professional naysayers.

There were other, more tangible reasons as to why SpaceX appeared to be singled out for particularly inspired questioning. Although both Dragon and Orbital Sciences Cygnus spacecraft were designed to be operated autonomously, the SpaceX decision to develop its own software to be run on commercial grade computers was a source of concern. Orbital Sciences by contrast, selected outsourced, radiation hardened, space grade mil-spec systems with a proven track record in orbit for Cygnus, a conservative choice which naturally elicited a lower level of concern.

Two months after the August meeting, former Gemini, Apollo 10 and Apollo-Soyuz Test Project pilot Thomas P. Stafford, chairman of the ISS oversight committee, included a summary of the exchanges in testimony before the House Space Science and Technology Committee. The inevitable result of the review team's comments, which the space media described as "dinging" SpaceX, led to a renewed focus on satisfying the committees' concerns and an extended sequence of testing and then retesting of every aspect of the software which would be controlling the Dragon whenever it was ultimately cleared for launch.

Less than three weeks later, events high in the atmosphere above the Siberian wilderness re-enforced the potential for disaster with even the most veteran space launch systems. On August 24, 2011, a Russian Progress freighter lifted off on a seemingly routine mission to ISS from the Baikonur Cosmodrome in Kazakhstan. Liftoff occurred normally, at 7:00 PM local time, with nominal first and second stage burns, but 325 seconds into the flight, as the third stage engine fired to give the craft the final push into orbit, sensors indicated pressures were falling off in the RD-0110 engine. Moments later, the stage lost thrust, sending the Progress capsule tumbling into the remote and rugged Altai region of southern Siberia. Despite several searches, no trace of the doomed freighter was found. The implications were potentially enormous, highlighting the vulnerable position of a space station depending on a single supply line for crew transport.

Because the Soyuz rocket which launches the Progress shares the same upper stage engine as the Soyuz crew transfer vessel, the immediate result was a grounding of both systems until the underlying problem could be analyzed and successfully corrected. With the Space Shuttle having completed its final flight; for the first time in its history, there was no way to conduct crew flights to the Station. Although not in any immediate danger, ISS partners faced the prospect of progressively abandoning the outpost as each of the two currently attached three man Soyuz return craft, which double as "lifeboats" reached the end of their 200 day on-orbit lifecycle.

The problem was accentuated by a quirk of orbital mechanics and the limitations of the Soyuz landing procedures. Unlike the Shuttle which enjoyed the benefit of two West Coast landing sites in addition to the primary Shuttle Landing Facility at the Kennedy Space Center, and true emergency backups across the globe, the Soyuz must return to Earth on the wide open fields of the Kazakh steppes, where its crews wait for the arrival of helicopter based search and recovery teams, a process which can take several hours. Because of these limitations, ISS rules call for Soyuz landings to take place in a window between an hour after sunrise and an hour before sunset. Both crews aboard the station however, were approaching respective 5 week blackout windows in which the station's orbit rendered daylight landings impossible. The window closed for the first crew on September 19, and for the second on November 19, after which the capsules would exceed their 200 day rated orbital life.

Besides the immediate implications for the ISS both in terms of resupply and crew exchange, the accident also potentially impinged on the SpaceX COTS 2/3 launch which was being pushed steadily to the right. The next scheduled Soyuz launch, originally intended for September, would include the first two astronauts, Don Petit, and Andre Kuipers, to receive intensive training for capturing and berthing the Dragon. For a moment, at least, it looked as though SpaceX might be ready before NASA.

Fortunately, the Russian Space Agency was able to trace the likely cause of the Progress accident to human error, specifically contamination in the fuel line to the gas generator, possibly by a piece of cloth, which restricted fuel flow to the gas generator, creating both an oxygen rich condition and reducing engine thrust. As a precautionary measure, all 16 engines already allotted for upcoming flights were recalled and tested, with no indication of a similar problem.

According to Russian officials, had such an incident taken place aboard a Soyuz with crew aboard, the emergency escape system would have pulled the crew out, but the ride, and in particular the unplanned landing well outside the normal range on steep mountain slopes, could have proven extremely dangerous, and quite possibly fatal. Following the rapid investigation, a subsequent Progress launch took place without incident in late October, and several weeks later on November 14, the crew which would eventually berth the SpaceX Dragon near the end of their mission, blasted off from Baikonur.

The incident served to highlight the inherent vulnerability of space launch systems with single failure points, the tenuous supply chain to the International Space Station, and the need for a full time emergency lifeboat. On a more positive note, NASA's perseverance in pushing for one final Shuttle launch, the STS-135 supply flight which had left the station with more than a year's supplies suddenly looked like a very prescient decision. For a Russian space program which had only months before been virtually bragging that it was now the sole means of access to the ISS, it was also a sobering reminder of the admonition that "those whom the Gods destroy, they first make prideful." As it turned out, the aging Russian space program would soon have even more to regret in the coming months, as it suffered a string of embarrassing failures.

Following the Progress flight 44 failure, the U.S. House Committee on Space, Science and Technology, Subcommittee on Space and Aeronautics held a hearing on lessons learned from the incident on October 14, 2011. As part of the testimony in that hearing and in response to a specific question, ISS Safety Panel Chair Tom Stafford included the review teams' summary of the August 9 SpaceX COTS review. Published in the public record, the cautious wording characterizing the software coding issue provoked a new round of criticism for the company's opponents in proxy fighting over the hotly contested Commercial Crew budget. Although ostensibly two separate programs, Congressional opponents of the Obama space policy, and SpaceX in particular, were hard at work slashing the budget requests for the Commercial Crew program, and forcing at least an additional year's delay in implementation. From this point forward, and despite the fact that traditional aerospace stalwarts Lockheed Martin and Boeing were also deeply involved with Commercial Crew bids, Boeing with the CST-100 Capsule and Lockheed Martin through ULA, with the Atlas V rocket selected by Boeing, Sierra Nevada and Blue Origin, every obstacle and delay SpaceX faced in mounting the next flight became a tool to attack the Commercial Crew program as a whole.

As it worked through the battery of simulations and inspections, NASA in mid-December finally granted formal approval for SpaceX to combine the next COTS 2 and 3 objectives. In return, SpaceX quietly dropped its plans to launch a pair of small commercial satellites for Orbcomm during the course of the same mission. With formal permission finally in hand, the launch date for the combined flight was set for February 9, 2012. Nevertheless, as the company approached the one year anniversary of its most recent launch, NASA Human Spaceflight chief Bill Gerstenmaier warned that although the upcoming SpaceX mission was now tentatively scheduled for early February, further delays should be anticipated.

The New Year dawned on somewhat of a mixed note, with the announcement on

January 7 that the next flight would be further delayed to allow for additional software testing. Days later, a press release informed the media that respected astronaut and Mission Assurance Chief Ken Bowersox had left the company shortly before the Christmas holidays. Coming after a handful of other high profile departures, the news inevitably led to more speculation that the company was experiencing difficulty and the work environment was becoming increasingly stressed.

As February and March went by, matured, and perhaps a bit more cynical now, SpaceX elected to keep quiet about the challenges it was facing in getting the Dragon certified to rendezvous with the Station as NASA pressed the company to conduct several more rounds of simulations. Meanwhile yet another problem cropped up on the Russian side of the supply line. While conducting a routine pressure check of the next Soyuz capsule due to launch to the station, engineers over pressurized the capsule, causing a rapid, explosive decompression which resulted in severe damage. Already running behind, the Russians would need to dip into the supply stream and sequester the next shell on the production line in its place. Finally, after another series of delays which Musk characterized as teaching the Dragon to not be too afraid and "runaway" on the basis of a false alert, NASA indicated at last that it was satisfied, and a new launch date was scheduled.

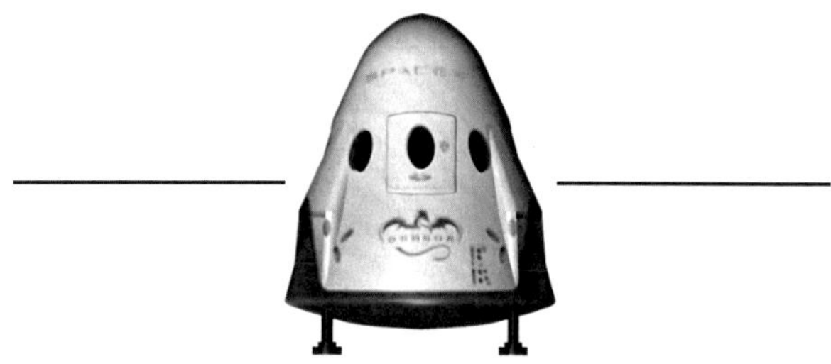

Dragon on the pad at SLC-40 prior to COTS 2/3 mission. Credit: Stewart Money

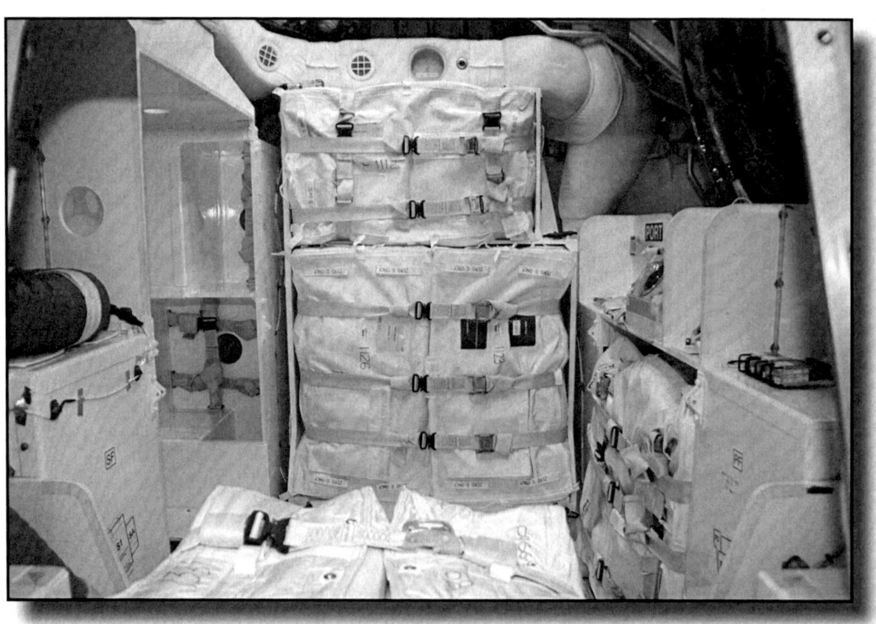

COTS 2/3 Dragon partially packed for its upcoming mission delivery. A robust air handling system is necessary for even a cargo vessel. Credit: NASA

Chapter XXII: COTS 2/3 Flight of the Dragon

Unlike the two previous flights of the Falcon 9, the COTS 2/3 mission would occur against the backdrop of an American space program fundamentally altered with the final flight of the Shuttle program and Space Shuttle Atlantis on STS-135 in July of 2011. The flight, which was added to the Shuttle manifest in fulfillment of a 2008 Obama campaign promise, and to the welcome relief of what remained of the once massive Shuttle workforce, carried one last load of supplies in the Multi-Purpose Logistics Module Rafaello, intended to keep the U.S. commitment to the station adequately stocked for at least a year. Enough time, hopefully that either SpaceX or Orbital Sciences could complete its COTS objectives and begin making supply runs. In the meantime, even though the Shuttle program's long good-bye had been underway since 2004 and the introduction of the Vision for Space Exploration, it struck the nation with surprising intensity, provoking scathing commentary from the media, often accompanied by interviews with astronauts from the Shuttle and even the Apollo program, who none too subtly began to equate the end of the Shuttle era with the death of the American space program overall. For an American public generally disinterested in space unless something significant occurred, the reaction to the final flight took on the air of "you don't know what you've got till it's gone." For most, the Shuttle was an example of American greatness, a symbol of the nation itself, even it was not all that clear what the Shuttle was actually doing most of the time.

The sense of loss, and of seeing a retreat from history was heightened both by the fact that the nation was still in the throes of two unpopular wars and a seemingly intractable recession. For many, it seemed that the country was no longer capable of doing great things, and the apparent loss of the Shuttle program was another in a long line of examples. Although it would be a stretch to suggest that the nation clamored for a new program of space exploration, (it didn't) the Obama administration's cancellation of Project Constellation in 2010, and the poorly rolled and rather questionable plan for following an undefined "flexible" path for exploration in the future, beginning with a trip to an asteroid, lent credence to the notion that the space program was being brought to a close with the end of Shuttle. For those that were paying attention, it also didn't help that the President elected to monopolize media coverage with a generally pointless press conference in the moments counting down to the final launch, all without giving even a nod of respect to the moment in history which was taking place at the Kennedy Space Center. As Shuttle Atlantis lifted off on a poignant final flight, those who suspected that the President's new plan was much more about kicking the can down the road, this brief little vignette offered a glimpse into what he really thought, or didn't, about the space program.

To a great extent, the critics were right. When Atlantis rolled to a stop after its final mission, NASA upper management was still arguing with the U.S. Senate over its mandate to produce a heavy lift booster comprised out of Shuttle components. As NASA had yet to produce a specific plan for a new booster, for all practical purposes there wasn't one. America did not have a plan, merely a plan to have a plan. Furthermore, even though the giant booster, dubbed SLS, for Space Launch System, or more derisively the Senate Launch System in mocking reference to the highly questionable collection of "experts" which had demanded it, there would be no specific plan of exploration forthcoming other than a flyby of the Moon, an event the agency had already accomplished in 1968, and even that was not scheduled to occur for nearly a decade. During that time, as nearly somewhere between $29 and $38 billion [85] was being plowed into building the SLS out of Shuttle era components, the nation would of course continue to have a manned space program in the form of NASA astronaut presence aboard the International Space Station, but with the significant caveat that supplies and then crews would be lofted on commercial rockets. The problem was that much like with the SLS, the proposed commercial crew transfer systems did not exist yet, and would not until 2015 at the earliest. Compared to the iconic winged spacecraft which had dominated America's and even the world's concept of space exploration for more than 30 years, even though it performed no actual exploration, one rocket by NASA ten years in the future and one or more undefined commercial vehicles offering LEO services sometime before that hardly sounded like much of a plan. In the meantime, the United States would have to buy launch services from Russia, which tactlessly and very publicly raised its prices to NASA almost the moment the Shuttle retired, adding insult to injury. Weeks later, Russian space officials were openly bragging about the ascendency of the Russian space infrastructure over the once mighty United States.

The first available opportunity to even begin reversing the sense of a retreat from history lay with SpaceX and the flight of the Dragon. Consequently, the COTS 2/3 flight, even though, as NASA officials repeatedly emphasized was still only a test, inevitably took on an atmosphere and an element of pressure which would not have been quite the same if the Shuttle was still flying. As for SpaceX, which was only looking at the third flight of the Falcon 9, the company was very much aware that in the historical context, most new vehicles suffered a catastrophic failure at some point within the first three flights. That awareness, combined with the sheer complexity of the mission itself, contributed to an urgency not felt since the all-important fourth flight of Falcon 1, when the company's continued existence depended on a favorable outcome. In this case the urgency was not so much a matter of betting the company, as SpaceX already had more commercial launches on its manifest than the NASA contracted missions, but one of losing the inside path to fast tracking its deepest ambitions.

According to the milestones as originally laid out, the second COTS flight called for SpaceX to bring a fully functional Dragon Capsule to within the vicinity of the station and demonstrate the ability to go through a series of carefully orchestrated maneuvers designed to ensure that Dragon and the SpaceX team directing it, could exercise the delicate control necessary to safely berth with the station on its next mission. After doing so, the Dragon would return to Earth and SpaceX would begin preparing for the next flight, COTS 3, which would see it berthed to ISS for the first time.

Even before the ultimately successful COTS 1 flight, SpaceX began lobbying NASA to combine the next two mission objectives into a single flight. After all, if the Dragon successfully completed the COTS 2 objectives, it had already demonstrated the command and control necessary to safely approach with the station's 200 meter

"Keep Out Sphere." The remaining challenge would be activation of the two short range target acquisition sensors via DragonEye. The agency was initially cool to the idea, partially due to the fact that the combination of two separate flight objectives would require the Dragon to consume close to its full complement of fuel. Following the COTS 1 flight however, NASA concurred, swayed by the argument that if the Dragon could successfully pass the COTS 2 objectives and demonstrate the ability to approach, retreat and hold on command, with the craft already positioned on the edge of the 2.5 km approach ellipsoid, there would be little risk in proceeding with the final set of objectives leading up to berthing. If a problem arose, the capsule would be commanded to retreat, and if it could not be readily addressed, SpaceX would proceed with a separate flight three after all. If on the other hand the flight was successful, then NASA could effectively eliminate at least part of the "gap" created with the decision to retire the Shuttle prior to any US ability to access the station and remove a major area of concern, and not to mention significant national embarrassment if the U.S. lost its supply chain to its $100 billion castle in the sky.

After seemingly endless reviews, the launch date was finally set for Saturday, May 21, at 3:41 AM EDT. On the two previous Falcon 9 launches, SpaceX had the luxury of launching when the vehicle was ready, and on the maiden flight, it missed the first opportunity. This time however, the company found itself facing an almost instantaneous launch window which offered no room for error, nor even for the briefest of holds. Unlike the Shuttle, which launched with substantial orbital maneuvering system fuel reserves, enough to create a rough 10 minute window, the Dragon carried only enough to allow for an instant launch.

Tanking operations commenced in the early minutes after midnight, and proceeded smoothly on a warm Florida night. Across the Indian River, all along US 1, and along the NASA causeway, where hundreds of thousands had gathered to say farewell to the Space Shuttle less than a year before, there was no sign America's next mission to the International Space Station would soon be underway, and only the tell-tale glow of floodlights on the Pad at SLC-40 gave any indication that once again, somebody was attempting to make history at one of the many launch pads which connected this small portion of the Florida coastline to the immensity of the universe.

The countdown procedure progressed smoothly and relentlessly down to the final seconds before ignition, and then with the briefest of flashes the 9 Merlin main engines roared to life…and immediately shut right back down. The launch window missed, SpaceX would be going nowhere today except back to the drawing board and on to a press conference to explain what happened.

Two hours later, by the time she joined COTS program manager Alan Lindenmoyer on the platform at the Kennedy Space Center's Media Press center, SpaceX President Gwynne Shotwell had a partial explanation. Everything had proceeded normally down to and through main engine ignition, but an abort was initiated when sensors indicated pressure spiking in the number 5, center mounted engine, with an undeniable trend towards a major problem. For once, it clearly was not a sensor issue, but an actual physical problem with the engine itself. As a number of invited guests began streaming away from KSC, including high school students whose experiments were contained in a NanoRacks package stowed aboard the Dragon, as well as some 50 randomly selected social media attendees NASA had invited as part of its newly introduced NASA Social campaign, SpaceX got to work safing the launch pad and getting a team of engineers out to the Falcon 9 to begin figuring out how, after testing each engine separately in Texas, firing the whole integrated thrust assembly in Texas, and again then briefly during the hot fire test only days before as it sat on the pad, something had still gone wrong to stop the launch when it counted the most.

For one thing, quite a bit had gone right. Once again, the Falcon 9 emergency detection section had functioned flawlessly, detecting a problem and safely shutting down all nine main engines after ignition occurred, in a process Shotwell labeled as "aborted with authority." While certainly not the outcome anybody wanted, the successful launch abort provided another piece of evidence that SpaceX had equipped the rocket with a state of the art diagnostic system which knew when to launch and when to hold up. With a decision on the next phase in the Commercial Crew competition expected sometime within the next two months, a less favorable outcome under an abort scenario, like almost everything about this mission, could have been devastating to the SpaceX entry.

In preparing for a new launch attempt, engineers faced the task of determining whether or not the rocket could be repaired on the pad, or if it would need to be wheeled back into the horizontal assembly building. If the engine could not be quickly repaired, a spare was waiting inside, but the operation would add at least two days to a launch window which was already closing. The very narrow window created by the need to conserve fuel meant that the Falcon 9's best opportunity for the next launch came in three day intervals, and flight planners were rapidly coming up against a new problem. Beginning in early June and lasting approximately three weeks, the International Space Station's orbit would place an attached Dragon spacecraft in a zone of maximum solar heating, one which exceeded safe design parameters. If the ship did not lift off sometime within the next three intervals, the launch would need to be postponed until late in the month.

Fortunately, an engine swap was not necessary. By mid-day Saturday, engineers were able to run a boroscope up inside the engine and positively confirm initial analysis as to the source of the problem. As part of standard procedure, in the immediate seconds prior to liftoff, liquid nitrogen is blasted through all the lines which will soon carry fuel and liquid oxygen in order to purge any possible debris. During this procedure, a safety check valve on the fuel line for the number 5 engine gas generator stuck in the open position. At the moment of ignition, as fuel began to flow to the gas generator to build the thrust to drive the turbo pump, the open valve allowed some of the fuel to return to the tank, changing the delicate balance between fuel and oxidizer in favor of the latter, leading to an oxygen "rich" fuel combustion causing a spike in temperature and pressure, and triggering the shutdown.

Repairing the engine was a comparatively simple matter of replacing the faulty valve. Despite the confined space of 9 engines located within the 12 foot circle which marked the dimensions of the thrust assembly, technicians replaced the valve within a matter of hours. Even this procedure as frustrating as it was, served to highlight the remarkable simplicity and work-a-day practicality of the Falcon 9 system. On almost any other launch vehicle, a similar repair would have required a return trip to the assembly building, and several days of laborious repair. For the ever problematical Shuttle, it would have easily been a week or more.

By Sunday afternoon, the next launch attempt was officially scheduled for two days later, this time at 3:21 AM Tuesday.

For the second time in the launch campaign, the midnight hours found the lights turned on at SLC-40, and fueling operations underway, and for this latest attempt, even if there were fewer spectators, and only a limited press attendance to record the event, the mosquitoes did their very best to make up the difference in attendance, particularly for the small crowd gathered at the water's edge of the turning basin behind the Kennedy Space Center's iconic countdown clock. This time however, attention was focused not towards launch pad 39A, but East, towards the dim glow of the spotlights cutting through the misty air above the pine scrub forest which marked the distance to the pad.

Once again, the countdown proceeded, accompanied if possible, by even more trepidation about the likelihood of yet another last second abort or worse. As the clock began to countdown the final ten minutes, the slow steady march of a space launch, which at first seems so controllable, suddenly began to speed up, taking on an air of inevitability and bringing everything into the moment, for better or worse. Like the final seconds of a football game, which seem to take forever when the clock steadily counts down and both teams line up for the field goal attempt which will mean either victory or defeat, agony or triumph, all of a sudden time speeds up, the whistle blows, the ball is snapped and it all hangs in the balance. And everyone is swept away in the moment.

This time, the night suddenly erupted in a fierce orange glow and a deep rumbling roar, as the Falcon 9 slowly, steadily clawed its way into the sky, gradually arcing out over the Atlantic Ocean towards the northeast, lofting the Dragon in pursuit of its prey circling overhead at over 17,000 mph. With so much riding on the outcome of the launch, and of the mission, every second the distinctive red orange glow, and watery sound of liquid oxygen and kerosene engines continued to bathe the Florida sky signaled a mounting optimism, and increasing relief, that this might work, that history was being made, and somehow, at long last, two hours before official sunrise, a new era was dawning in mankind's eternal quest to expand the boundaries of human existence.

As it had twice before, the Falcon 9 completed its first stage burn and stage separation without incident. Second stage ignition began cleanly, and forty seconds later, the Dragon's disposable nose cap ejected on cue. With the second stage rapidly accelerating towards space, attention turned towards the next obstacle, achieving orbit and the nearly instantaneous deployment of solar panels. If something went wrong, the craft might just as well have never left the ground.

The first step in the process is a successful release of the protective covers, immediately followed by deployment of the arrays themselves. Both went off without a hitch. Ten minutes after it was sitting on top of the rocket overlooking the nighttime roll of surf on the Cape Canaveral shores, the Dragon spacecraft was in orbit, its solar panels deployed and soaking in the bright, unfiltered sunlight of space. Fully powered and operating nominally, it was time to begin the pursuit of the Station in earnest.

Back at KSC, NASA Administrator Charles Bolden stepped up to a microphone in the pre-dawn still. Outside the press center at KSC, with the countdown clock in the background now counting forward every second of mounting mission success, the former Shuttle astronaut congratulated SpaceX on a successful start to what would be a very demanding mission. After giving sincere and clearly enthusiastic congratulations, once again and not for the last time that day, a high NASA official repeated the mantra that through COTS, SpaceX was helping free up NASA resources to do what it does best "and go exploring." If he was aware of the irony that his agency had just witnessed the launch of a space transportation system which was designed and built for a fraction of the cost of the yet to be completed Orion capsule housed only several hundred yards away, it did not show.

Historic first private launch to ISS, the COTS 2/3 mission. Credit: SpaceX

NASA Administrator Charles Bolden at a pre-dawn press conference following news that the first SpaceX Dragon mission to ISS is safely deployed in low Earth orbit. Credit: NASA

Back in space, and before it began to play catch up, during the remaining course of the first day of the mission, SpaceX controllers in Hawthorne put the craft through a series of exercises to satisfy COTS 2 objectives and to confirm that once the command was given to go find the Station, the Dragon would behave politely. First up was a demonstration of the absolute GPS system to confirm the craft knew exactly where it was in space. Next, the Dragon opened its GNC bay door, deploying the rendezvous sensors and exposing the hard attachment point. Once again, if the small hatch failed to open on command for any reason, the mission would be over. With sensors now exposed, SpaceX ran the first of what would be many tests on the DragonEye Lidar array and Thermal Imager, which would be used together to independently confirm Dragon's final approach to ISS. Next, the capsule demonstrated a full-on abort and the ability to perform a continuous burn, necessary to protect the station in the event of an impending catastrophe. Following that, the system demonstrated a pulsed abort, which could be called for under a less dire scenario in which a series of very brief firings of the Draco thrusters moved the Dragon out of harm's way. The latter was necessary because unlike most other spacecraft, including the yet to be flown Orbital Sciences Cygnus, the Dragon was not equipped with conventional small thrusters, but instead depended on micro-bursts from the much larger 90 lb. thrust Draco motors to achieve the same result. Being unconventional, this solution had proved to be a specific area of concern for the two oversight boards which ultimately signed off on the mission. Among the easiest of the assignments, the Dragon was tasked with demonstrating the ability to do nothing at all, and simply hang disabled in space, called free drift.

With those objectives completed, the spacecraft began a series of co-elliptic burns to achieve a circular orbit, together with height adjustment burns to bring it to the same orbital altitude as ISS.

As the previous day's maneuvers brought the Dragon within range of the Station, flight operations moved into a new phase with the participation of the Johnson Space Center's mission control team and GO/NO-GO authority over each new step in the dance which would bring it closer to the Station. After the final height adjustment brought the craft to a position 2.5 km under the station, Johnson issued the first GO/NO-GO. Once within range, it was time to test the relative GPS system to make sure that the Dragon knew where it was not just in space, but in relation to ISS itself, with information between the two spacecraft transmitted via the CUCU aboard ISS lofted by Space Shuttle Atlantis, and now installed in the Cupola where crewmembers could carefully monitor the Dragon's progress and issue commands instantly. The communication link verified, the next step was another height adjustment burn to take the capsule away from the Station, followed by yet another Go-No-Go to begin the long looping ellipse above the station and then slipping again behind and below it. A day's work completed, the Dragon had in effect buzzed the Station in order to place itself right back where it had begun the previous day's exercises.

As day four of the SpaceX mission to ISS dawned, which according to international convention occurs at UDT, or Greenwich Mean Time, six hours ahead of Mission Control in Houston, SpaceX had so far demonstrated at a safe distance that the Dragon space craft could communicate with ISS, knew where it was, knew where ISS was, and could safely perform all the maneuvers to bring it within close proximity. The next and final phase was designed to ensure that the two range finding systems, LIDAR and thermal imaging could precisely guide the craft to the specified capture point within range of the Canadarm2 remote manipulator arm. Once again, the day's plan called for a series of retreats, holds and advances, with Mission Control at JSC again commanding the Go-No-Go decisions. This time the maneuvers would be performed along the R-Bar, an imaginary radial line which connects ISS to the center of the Earth.

COTS 2/3 Dragon making its approach to ISS. Credit: NASA

First, Dragon performed a height adjustment burn to bring it back to a point along the R-Bar 2.5 km below the Station, and after demonstrating continued communications with astronauts via the CUCU unit, received a GO to decrease the distance to 1.2 km for another planned hold. Following the go command, the next burn brought the spacecraft for the first time within the station's protective approach ellipsoid. At this point, the Dragon's Lidar system came fully into play, and demonstrating it was up to the task, JSC issued the command to approach again, this time within 250 feet for the next R-BAR demonstration. At this point, SpaceX ran into the first and only difficulty of the approach procedure when the thermal imager picked up a confusing signal from the nearby Japanese KIBO Module. While the Dragon briefly retreated and held station, engineers in Hawthorne adjusted sensor input parameters to account for the reflection, and after conferring with JSC, began the approach back to 250 meters. This time, with everything clear, Dragon performed the final set of R-BAR demonstrations and received the next to last command prior to berthing, a hold at 30 meters. Moments later, JSC gave the final okay, and the brilliant white spacecraft edged ever so much closer, coming to an all stop at the designated capture point 10 meters below the Harmony module.

At the COTS 2/3 post launch press conference, NASA Associate Administrator William Gerstenmaier, Commercial Crew and Cargo Program Manager Alan Lindenmoyer and SpaceX President Gwynne Shotwell can finally relax. Lindemoyer's contribution to the program was indispensable. Credit: NASA/Kim Shiflett

Perched in station's Cupola, the Earth facing glass enclosed node which also serves as the American side's control tower for berthing, astronauts Andrew Kuipers and Don Pettit deftly moved the robotic arm into position to grasp the Dragons' attachment clamp, and moments later, succeeded on the first attempt.

With the Dragon successfully maneuvered into its final berthing position, and correctly aligned for contact, Petit visually inspected the spacecraft's seal to make sure it had not suffered damage or contamination during the ejection of the protective nose cap, and satisfied, secured Dragon against the active side of the Common Berthing Mechanism, engaging 16 powered bolts to lock the two craft together. Donning protective masks in the event an unexpected gas had leaked into the capsule's interior, Petit opened the hatch and for the first time ever, a private spacecraft stood ready to accommodate astronauts in Earth orbit. With something a little less than one giant leap, the Dragon became for the moment at least, a crewed spacecraft.

Astronauts spent the next 5 days gradually unpacking the "non-critical" supplies carried on this first cargo run; 1,014 pounds of food, batteries, power cables and spare cargo bags, along with a NanoRacks Cubelab experiment, all of which had been sealed in Dragon back in Florida over 10 days ago. After repeated admiring visits by all six of the station's crew, the approaching solar heating max meant it was time for astronauts to begin packing the craft for its return to Earth, and the fulfillment of the remaining COTS objectives.

Dragon captured by Canadarm 2 on its first visit to Station. Credit: NASA

On Thursday, May 31st, the final day of the mission, after carefully maneuvering Dragon to the 10 meter release point, the ISS crew released the Canadarm2's grip on the attachment point, allowing SpaceX, once more in command of its craft, to begin the first of a series of three short burns to remove the Dragon from the vicinity of the station and send it on its way to the point of re-entry. Following a 10 minute de-orbit burn, which saw the release of the Dragon trunk and solar array to be consumed in the fire of re-entry, the Dragon began its plummet back through the atmosphere, this time laden by more than a wheel of cheese. In contrast to the previous landing, weather conditions were slightly more challenging, with higher winds and modest seas. With a Navy P-3 aircraft flying steady circles around the designated landing area, and capturing the scene with high resolution cameras, right on schedule, the descending spacecraft deployed its two drogue parachutes, and 90 seconds later the three 116 foot orange and white mains followed suit, bringing the craft to a splashdown in the Pacific, where it even surfed for a bit as the steady wind caught two of the mains and pulled the Dragon along the wave tops. They finally collapsed, leaving the scorched spaceship floating in Pacific swells, patiently awaiting the arrival of the nearby recovery team. Once again, had a crew been aboard, they would have had a very nice ride.

Four days later the Dragon was back in port and on its way to Texas, and a place in history. More importantly, less than a year after it temporarily abandoned the ability to reach ISS on its own, the United States was back in business with a space launch system which had shattered all expectations. The event marked an impressive number of space firsts. Among these were the first privately developed spacecraft to dock with the station, the first shirt sleeve operation inside a privately developed spacecraft, and most important for an ISS program just entering a new phase of research, the first and only meaningful return capability in the post Shuttle era. NASA might not be back, but it was unmistakably on the right path. With regular CRS missions set to begin by the end of summer, all eyes turned towards the next challenge, Commercial Crew and the resumption of American manned spaceflight.

ISS crew gathers inside the SpaceX COTS 2/3 Dragon before it departs the Station. Pictured clockwise (from lower left) are Russian cosmonauts Gennady Padalka and Sergei Revin, European Space Agency astronaut André Kuipers, Russian cosmonaut Oleg Kononenko, Commander, and NASA astronauts Don Pettit and Joe Acaba. Credit: NASA

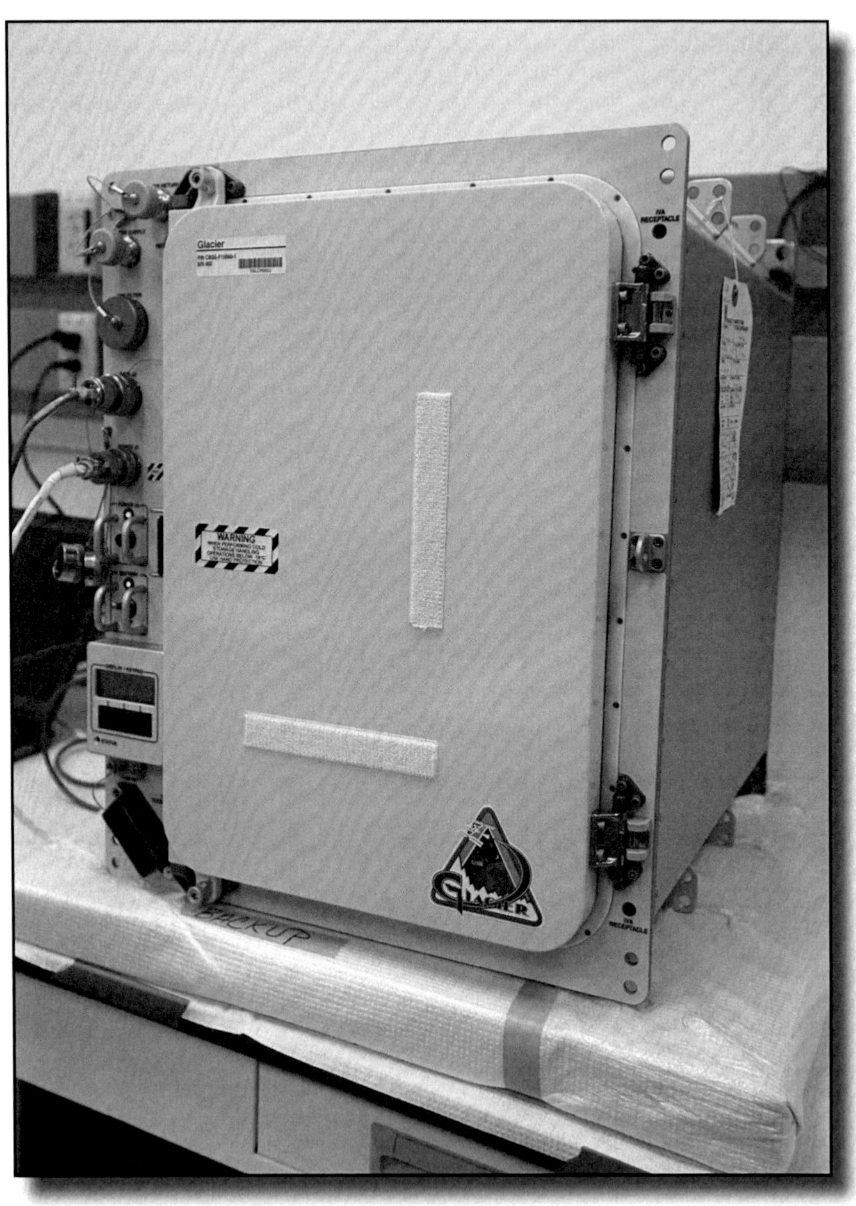

A GLACIER freezer, one of the most critical links in the ISS research chain. The Dragon spacecraft provides continuous power while it is aboard. Credit: NASA

Chapter XXIII: CRS Missions 1 & 2

After the success of the COTS 2/3 mission, both NASA and SpaceX were eager to begin the 12 scheduled resupply flights to ISS under the separate, $1.6 billion dollar Commercial Resupply Contract. With Station construction now complete and more astronaut work hours available for other tasks, it was at last time to begin ramping up the pace of scientific research, a goal which depended on re-establishing regular two way traffic as soon as possible. With very little in the way of issues to resolve following the Dragon's first visit to ISS, NASA set the next flight for early October 2012.

If there is a noteworthy limitation on the cargo version of the Dragon, it is not the total mass that in can carry into space, but rather the volume. Quite simply, although its comparatively steep profile offers a higher percentage of usable pressurized volume than other recoverable capsules such as the Boeing CST-100 and NASA's Orion, when compared against the ATV and HTV, and certainly the Shuttle's MPLM, the Dragon is not efficient at transporting low mass, high volume materials. Furthermore, the need to carefully package cargo to undergo the rigors of both launch and recovery means that even for the volume available, the relative density of cargo carried in the pressurized section is low. The somewhat ironic result is that cargo service using the Dragon results in an astronomical price per pound, even though the mission price, at $133 million, is a relative bargain.

For the CRS-1 flight, the cargo being placed aboard Dragon had a total mass of 1,995 lbs., of which 1,113 lbs. was packaging, leaving a usable payload of 882 lbs. Low density however, doesn't mean low value, and science is rarely measured by the pound. In contrast to the COTS 2/3 mission, this time, in addition to routine supplies, NASA tasked SpaceX with both transporting scientifically significant payloads to ISS, and bringing others back down.

Reflecting the cooperative arrangement between all parties, the Dragon was packed with supplies for the United States, Japan and the European Space Agencies, and scheduled to return material from all partners including Russia. Altogether, the CRS-1 launch included 260 lbs. of crew supplies, part of which was intended to do double duty in an ongoing experiment exploring the effect of low sodium diets on reducing bone loss in space. The 390 lbs. of utilization payload included a GLACIER freezer for ultra-cold storage samples, a Fluid Integrated Rack for hosting investigations on the behavior of liquids in zero gee, as well as racks headed for both the Japanese Kibo and ESA Columbus modules. In addition NASA also sent along the equivalent of a Saturday morning trip to the hardware store, complete with air filters, replacement electrical components and even a fan for providing fresh air to the next European ATV scheduled to arrive.

And, in a rapid return to some of the same kind of research hosted by the Shuttle, a number of experiments would make a round trip flight; carried to space aboard Dragon where the relevant experiments could be performed inside ISS and then sent back to Earth on the return trip for timely analysis by the principle researcher.

After being nixed by NASA as a secondary payload for the COTS 2/3 mission, the CRS-1 mission also provided SpaceX the first occasion to launch a long delayed Orbcomm satellite originally slated to fly on Falcon 1, then Falcon 5, and now finally making its way to space aboard the Falcon 9. After depositing the ISS bound Dragon in low Earth orbit to independently rendezvous with the Station, the plan called for the Falcon second stage to re-ignite for a new burn to carry the Orbcomm satellite, labeled OG2, to an orbit at 750 kilometers, and the beginning of a series of check out tests meant to verify the prototype for a fleet of spacecraft to be deployed over the coming years.

SpaceX employees gather to cheer on CRS-1 launch at Mission Control / Hawthorne. COTS 1 Dragon capsule overhead. Credit: SpaceX

COTS 2/3 Launch of a new era. Credit: Stewart Money

The Wet Dress Rehearsal took place without incident on August 31st, followed by an equally uneventful static fire on Monday, October 1st marking the beginning of launch week. Following a Launch Readiness Review on Friday, October 5th, NASA gave the mission, scheduled for a Sunday evening liftoff, a green light. As launch day dawned, SpaceX began loading the last minute, powered cargo aboard Dragon with the Falcon 9 retracted in the horizontal position.

Going back to the vertical once loading was completed, launch controllers powered up the Falcon and Dragon at T-7.5 hours. Four hours later, fuel loading commenced with liquid oxygen first, followed by RP-1. After a Launch Readiness Poll at T-13 minutes the terminal countdown began at T-10 minutes. At T-2.5 minutes, SpaceX reported "go" and 30 seconds later the Air Force 45th Space Wing range control followed suit.

As the giant NASA clock counted down the final seconds to liftoff, the Falcon 9 was poised to complete its fourth consecutive launch, and in doing so, fly safely out of the three launch threshold which Aerospace Corporation statistics suggested was the most likely time for a failure in design to manifest itself. With no last minute drama, the engine ignition sequence began at T-3 seconds with a brief green flash of the Tea-Tab ignition fluid and the booster roared to life.

Lighting up the Florida night as it had done once before, the Falcon 9 made an on-time liftoff for CRS-1, accelerating north and east over the Atlantic Ocean. Seventy seconds into the flight, the booster went supersonic, with Max-Q approaching ten seconds later. Just as the first stage was approaching the edge of visibility with the naked eye however, a brief, bright, flare of the orange exhaust at almost the split second of Max-Q caught the attention of the small crowd of journalists and visitors gathered at the NASA causeway to witness the event. Whatever the source, the flare faded quickly, and as the booster continued to climb into the night on a steady course, it looked as though all was well.

What the human eye could barely see, the high resolution tracking cameras which record every launch from Cape Canaveral were able to render in stark detail, and what they showed was anything but normal. At the 79 second mark, a burst of flame erupted from the aft end of the Falcon 9, disturbingly accompanied by a shower of debris. Clearly, something had gone seriously wrong. Surprisingly, the flight itself appeared to be going as planned, with on-board cameras dutifully noting both first and second stage separation. Nine minutes after launch, the Dragon was in orbit once again with its solar panels deployed.

Though it would take some time for SpaceX to confirm precisely what happened, initial telemetry indicated that the number one engine suffered a chamber breach and abruptly terminated thrust. With data still flowing in from the engine though, it was clear that despite appearances, the unit had not suffered a catastrophic failure. What appeared to be pieces of the engine flying off were in fact from the protective engine fairing, torn away by the sudden change in pressure. Amazingly, the Falcon 9 itself seemed almost indifferent to the incident, shaking it off with no apparent ill effect. In fact the engine controller and flight computer had done exactly what they were designed to do in the event of a major problem; instantly shut down the afflicted engine and recalculate the trajectory and firing duration based on the new parameters and complete the burn. Validating SpaceX claims that its vehicle could do precisely that, the second stage burned slightly longer than planned, and placed the Dragon precisely in the pre-determined orbit and en route to another rendezvous and docking with the International Space Station.

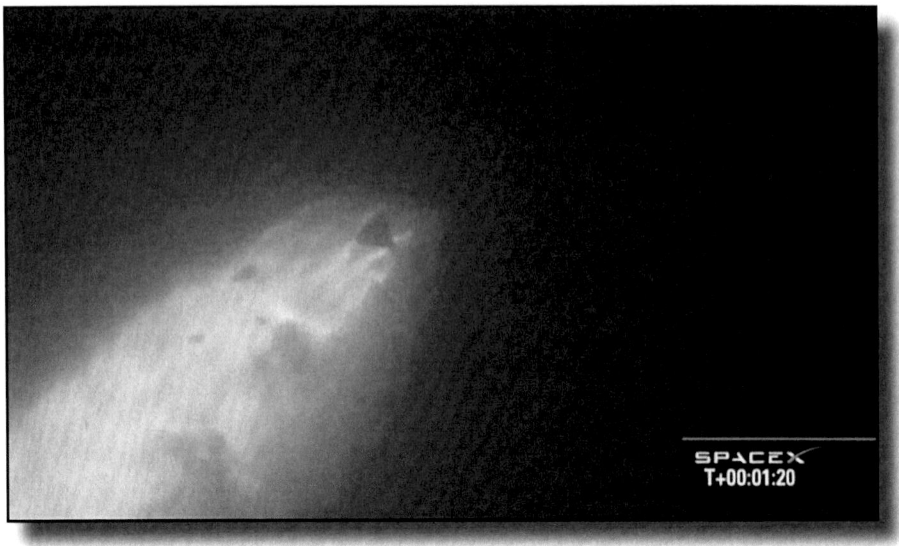

An unplanned demonstration of the Falcon 9's engine out capability. Video capture of Merlin 1-C chamber failure and fairing breaking away during CRS-1 launch. Credit: SpaceX

CRS-1 Dragon. Credit: NASA

Although the Dragon continued to proceed to the station unaffected, there had been a significant casualty. As a result of the extended second stage burn, iron-clad safety gating protocols negotiated with NASA before the launch required that the Falcon 9 second stage eject the Orbcomm satellite in the initial, and very low, parking orbit. With the OG2 satellite scheduled to be released in the same orbital plane as ISS, but at a much higher altitude, NASA required that an upper stage restart could only take place if there was an over 99% probability that it would result in a completely successful second stage burn. Due to the modest LOX depletion from the slightly extended burn to orbit, the probability dropped to 95%, meaning that no restart would be permitted. Consequently the Orbcomm satellite was no longer destined for a higher orbit, but instead for a quick and untimely demise in the upper atmosphere. Even as the Dragon set off for the station, the OG2 orbit began degrading rapidly.

Back aboard Dragon however, everything was going as planned, and this time, freed from the sequential test requirements of the COTS 2/3 mission, the spacecraft made its way to station's R-bar approach position the following day. Following an abbreviated version of the three phase "GO/NO-GO" final approach protocol with Mission Control in Houston, the Dragon arrived for an on-time capture by Expedition 33 Commander Sunita Williams and JAXA astronaut Akihiko Hoshide. Grappling took place at 6:56 AM ET, and barely two hours later the Dragon was berthed, beginning an 18 day stay at ISS.

A sparse crowd reacts to lift-off the CRS 2/3 mission at KSC press center. Credit: SpaceX

After several days of unloading cargo and experiments, the crew began the process of loading 1,673 lbs. of return items including cargo from each of the stations four segments; US, ESA, Japanese and Russian, as well as ISS equipment for analysis and repair. Some of the most important material came from the astronauts themselves; 112 blood and 384 urine samples taken over the previous year as part of an ongoing effort to understand bone mineral density loss and other effects of micro-gravity on the human body. Once safely recovered by a team at the splashdown site, the frozen samples would be unloaded and transported to waiting labs for study. With direct implications for treating the global problem of bone density loss due to aging, the samples being transported back to Earth represented one of the most clear cut examples of how bio-medical research being conducted on ISS is quietly bridging the gap between space exploration and everyday life. As Commander Williams put it, "literally and figuratively there's a piece of us on that spacecraft going home to Earth."

On departure day, as the Station flew high over Burma, ISS astronauts released the SpaceX CRS-1 Dragon capsule at 8:29 AM CDT. Following three brief burns to safely remove the spacecraft from the International Space Station's "keep out sphere" the craft rotated, and began heading back to Earth for an estimated landing at 2:20 PM CDT. The deorbit burn began at 1:28 PM, with the Dragon trunk and solar panels jettisoned moments before atmospheric interface.

First Dragon to go to ISS safely secured aboard recovery vessel for the voyage home. Credit: SpaceX

For only the third time, the Dragon once again made a picture perfect landing, but in a sign of how quickly events were expected to become routine, this time there was no P3 aircraft circling the landing zone in order to capture aerial footage. There was hardly any need. Demonstrating a degree of re-entry control already well ahead of what the Soyuz could accomplish, the Dragon splashed down at 3:22 p.m. within clear view of the waiting recovery boat. Only a few minutes later, the dive team was already in the water performing a preliminary inspection before it could be hoisted aboard.

SpaceX was fortunate the landing was so accurate. As the spacecraft plunged into the Pacific, salt water entered one of the submerged equipment bays, promptly shorting out a power management unit and killing all three coolant pumps as well as electrical power to the returning GLACIER freezer inside the pressure vessel, which remained water tight. Without power however, the samples, which were stored at -95 Fahrenheit, immediately began to warm, potentially threatening their scientific value. Fortunately, the combination of cool Pacific water and cabin circulation fans which did remain powered, minimized the temperature climb, and the recovery crew was able to restore power and restart the freezer only moments after the Dragon was lifted out of the water. As a result, the temperature never climbed above a comparatively balmy -65. Two days later, with time critical cargo already delivered to NASA and on its way to destinations across the U.S., the Dragon was back in port and headed towards McGregor Texas for complete unloading and fault analysis.

The problems encountered upon landing were not immediately publicized, and in fact only became part of the public record during the course of a routine update ISS Program Manager Mike Suffredini provided to a Congressional oversight committee several weeks after the mission's completion. As it turned out, there were a few other problems as well. In addition to the water intrusion, Suffredini's update revealed several other issues which occurred while the Dragon was in orbit. While attached to the Station, one of three redundant flight computers, Computer "B" de-synched from the other two, presumably due to a radiation hit. Although the computer was successfully rebooted, SpaceX, after discussion with, and at the suggestion of NASA, followed the easiest course of action and did not attempt to re-synch the computer. The decision was completely within the logical parameters of the Dragon's triple string redundant voting architecture which requires two of three computer units, each of which consists of two computers, to agree on each course of action. There were several other radiation events as well; one of three GPS units suffered a fault, but was successfully rebooted. Two other items, the Propulsion and Trunk Computer as well as an Ethernet switch also took suspected hits, but were all rebooted successfully. Finally, what had been described as an undisclosed problem with a set of Draco thrusters was presented as an issue with two pressure transducer sensors and one injection resistance temperature detector. The Dragon's onboard fault detection, isolation and recovery system handled these issues without difficulty as well. Despite the various glitches, the first CRS mission was in the books, and with Orbital Sciences still almost a year away from making its first test flight to the Station, SpaceX had come through for NASA with the completion of two supply flights to ISS in 2012, an early requirement which the agency told the GAO was an absolute necessity to keep the station's scientific output on pace, as well as forestall any "gradual degradation" if the delays went further.

The news was not as good for Orbcomm however. Although controllers were able to complete some basic functions with the prototype spacecraft, including deploying the solar panels and antennas and performing elementary system checks, they were not able to fully test the satellite before it re-entered the atmosphere. As a result, Orbcomm filed an insurance claim for $10 million, declaring the small satellite a total loss. As might be expected, critics seized on the Orbcomm loss as evidence of a flight failure, but SpaceX firmly rebuffed the characterization, observing that the Orbcomm element, however important, was a secondary aspect of the mission, and thus, the Falcon 9's launch record was intact, and now standing at 4 and 0.

Depending on one's perspective, the CRS-1 mission could be characterized as another outstanding success, albeit one with a few anomalies to address, or a series of near misses which warranted serious concern. Either way, the company was determined that the next flight, scheduled for early 2013, would be much smoother. Instead it offered the most intense drama yet.

CRS-2

With SpaceX already fully engaged in testing the new Merlin 1D engine for its introduction on flight number 6 with the debut of the Falcon 9 V1.1, the engine failure on flight 4 left the company no choice but to conduct an exhaustive examination of the Merlin 1C series it had been hoping to ease into retirement with its final flight. After an initial joint examination which failed to yield any clear culprit, SpaceX remained uncharacteristically quiet about what caused the engine shutdown, and what was being done to prevent it from happening again, even as preparations for the next mission got underway. In fact, the only information released until immediately before the CRS-2 launch, was that the particular engine which failed had already accumulated a higher than normal total runtime through the course of testing, suggesting that the most obvious solution, albeit an imperfect one, was to limit run time for each of the 10 flight engines to be used on the next launch.

Finally however, at the pre-launch press conference for the CRS-2 flight, SpaceX President Gwynne Shotwell revealed both the source of the problem and the reason for the near embargo on discussing it.

The chamber breach had occurred due to the combination of the extended runtime previously cited, which aggravated an undetected material flaw and weak spot in the overworked combustion chamber. The solution to preventing a recurrence was a non-destructive metallurgical analysis of each of the engines for the upcoming flight. That was as far as it went. Citing ITAR regulations and the nation's continuing quest to prevent China from gaining any further insight into methods of metallurgical analysis, she declined further comment into the incident. Apparently NASA was satisfied however, and for the mission immediately at hand, that was all that mattered. As for future flights, the upgraded Merlin 1D featured an entirely different chamber design which should not prove susceptible to the same type of failure.

The CRS-2 flight was scheduled for Friday, March 1st, with a daytime liftoff at 10:10 AM EST, and once again it would be lightly burdened with approximately 1,200 lbs. but in a clear sign that NASA was ramping up the pace of research, the Dragon carried materials for 160 investigations, including 50 which were entirely new. Also, for the first time the Dragon would fly with unpressurized cargo secured in the trunk, a pair of external truss beams intended to be mounted to the Station's outer hull.

With perhaps a heightened sense of concern following the previous launch, the final first generation Falcon 9 sailed smoothly through pre-launch preparations and countdown on its way to a flawless liftoff under partly cloudy skies. As it passed through the first stage burn and Main Engine Cutoff without any incident, the sense of relief was palpable. Stage separation and second stage ignition went just as smoothly, and just over 10 minutes after main engine ignition, the Dragon was safely in orbit and separated from the second stage. That's when the trouble began.

Late loading of Dragon just prior to CRS-2 mission. Credit: Stewart Money

The cargo version of the Dragon rides to orbit with its on-board thruster systems in an unpressurized state. One of the first actions on release from the second stage is a pressurization of the 8 tanks, four fuel and four oxidizer, and the 18 Draco thrusters, arranged in four pods, from the dual high pressure helium supply tanks. This time however, as the flight computer issued the command to pressurize the tanks, only the number one tank came up to operating levels. Tanks two and four remained unpressurized, while number three showed a gradual but inadequate buildup of pressure. In an instant, SpaceX was facing a serious problem. Somehow, despite the built-in redundancies, the Dragon had only one out of four Draco pods at its disposal, and NASA rules required that at least three be working before it could approach the Station. To make matters worse, as flight controllers started to troubleshoot the problem, the Dragon gradually began to experience difficulties with attitude control. With the reaction control system disabled, SpaceX originally refrained from deploying the solar panels, but with the temperature steadily dropping, engineers were afraid that the panel might be frozen in place before the thruster problem could even be addressed. Reluctantly, mission controllers issued the command to unfurl the solar panels, hoping that in addition to buying time beyond the craft's 15 hour battery life, they might act much like a skater who extends her arms to slow down from a spin. The maneuver worked, but the stricken spacecraft was still tumbling sufficiently that maintaining a constant communications lock became impossible.

With NASA Administrator Charles Bolden in attendance at Hawthorne Mission Control, the flight control team was initially stunned by the nature of the problem. Ordinarily, a similar situation could be caused by either frozen or stuck valves, most likely caused by either ice or physical debris in the lines. What was truly perplexing however, was the fact that three of the four tanks were out of commission, which really should not have been possible. Originally believing the culprit to be ice lodged somewhere in the lines, the team tried cycling the check valves between pressure and oxidizer tanks several times, but without result. What appeared for a moment to be a temporary problem suddenly loomed as a possible mission ending disaster. Deducing the problem must lay in the valves themselves, SpaceX software engineers quickly wrote a new program that would sent a series of commands to "hammer" the valves by releasing high pressure slugs of helium gas, hoping to literally force them open. Due to the slow tumble however, mission control was having difficulty in transmitting the new commands back up to Dragon. With the orbit's low perigee at 199 km threatening to degrade further, SpaceX was facing the prospect of an uncontrolled re-entry which would have both destroyed the Dragon, and posed a credible threat of debris over an undefined landing zone.

Falcon 9 clears the tower on CRS-2 launch. March 1 2013.A tense drama unfolds only moments later when the Dragon thruster system fails to fully pressurize. Credit: Stewart Money

Fortunately, cross agency cooperation saved the day. Reaching out to the Air Force, SpaceX finally gained a better uplink via the U.S. Air Force's more powerful satellite transmitters, allowing mission control to send the hastily written program which appeared to be the best, and maybe only, chance to save the mission. With the code delivered, the Dragon began releasing high pressure blasts to its uncooperative oxidizer tanks. At first, nothing. Finally, on the fourth try, the pressure reading from one of the three LOX tanks came up, and was then followed by the other two.

The repair had apparently worked, and moments later, the spacecraft was once again in stable flight. Before the Dragon could be cleared to begin its burn to match orbits with ISS, controllers would need to satisfy NASA that everything really was back to normal. Following a brief four minute burn to stabilize the orbit, SpaceX and NASA set to work on a long day of carefully testing the spacecraft's systems. By the next morning with Dragon responding perfectly, SpaceX was cleared to begin its pursuit of the station in earnest, having lost only a day.

Having overcome its thruster issues, the CRS-2 Dragon is photographed in free flight.

Credit: NASA

Astronauts pull Dragon in for berthing on CRS-2. Credit: NASA

Dragon berthed to ISS, CRS-2. Credit: NASA

After the high drama of the first few hours in orbit, the rest of the CRS-2 mission to the International Space Station proceeded without incident. Once captured by the robotic arm at 4:21 AM CST, Mission Control in Houston temporarily assumed control of the arm, and moved the Dragon into berthing position via ground command. Once berthed, astronauts opened the hatch at 12:14 pm CST on Sunday March 3rd, and began unloading the fully packed spacecraft. Dragon even got to make up the lost day in space when high seas in the recovery zone prompted mission managers to delay departure by a day to March 26th.

As before, Dragon inched its way away from the station carrying more cargo than it had hauled up. Following a nominal reentry burn and flawless descent, which once again occurred on target, the Dragon splashed back down into the Pacific, but this time, with re-designed seals, there was no water intrusion into the equipment bay. As for the valve issue which had nearly doomed the mission, a subsequent analysis revealed that unbeknownst to the company, the vendor which supplied the valve had made a subtle, almost invisible change to its design, and three of the new valves had been installed in the CRS-2 Dragon, leading to the failure. In the period before launch, SpaceX had performed a low pressure check of the Dragon's thruster system, but had not performed high pressure checks which would have revealed the problem.

If there can be such a thing as a useful failure, the difficulties SpaceX encountered on what seemingly should have been two routine flights to ISS may have been just that. With the company preparing to introduce a radically upgraded engine in the new Merlin 1D and what was arguably a new booster in the Falcon 9 v1.1, perhaps the most dangerous threat it faced was complacency and a sense that it was on the way to mastering the art of routine, safe spaceflight. The first two CRS flights were a bracing reminder that the margin for failure in space operations is so narrow that it is barely detectable, and the only insurance is relentless testing and analysis. At the same time, and particularly on the CRS-2 flight, the deft response in analyzing and then solving a difficult problem against the clock and under the microscope of a public mission only served to further burnish SpaceX's credentials as a truly remarkable organization which undeniably reminded observers of what NASA itself had once been.

An era came to an end with the conclusion of the second CRS flight to ISS on March 31st, 2013, as the original series Falcon 9 rocket made its fifth and final flight. Although a story understandably lost in the narrative of the Dragon thruster problem and the rapid fire sequence of events which salvaged the mission, the CRS-2 launch marked in some ways the end of the second era of SpaceX and the beginning of a new, even more challenging era which is still taking place today.

Beginning with the formal introduction of the COTS program which resulted in the development of both the Falcon 9 launch vehicle and the Dragon spacecraft, SpaceX made incredible, almost hard to believe strides in successfully introducing both systems. In doing so at a total NASA investment of $396 million including the additional risk reduction funding allocated in 2010, the company answered the challenge posed by Mike Griffin in 2005, and provided undeniable existence proof that new commercial contractors could develop and deploy a highly complicated space transportation system for a fraction of the cost of either traditional NASA procurements such as Ares-1, or semi-commercial efforts such as those which took place with Boeing and Lockheed Martin under the USAF Evolved Expendable Launch Vehicle Program.

Nevertheless, even though SpaceX beat the historical odds which suggested a high probability of failure within the first three launches, and instead recorded five successful flights for the Falcon 9, including four safe recoveries for the Dragon spacecraft, it was in many ways still a pathfinder system, developed within a unique and mostly protected environment. Expanding its scope beyond the 10 remaining Commercial Resupply missions to ISS required new developments to successfully enter three new areas, NASA's Commercial Crew program, the Geostationary communications satellite launch business, and the untapped market offered by the Department of Defense and the EELV launch business. Though different, what all three had in common was that they demanded a level of expertise and proficiency which SpaceX was still trying to develop, all while presenting experienced and deeply entrenched competitors, each of whom was determined to make SpaceX earn every step it took.

The new era begins with the introduction of the Commercial Crew program, and ends with the introduction of a bigger, better launch vehicle, the Falcon 9 v1.1.

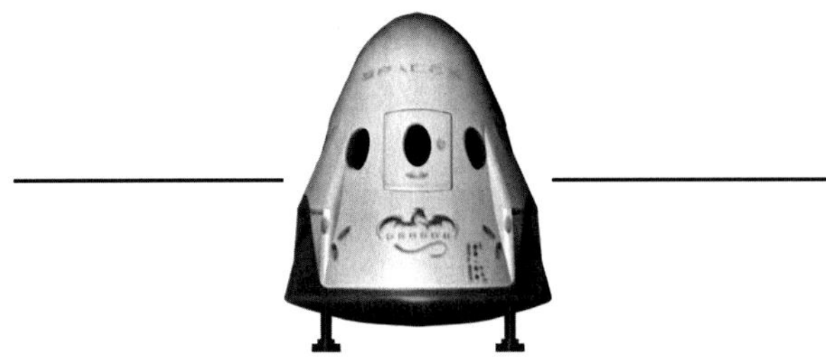

Chapter XXIV: Commercial Crew

On June 21, 2004, Burt Rutan's all composite, airplane launched SpaceShipOne earned world-wide publicity and a coveted place in history by becoming the first privately developed manned vehicle to achieve suborbital spaceflight. In light of the accomplishment, space enthusiasts could perhaps be forgiven for believing that a new age of commercial spaceflight was taking off. As the spacecraft was being towed back down the runway after a successful landing, newly minted astronaut, a beaming Mike Melville stood triumphantly astride it, holding up a sign which read SpaceShipOne–GovernmentZero. The good natured jibe concealed the fact that while the team behind SpaceShipOne had clearly recorded one of the red letter dates in space history, there was a world of difference between the suborbital flight just completed, and the vastly more challenging aspect of reaching orbit. Although it had pierced the Karman line and the boundary of space at altitude of 100 km (62 miles), and achieved several minutes of microgravity, SpaceShipOne, achieving a maximum speed of 2,150 mph, in fact came nowhere close to reaching the minimal velocity of 17,500 mph required to attain orbit.

Unlike Project Mercury's first brief foray into suborbital space with Alan Shepard aboard a Redstone rocket on May, 5th 1961, a feat which the X-Prize winner had just replicated albeit in a very different vehicle, there was no alternate delivery system comparable to the Atlas booster waiting in the wings to take the next step any time soon. At the end of the day, the prize of launching the first privately funded manned spacecraft into Earth orbit was still very much anyone's to take, and with it, one of the most enduring accomplishments in the history of the space age.

Just six years later, as the first Falcon 9 hurtled through the upper atmosphere on the way to orbit in June 2010, Elon Musk and SpaceX, with indispensable help from NASA and the American taxpayer through the COTS program, had taken a critical, and in many ways the biggest single step in producing a launch vehicle capable of propelling a manned spacecraft to orbit, but both financially and chronologically, they were less than half-way towards reaching the goal which seemed so achievable when SpaceShipOne rocketed into the heavens high over the Mojave desert.

While the first vehicle Elon Musk and SpaceX introduced to the public was the decidedly modest Falcon 1, he had begun the company with the shamelessly immodest goal of creating a transportation system fully capable of providing for the settlement of Mars with a purpose no less than enabling mankind to become a multi-planet species. Following that path, there was never any question that if his company progressed beyond the Falcon 1, the next launch vehicle would be capable of launching people into space. Even as his engineers struggled with the first generation, ablatively cooled Merlin engine, and the very rudimentary elements of attaining orbit, the company was quietly planning a launch vehicle and manned spacecraft which would comprise its first end to end space transportation system.

Although from a purely financial standpoint, as detractors were always eager to point out, SpaceX products could not be considered "privately developed" to the degree that applied to SpaceShipOne and its White Knight carrier, there was also no doubt that the Falcon 9 represented something new under the sun, the product of a collaborative relationship between NASA and SpaceX, in which the latter had invested more than 60% of the total expense in a design which was strictly its own.

Perhaps even more surprising, despite a nearly flawless launch record in the USAF EELV program which had seen more than 30 Atlas V and Delta IV rockets successfully deliver payloads to orbit since 2002, eight years later, it was the Falcon 9, with precisely one successful launch to its name, rather than either of the more proven boosters which was quickly developing the inside track to becoming the first new American crew launch vehicle since the Shuttle and the first real step towards affordable space transportation in the history of the space age.

The sequence of events which took SpaceX from a tiny startup in 2002 to the cusp of replacing the role of the Shuttle in barely more than a decade is a story as much about politics and luck as it is about determination and technology, and one which will be studied for years to come. Although SpaceX and the unique leadership supplied by Elon Musk take center stage, it is also indisputably about NASA, and a sea-change in the way some elements within the agency had come to view the potential role of commercial companies in reaching low Earth orbit, and what they accomplished to bring it about. Though bolstered by external developments such as the X-Prize, it also came from a realization that if the agency wanted to resume Apollo's legacy of exploration into deep space, it had little choice but to change how it got to orbit in the first place.

What emerged was a curious mix in which very different groups of stakeholders saw in the developing Falcon/Dragon system a means to advance their goals. For NASA, the introduction of the Vision for Space Exploration created what was for many a difficult dilemma. In order to go forward, the agency would have to leave behind the one iconic symbol which had stood at the centerpiece of the manned space program for the last 30 years. And although NASA could now begin planning missions to the Moon, and perhaps beyond, there was no getting around the fact that the new crew capsule which would carry astronauts to space and back seemed an obvious and depressing retreat from the remarkably advanced spacecraft which was the Shuttle. Although horrendously expensive, the Shuttle offered an astounding array of features; enormous cargo capacity, research deck, manipulator arm, airlock, and extended habitable space in the form of Spacelab, all of which would be painfully absent in the capsule which seemed destined to replace it. Furthermore, in pursuing an architecture capable of supporting deep space missions with a combination of fully expendable launch vehicles, the agency was also turning its back on what was once presumed to be the proper legacy of the Shuttle, a fully reusable next generation launch vehicle such as the ambitious X-33/Venturestar program, or even the more restrained goal of an incrementally reusable Second Generation RLV program which NASA was just beginning to pursue when it was supplanted by the Vision for Space Exploration. Either way, the goal of achieving a breakthrough in the costs of reaching low Earth orbit through direct technology development was cast aside.

For space advocates outside the NASA/contractor community, this meant private developments such as the X-Prize took on added importance, and it would be up to the nascent NewSpace community to carry the torch. When SpaceX emerged on the scene with its small rocket but clearly expressed plans for establishing reusability, for anyone willing to take a leap of faith, there was a new reason for hope. Four years later, as the company was tapped for the Commercial Orbital

Transportation Services program and tasked with developing a system which could be readily upgraded for a crew transport function, what had previously seemed very improbable, was quickly becoming possible. And although developing a crew transport capability for NASA did not necessarily translate into a system which could open up space for all, it was, and still is, a critically important milestone along the way. Like the successful launch of the first privately developed liquid fueled rocket in the Falcon 1, it held the potential of tearing down the barrier in perception of what could be done. The shortest path to becoming reality however, extending the experiment begun in COTS and CRS to crewed spaceflight, required NASA to take a step it was very reluctant to take. It was also one which a powerful collection of elected representatives, entrenched contractors, and even storied astronauts were determined to prevent.

When NASA Administrator Sean O'Keefe was helping to craft the proposal which would become the Vision for Space Exploration, the creation of a gap in U.S. manned spaceflight between the retirement of the Shuttle and the introduction of its successor was accepted as a necessary and manageable consequence of such a significant change in U.S. space policy. At the same time, although lost in the bigger story of NASA's planned return to the Moon, the VSE called for an expanded role for the commercial sector in providing cargo, and very importantly, crew transport services to LEO. The Aldridge Commission, convened to recommend options for implementing the VSE, lent credence to the call for separating the conceptualization of outer space as being between two different destinations, low Earth orbit and everything beyond, but like many of the recommendations in the report itself, the call for pursuing commercial alternatives for low Earth orbit crew transfer was initially swept aside in the implementation of Project Constellation.

Significantly however, the very language of the legislation writing the VSE into law, the NASA Authorization Act of 2005, did call for the use of commercial transportation services to Leo to "the greatest extent practical." [86]

> (2) CONSULTATION AND COORDINATION. —In carrying out the programs of NASA, the Administrator shall—
>
> (A) consult and coordinate to the extent appropriate with other relevant Federal agencies, including through the National Science and Technology Council;
>
> (B) work closely with the private sector, including by—
>
> (i) encouraging the work of entrepreneurs who are seeking to develop new means to launch satellites, crew, or cargo;
>
> (ii) contracting with the private sector for crew and cargo services, including to the International Space Station, to the extent practicable;

(iii) using commercially available products (including software) and services to the extent practicable to support all NASA activities; and

(iv) encouraging commercial use and development of space to the greatest extent practicable; and

(C) involve other nations to the extent appropriate.

In the immediate period after the introduction of VSE, and before the emergence of Exploration Space Architecture Study under Mike Griffin, NASA looked to the recently introduced Atlas V and Delta IV Evolved Expendable Launch vehicles as likely alternatives to replace the Shuttle's basic crew transport capacity to LEO, and perhaps in the case of the Delta IV heavy, to lunar space.

Mike Griffin however, had arrived with a mandate to speed up the deliberate process his predecessor was pursuing, and wasted little time in commissioning the Exploration Space Architecture Study to support his very different concept of a crew launch system derived not from the EELV class, but instead comprised of components from the Shuttle system. Ostensibly, the use of already existing "man rated" components offered the promise of a quick introduction and low risk development of first a new launch vehicle for crew, the Ares-1, followed by a super heavy lift booster, the Ares-V, with the supposed benefit being the commonality between the two components.

Had the original engineering been sound, and not resulted in major design changes to the Ares-1, the program might have worked, and resulted in a crew and cargo launch capacity to ISS only a few manageable years after the Shuttle was retired, reducing the "gap" in providing a U.S. crew launch capability. Instead, the Ares-1 program encountered one problem after another, beginning with the realization that it would be exceedingly difficult to "air start" the single Space Shuttle Main Engine powering the second stage. This revelation resulted in the decision to substitute a new second stage engine, one based on the Apollo program's own J-2 engine. The problem was this engine, the J-2X, did not actually exist, and would require its own development program. Furthermore, this change meant that the existing SRB based first stage would now be insufficient for the task at hand, leading to yet another significant change; adding a fifth segment to the SRB. With each change, the "gap" gradually became wider. As a result of the chaos, the United States was now looking at 5 or more years of purchasing Soyuz flights from Russia as NASA's rusty engineering base tried to find its feet. There was however, an alternate possibility already under study.

In addition to its commitment to the Ares-1 and Project Constellation Architecture as a means not only for conducting a return to the Moon, but also supplying crew transport services to ISS, the agency had been required by law to pursue a commercial track as well, which it did so in establishing the Commercial Crew and Cargo Program Office(C3PO) at Johnson Space Center. Under its auspices, and the direction of Alan Lindenmoyer, NASA made astounding progress in developing commercial cargo capacity through the COTS program. One might have thought that on the basis of the undeniable success of NASA and SpaceX in conducting the first Station resupply in the COTS 2/3 mission less than a year after Atlantis came to a wheel stop for the final time, Congress would have pumped money into the nascent Commercial Crew program to regain American crew launch capacity as soon as possible. Instead they did the opposite.

While COTS as originally outlined included part "D" which called for commercial crew transportation, for which SpaceX had originally proposed a series of milestones, NASA under the direction of Mike Griffin, although supportive of the COTS cargo initiative, was reluctant to explore the fourth option, and never requested funding to pursue it. Nevertheless, C3PO continued to make progress in outlining how a Commercial Crew program might eventually take shape, and as the Ares-1 delays mounted, it found a strong advocate in incoming NASA Deputy Administrator Lori Garver.

Following a series of well publicized difficulties, which included reports that the Ares-1 might shake so violently on ascent that it posed a risk to the crew, as well as the release of GAO reports in 2007 and 2008, which called into question the financial basis of the program and shortfalls in required funding, the newly elected Obama Administration commissioned the Augustine Committee in early 2009 to review Project Constellation and recommend options. As it became increasingly clear that the panel would include cancelling the Ares-1 as one of the possible courses of action being presented, a group of Constellation contractors, led by ATK, began a PR campaign to preserve the imperiled program by highlighting the purported safety hazards of the alternative course of allowing private industry to develop launch solutions. The overly simplistic line of reasoning was that because NASA was not designing the systems, they could not be certified as safe.

Given the history of the Solid Rocket Boosters in the Shuttle program and questions being raised about the 5 segment first stage for Ares-1, it was a brazen, and some would say shameless approach, but it was effective. Before the final Augustine report was even delivered, supporters submitted letters from concerned astronauts, many of who now happened to be employed by the Project Constellation's prime contractors, calling on Congress to preserve the status quo. In a preview of what was to come, in January 2010, the Astronaut Safety Advisory Panel, led by Admiral Joseph Dyer, used the occasion of its annual report to Congress to preemptively advise against a possible commercial crew program on the same grounds. Again, it was an odd point to make, considering that the panel had been relatively sanguine regarding the well documented risks posed by the Ares-1 booster. Notably, the ASAP report called out SpaceX as an example of an unproven and unknown system which "did not meet NASA safety standards for piloted vehicles," going on to question the company's honesty by adding "despite some claims and beliefs to the contrary." [87]

Outraged, Elon Musk released a rebuttal pointing out that the ASAP had had minimal interaction with his company, and could not possibly be in position to comment on a system which it knew next to nothing about, but had been designed to exceed all of NASA's published human ratings requirements. Though years would pass before any system developed under Commercial Crew was ready to fly, a time frame in which NASA would have ample opportunity to thoroughly evaluate any proposal it was considering, the ASAP report was a clear indication that unlike the relatively calm stand-up of the COTS program, the battle lines against Commercial Crew were already being drawn before the Congress had allocated any meaningful funding. As always, the one issue left unaddressed by the aerospace establishment fighting to protect the status quo, was the undeniable math which underlay the Augustine Committee's conclusion. Absent a significant increase in NASA funding, the agency would not have the resources to follow through on its plans. While not a prescription or an excuse for taking unnecessary risks, it was strong admonition that something had to change.

When President Barack Obama signaled a major change in U.S. space policy with the call to end Project Constellation and its goal of a return to the Moon in his speech at the Kennedy Space Center on April 15, 2010, he formally assigned the task of regaining domestic launch capability to low Earth orbit to commercial industry and the pilot program which had been underway in the C3PO office for the last year. Beginning in 2009 as part of the American Recovery and Reinvestment Act, the Obama administration formally initiated the Commercial Crew program office to study ways of implementing what would have been COTS D with an allocation of $50 million. The purpose was to study sub-systems which could contribute to a future commercial crew program. Conducted under Space Act Agreements, and called Commercial Crew Development Phase I, or CCDev 1, NASA received bids from more than 30 companies, including SpaceX, for sub-system research. It chose five:[88]

Boeing, Blue Origin, Sierra Nevada, United Launch Alliance, Paragon Space Development.

The bitter political obstructionism which would afflict the Commercial Crew program from the outset did not take long to develop. NASA's total allocation under the Recovery act was $1 billion, and from that amount the agency originally intended to direct $150 million towards jumpstarting the new program. The effort was almost immediately sidetracked when Republican Senator Richard Shelby of Alabama successfully diverted $100 million on top of its existing funding to the vastly over budget and soon to be cancelled Project Constellation. An allocation NASA neither requested nor wanted, this wasn't just bringing home the bacon, it was rendered pork fat fondant over a plateful of deep fried souse meat. Left to starve by contrast, the eventual re-introduction of an American crew launch capacity via commercial providers began a steady slide into the future which made Russia the undeniable long term beneficiary.

Among the CCDev-1 recipients, Boeing was awarded $18 million for development work on its proposed CST-100 spacecraft, a modestly upsized version of the historic Apollo capsule, equipped with a small expendable service module and an innovative pusher escape system. Even though Boeing had developed the Delta IV launch vehicle and was a 50/50 partner in United Launch Alliance, the CST-100 capsule was designed from the outset to be "launcher agnostic," capable of being carried to space aboard the Delta IV, Atlas V, Falcon 9, or even ATK's proposed Liberty launch vehicle. After a surprisingly long period of deliberation, Boeing ultimately announced in August of 2011, that if the CST-100 proceeded to development, the launch vehicle for the immediate future would be the ULA Atlas V.

Also offering a pusher escape system was Blue Origin, the secretive company funded by Amazon.com founder and CEO Jeff Bezos. Already working on a reusable suborbital launch vehicle program intended for tourism flights under the name New Shepard, Blue Origin's CCDev-1 proposal was focused on demonstrating a pusher abort system for the terribly named "biconic space vehicle," a reusable capsule also intended to be launched on the ULA Atlas V. For Blue Origin, which like SpaceX was pursuing a breakthrough in lowering the costs of space transportation, the Atlas V was selected as a temporary solution while the company gradually worked through the development of its own two-stage reusable launch vehicle.

Sierra Nevada Corporation was also selected to further the development of its long sought Dream Chaser spaceplane, which the company had previously submitted to NASA under the COTS program. With its sight lines clearly evocative of the Space Shuttle, although on a smaller scale, and the only winged entry to receive funding, the Dream Chaser, the most technologically demanding entry in the field, would also hold a strong emotional advantage for an agency in which many were still mourning the passing of the Shuttle.

With three selectees sharing in common the Atlas V as their chosen launch vehicle, a fourth award, made to ULA to enhance the Emergency Detection System aboard the booster, provided a much needed common component. While Boeing might have been expected to specify its own Delta IV launch vehicle for the CST-100 rather than the Atlas V, the Delta IV is slightly underpowered in its most basic configuration without the addition of small GEM (graphite epoxy motor) strap-on boosters, which cannot be readily "man-rated." Although the Atlas V also frequently employs supplementary boosters as well, its basic version, the 401, supplies the necessary first stage thrust without them, although it comes at the substantial cost of adding a second RL-10 engine to the Centaur upper stage, making it a 402 designation. Combined with the higher cost of the Delta IV compared to the Atlas V, it has set up one of the unwitting ironies of the Commercial Crew program. Among the finalists which entered the eventual third round, only the SpaceX Falcon 9 would actually offer a completely independent alternative to Russia's continued pre-eminence over American manned spaceflight program which extended from the Soyuz to ULA's reliance on the Russian built RD-180 main engine for the Atlas V.

The one name curiously absent from the CCDev 1 winner's list was SpaceX, which did submit a bid, but was discounted for asking for too much funding to match the limited list of milestones proposed. Although following a parallel track through the COTS program, SpaceX, which was keenly focused on winning Commercial Crew, would not make that mistake again when the next round got underway. The fifth selection was to Paragon Space Development Corporation for work in advancing its area of expertise, the modular Commercial Crew Transport-Air Revitalization System, (CCT-ARS) which was expected to be the likely supplier of life support capability to whichever teams were the eventual winners. Already working with SpaceX as a vendor for the Dragon capsule, the Paragon bid was something of a proxy for the company in this initial round.

After the initial CCDev-1 awards in early 2010, NASA issued a call for a second round of solicitations that same October, to be funded as part of the FY 2011 budget appropriations. Following the restricted funding allocated in the first round, NASA hoped it could now start the program in earnest, beginning with a budget request of $500 million, and increasing to $1.4 billion per year for the next two years, followed by two more years at $1.3 billion. Once again, these ambitions were stifled when the U.S. Senate, led by Richard Shelby, along with participation from Kay Bailey Hutchison of Texas, and Bill Nelson of Florida, allocated only $312 million, with $270 million going to the actual development contracts. As before however, the trio of senators succeeded in showering money on the remnants of Project Constellation instead. It was an unmistakable signal that Congress was prepared to protect the status quo no matter what the cost, even if it could be measured in the loss of American prestige and the transfer of hundreds of millions of dollars to Russia to pay for rides to space aboard Soyuz and the production of even more RD-180 engines in the suburbs of Moscow.

Making the best of the situation, NASA divided the appropriation between four companies in the next round in awards announced on April 18, 2011. This round, CCDev-2, and also operating under the flexibility of Space Act Agreements, sought for the first time to begin developing complete systems rather than operating at a component level. Specifically:

> "The goals of the CCDev 2 investments are to advance orbital commercial CTS concepts and enable significant progress on maturing the design and development of elements of the system, such as launch vehicles and spacecraft, with the overall objective of accelerating the availability of U.S. CTS capabilities while ensuring crew and passenger safety."
>
> Notably, and in keeping with the philosophy behind COTS, C3PO was hoping the program would help set the stage for broader commercial development and manned spaceflight to LEO:
>
> "Through this CCDev 2 activity, NASA may be able to continue to spur economic growth as capabilities for new space markets are created, and reduce the gap in U.S. human spaceflight capability."

Following the process established in the COTS program development, the relationship between NASA and the contractors took the shape of collaboration rather than demand, with large scale issues discussed in a series of workshops in which officials from the Commercial Crew program office and the industry partners could discuss the top level requirements, and then through question and answer sessions, mutually agree on ways to meet those requirements without the agency specifying the vehicle particulars. As the proposals took shape, the day to day work of designing the systems and subsystems took place in a steady exchange of information between NASA and the contractor teams guided by a NASA assigned integrating team leader, who again following the model developed in the COTS program worked closely with their respective industry partner.

The top level requirements called for the respective spacecraft to perform three primary functions:

Deliver and return four crew and their equipment to ISS

Provide Assured Crew Return in the event of an emergency

Serve as a 24-Hour Safe Haven in the event of an emergency

Most significantly, any system selected would have to provide a higher level of safety than existing crew transportation systems, Shuttle and the Soyuz. The launch rate required was a minimum of two launches per year, performing the ordinary crew rotations at 180 day intervals, but with an ability to linger as long as 210 days. In addition to transporting four crew members, each flight also needed to be capable of hauling a nominal amount of cargo, 220.5 lbs. in 8 cubic feet of storage, as well as a GLACIER freezer, the primary means of transporting biological samples to and from the station. [89]

Additionally, each spacecraft was required to provide a 24 hour linger in orbit capability both prior to arriving at ISS, as well as after departing the station en route to reentry and landing. Finally, any crew transport system must provide the capacity to function independently as a safe haven for up to 24 hours, with its hatch closed, but remaining docked to ISS. This latter contingency was part of the emergency protocol for the station which stated that in the event of a significant problem aboard ISS, the entire crew needed to be able to retreat to the comparative safety of their transport craft, not necessarily to evacuate, but to provide a safe environment from which to troubleshoot the issue without exposing crew members to any more risk than absolutely necessary.

From a broader perspective, NASA structured the Commercial Crew program, not just to provide a means of astronaut transport to ISS at lower overall investment than it might otherwise have had to make as evidenced by the Ares-1 program, but also with the explicit purpose of building in competition between multiple providers from outset. Encouraged by the success of the COTS program to date, as well as its own experience at KSC in competitively procuring launch vehicles through the National Launch Services Contract, NASA's rationale for the program depended on creating at least two fully functional providers, a calculation which necessarily led to a higher level of initial funding than might have been the case with a single provider. On the other hand, had the agency taken that path, it quite likely would have been through conventional, and far more costly FAR contracts from the outset. Based on the overwhelming preponderance of historical and concurrent evidence, it is not difficult to imagine that the overall development costs would have been greater, and the operational costs higher still.

Just as COTS sought to use the ISS and NASA as anchor tenants in what was hoped to be a growing market in LEO cargo delivery, the Commercial Crew program hoped the ultimate result would be crew transport services to other destinations in LEO as well. NASA's newfound enthusiasm for commercial operations in LEO was remarkable really, and a far cry from the days when the agency attempted to prevent Dennis Tito from journeying to ISS aboard the Russian Soyuz, and any commercial proposal regarding the Shuttle was out of the question. Now NASA was actively supporting the underpinnings of private, commercial space transportation. Although there was no other destination presently in LEO, a strong possibility existed in the proposal by Bigelow Aerospace to build an inflatable space station to be marketed to individuals, corporations and sovereign nations. Here was the classic chicken and egg dilemma writ large. Absent a crew transport capability, an inflatable Bigelow station would remain as a viewgraph only, but until such the time such a station was in orbit, the "market" didn't exist. NASA was convinced however that possibility was real, and one of the factors weighing favorably on Boeing's proposal for Commercial Crew was the inclusion of work performed with Bigelow to hammer out a preliminary crew transport service based on the CST-100. Boeing was careful to point out however, that if it was not selected for development under the NASA program, its business case for Bigelow could not close.

Along with prior award winners, each of which (except Paragon) was selected to continue work on their proposals from CCDev-1, SpaceX now joined the winning entrants with an award of $75 million to develop an integrated launch abort system for the crew version of the Dragon spacecraft which it named DragonRider. Among the other winners, Blue Origin received $22 million for work which would now include testing the thrust chambers for the BE-3 hydrogen/oxygen engines intended to power its two stage reusable launch vehicle. This segment of the Blue Origin award highlighted a little recognized aspect of the Commercial Crew program, namely that its mandate was to assist in maturing technology which could expand space commerce, even if it was not on the immediate path of the first generation crew launch vehicle NASA was seeking. Besides work on the BE-3 engine, other milestones offered by Blue Origin were directly tied to its crew vehicle, and focused on the escape system, with the most significant milestone being an escape demonstration test for a boilerplate version of the bi-conic space vehicle.

Boeing received an additional $92.3 million for the CST-100, an amount which would include a program of parachute drop tests, and Sierra Nevada Corporation received $80 million for the Dream Chaser and a steadily advancing program intended to culminate in captive carry flights of the atmospheric test article.

In addition to the funded winners, there were also several unfunded winners with interesting potential. One was Excalibur Almaz, a space tourism company based on the Isle of Man, and in possession of a rather unique set of functional Cold War space hardware, a fleet of six refurbished and reusable Soviet era space capsules and two partially completed small space stations. The overall program was named Almaz, while the re-entry capsules, named Vozvrashaemiy Apparat (VA) in the Soviet Union, and designated Merkur in the West, were part of a two vehicle system designated as the TKS spacecraft which also included an independently flown combination service and resupply module named the Functional Cargo Block. A very flexible system, the VA capsule was designed to launch aboard a Proton booster attached to FGB giving it independent free flight and maneuverability, or alternatively, attached directly to an Almaz space station.

While the FGB's went on to comprise core components of the Mir Space Station, and even the first element of ISS, the Zarya module, as well as a subsequent module added in 2012, the VA capsules saw extensive testing in a rather bizarre program which sought to maximize its investment by launching the capsules two at a time aboard the Proton on several occasions. The top secret program was ultimately cancelled in favor of the much more public Soyuz. Although none of the flights had been manned, they just as easily could have been, and two of the spacecraft were re-flown to demonstrate reusability, with one being flown and recovered three times. Equally surprising, Excalibur Almaz has in its possession two partially outfitted, single module space stations, whose predecessors had seen use in orbit as Salyut's 2, 3, and 5. [90]

All of the hardware was acquired by American businessman Art Dula, in hopes of combining the assets to enable a program of Earth and Lunar orbital tourism. Though the potential in-space components of Excalibur Almaz are very real, what the company lacks is a readily available launch vehicle and the funds to fully renovate its systems. As part of its unfunded Space Act Agreement with NASA, Excalibur Almaz envisions using an Atlas V to launch the recoverable Almaz capsules for crew transportation.

There is a bizarre side story related to Excalibur Almaz which has nothing to do with Commercial Crew but is simply too rich to be ignored. Reportedly, at least one Almaz station, Salyut 3, was equipped with a fixed 23 mm rapid fire cannon, sighted down the long axis of the station, which required cosmonauts to aim the cannon by maneuvering the entire station to align with the desired target, presumably American spy satellites which got too close. According to some reports, on at least one occasion, the cannon was actually fired in orbit, making the Almaz the world's first "fully armed and operational battle station!" albeit one a little bit undersized to be considered a proper Death Star.

Also unfunded, but still selected, was a somewhat brazen proposal by American aerospace giant ATK, maker of the Shuttle Solid Rocket Boosters, to revive the Ares-1 launch vehicle which NASA had just terminated, and offer it back to the agency for Commercial Crew, ostensibly at a price which was a fraction of previous estimates. The new vehicle, now rebranded Liberty and sporting a flashy red, white and blue paint scheme was intended to boost a composite version of the Orion spacecraft designed by Lockheed Martin. In a major design change, and one which reflected somewhat poorly on NASA's Ares-1 approach, the Liberty rocket proposal incorporated an entirely different second stage, to be provided by Astrium, the wholly owned subsidiary of European aerospace giant EADS, and derived from the Ariane V's first stage. Significantly, the new second stage differed from the Ares-1 proposal in utilizing a forward rather than an aft mounted LOX tank, a simple solution which would have helped to dampen the oscillation issue which plagued the Ares-1.

Among those entries not funded and not selected was also one offered by United Space Alliance, the Boeing/Lockheed Martin joint venture which served as the primary civilian contractor for the Space Shuttle program. The USA proposal was to privately operate the two newer Space Shuttles; Discovery and Endeavor, for two flights per year from 2013 to 2017. A last gasp, NASA was not about to revisit its painful decision to retire the fleet entirely, but the idea at least had the merit of retaining for a period, a hard to replace capability which would soon be gone forever. Also dismissed, a proposal by COTS competitor Orbital Sciences to produce a space plane of its own called Prometheus, to be developed by a consortium of companies led by OSC, with the airframe built by Grumman, participation by Virgin Galactic and launch provided by the ever available Atlas V.

For SpaceX the award breakdown and first set of milestones in the Commercial Crew program were as follows:[91]

1	CCDev 2 Kickoff	04/2011	$10M
2	Launch Abort System (LAS) Propulsion Conceptual Design Review (ConDR)	07/2011	$5M
3	Design Status Review	08/2011	$15M
4	LAS Propulsion Components Preliminary Design Review	09/2011	$10M
5	Crew Accommodation Concept Prototype And In-Situ Trial 1	10/2011	Internally Funded
6	DSR 2	12/2011	$15M

7	Crew Accommodation Concept Delta Prototype and In-Situ Trial 2	01/2012	Internally Funded
8	LAS Propulsion Component Test Articles Complete	04/2012	$5M
9	LAS Propulsion Component Initial Test Cycle	05/2012	$5M
10	Concept Baseline Review (CBR)	05/2012	$10M

Although the Falcon 9 rocket which would launch the Commercial Crew Dragon was in for a major upgrade, SpaceX's CCDev-2 milestones were not focused on the booster, but instead oriented around a major point of concern, the company's novel crew escape system.

In both the Mercury and Apollo programs, NASA sought to provide astronaut safety by including a solid rocket based "tractor" escape system which would pull astronauts to safety in the event of a vehicle failure. Project Gemini was the exception, including instead aircraft style ejection seats for its two occupants. For its first four test flights, the Space Shuttle was also equipped with ejection seats derived from the SR-71 Blackbird, a prospect one astronaut reportedly described as "avoiding death by committing suicide." The seats offered a limited chance of survival while the solid rocket boosters were still burning, but at the point of SRB cut-off, the orbiter was already too high and too fast to offer a realistic hope of escape. After the first four flights, which were conducted by two man crews, the ejection seats were removed, as the location of additional crew located on the Mid-Deck made any option for ejection problematic. From that point on until the end of the Shuttle program, anyone boarding the Shuttle did so in the knowledge that there was no chance for survival if a major malfunction occurred in the first few moments of flight while the SRB's were still burning. Even then, in the moments immediately following SRB termination and separation, a return to launch site abort was still a very risky scenario. Consequently, when it finally became time to begin planning a replacement for the orbiter fleet, NASA placed understandable emphasis on devising a far more reliable system which offered survivable options in all abort modes. For the very heavy Orion capsule, that meant developing a somewhat elaborate, tower based escape system.

Significantly, each of the CCDev participants, left to their own ingenuity in developing crew proposals recognized the inherent inefficiency of disposable escape towers. Instead, Boeing, Sierra Nevada Corporation, Blue Origin and SpaceX all proposed a pusher type escape system as the means to propel astronauts to safety in the event of a launch vehicle failure. Boeing and Blue Origin both offered systems which included existing and proven components as part of their proposals. Boeing for example, based its escape system on the Rocketdyne RS-88 Bantam ethanol powered liquid rocket engine originally developed for NASA in the 1990's.

SpaceX, as usual, was not constrained by convention. The basis of its crew escape system would be a new, radically upgraded version of its home grown Draco thrusters. At approximately 16,000 lbs. thrust, more than twice that of the Falcon 1 second stage Kestrel engine, and 200 times as powerful as its standard Draco thruster, the new unit, named appropriately enough SuperDraco, is capable of going from zero to full thrust in 100 milliseconds. Arrayed around the perimeter of the

Dragon capsule, 8 SuperDracos firing at full thrust would be more than enough to blast the spacecraft safely free of a booster in trouble, either on the pad or at any stage in flight. In doing so, SpaceX was offering powered safe escape all the way to orbit, integral to the capsule itself, a significant improvement over tractor based systems such as those found on the Soyuz and Orion, which both eject the escape tower while still in flight, and before reaching orbit in order to save weight.

The escape options offered by the SuperDraco system in nominal flights were impressive, but it was in circumstances not involving stricken launch vehicles which showcased the remarkable, almost stunning advance being offered by SpaceX. Quite simply, the massive burst of power contained in the SuperDraco escape system, could, if carefully applied, instead be used to slow the Dragon's descent without the aid of parachutes at all, leading to a precision landing on solid ground. Furthermore, as the company soon made readily apparent, solid ground did not necessarily mean planet Earth. It could in fact mean the Moon and Mars as well. This was the vision of space travel presented since the earliest days of science fiction; an image as "God and Robert Heinlein would have wanted," and it was one which SpaceX could use the opportunity of NASA's Commercial Crew Development program to perfect. Very significantly however, the DragonRider capsule which SpaceX officially proposed to NASA specified an updated and rapidly deployable version of the dual drogue, triple parachute landing system as its primary return hardware, a capability already successfully demonstrated on its first return from orbit. The long term goal was clear however, bringing a Dragon down to a pinpoint landing wherever required with the parachute system maintained as a fully independent backup system in all cases. In effect SpaceX was offering to NASA and its astronauts the highest level of safety ever built into a space return architecture and something no other system proposed or in development could match.

Taken as a whole, the four entries chosen for CCDev-2 represented what could only be considered a rational and measured approach to the challenge of re-gaining the capacity for independent U.S. human spaceflight. With the exception of Blue Origin, whose designs and ambitions remained veiled, none of the other three entrants would seem to be "risky" on the face of it. In the first place, the Atlas V selected by everyone but SpaceX held a near perfect launch record, and it was difficult to base any credible argument against the Commercial Crew program on the Atlas V vehicle, at least on a technical level. Much the same could be said for Boeing and its CST-100 space capsule, which is comprised mostly of flight proven, heritage systems. Again, given Boeing's long and extensive history, undeniable expertise in building most of the U.S. components of the very station to which the capsule would be travelling, arguments against this combination based on safety were difficult to take seriously. For a Congress which was clearly reticent to provide adequate funding for Commercial Crew as a whole, there was literally nothing to suggest anything other than an assumption that until proven otherwise, that the Boeing CST-100 and the Atlas V was capable of providing the services required. The real question was whether or not SpaceX was offering something better still; an all U.S. built solution which was both more affordable and far more capable. As for Sierra Nevada, in sharing the same Atlas V launcher as the CST-100, its issues boiled down to the increased cost and complexity of developing a spaceplane which faces a higher bar in both the launch abort and re-entry regimes of flight. Even so, based on both NASA's HL-20, and the flight history of its Russian forerunner, the aerodynamics for Dream

Chaser have already received extensive scrutiny. Furthermore, fellow competitor Boeing was concurrently flying the unmanned X-37B spaceplane to orbit and back, also aboard the Atlas V. The issue with SNC then, was whether or not the company had the management, technical expertise and financial ability to execute the design.

Despite the broad portfolio, complete with both ultra-safe and potential high payoff options, Congress still appeared to be intent on killing the program by starving it to death, and if that failed, changing the contractual basis of Commercial Crew in advance of the next round, in which solicitations would be due in March of 2012. Almost inexplicably, the opponents working to undo Commercial Crew, were once again primarily Senate Republicans, but aided this time by the House of Representatives and Frank Wolf of Virginia, Chairman of the Committee on Appropriations Subcommittee on Commerce, Justice, Science, and Related Agencies. NASA is one of those related agencies.

As the final flights of the Shuttle program wound down to the launch of Atlantis in July 2011, the U.S. Congress, led by the Senate was still pushing for a rapid start to the new heavy lift rocket called for in the NASA Authorization Act of 2010. NASA, following the course the President had outlined in his Kennedy Space Center address, was initially very resistant to the program on the grounds that there was far more to be gained from investing in new technologies, such as main engine propulsion where the nation was clearly lagging, than a continuation of Shuttle era components, whose roots lay in the late 1960's. With no answer to the affordability questions raised by the Augustine Committee, other architecture options such as refillable fuel depots serviced by more affordable EELV class vehicles appeared worthy of careful consideration prior to committing the United States to another long term, large booster development program. The U.S. Senate however, which would ultimately set a record of four straight years without passing a budget, would not be deterred, and instead wrote into law the specific requirements for the booster it wanted, even going as far as to suggest issuing subpoenas for immediate delivery of SLS design documents to NASA Administrator Charles Bolden if the agency did not comply.

> PUBLIC LAW 111–267—OCT.11, 2010
>
> (1) IN GENERAL. —The Space Launch System developed pursuant to subsection (b) shall be designed to have, at a minimum, the following:
>
> (A) The initial capability of the core elements, without an upper stage, of lifting payloads weighing between 70 tons and 100 tons into low-Earth orbit in preparation for transit for missions beyond low-Earth orbit.
>
> (B) The capability to carry an integrated upper Earth departure stage bringing the total lift capability of the Space Launch System to 130 tons or more.
>
> (C) The capability to lift the multipurpose crew vehicle.
>
> (D) The capability to serve as a backup system for

supplying and supporting ISS cargo requirements or crew delivery requirements not otherwise met by available commercial or partner-supplied vehicles.

SEC.304. UTILIZATION OF EXISTING WORKFORCE AND ASSETS IN DEVELOPMENT OF SPACE LAUNCH SYSTEM AND MULTIPURPOSE CREW VEHICLE.

(a) IN GENERAL. —In developing the Space Launch System pursuant to section 302 and the multi-purpose crew vehicle pursuant to section 303, the Administrator shall, to the extent practicable utilize—

(1) existing contracts, investments, workforce, industrial base, and capabilities from the Space Shuttle and Orion and Ares-1 projects, including—

(B) Space Shuttle-derived components and Ares-1 components that use existing United States propulsion systems, including liquid fuel engines, external tank or tank related capability, and solid rocket motor engines; and

(2) associated testing facilities, either in being or under construction as of the date of enactment of this Act.

The Congressional language made it perfectly clear that the priorities were preserving Shuttle workforce jobs above anything else, however the ostensible justification for the new booster under section D as backup for cargo and crew services to ISS was pure farce. Leaving aside the obvious fact that it was precisely the funding being diverted from Commercial Crew to the new booster which threatened the former program in the first place, on the issue of both cost and schedule the provision was absurd.

Even though it was now well on the way to getting its rocket, Congress, apparently threatened by the prospect of the Commercial Crew, and SpaceX's surprising success in the COTS 1 mission in December 2010, took advantage of the long break before the COTS 2/3 mission in 2012 to attempt to further derail the program as much as possible.

Amid solid progress in the CCDev-2 phase of the program in which each of the four winners completed their objectives either on time, or very close to it, the next phase, originally termed CCDev-3, was changed to Commercial Crew Integrated Capacity or CCiCap. Very significantly, NASA initially appeared to cave in to pressure and announced in July 2011, that the next round would be conducted under Firm Fixed Price Contracts and conventional FAR procedures. Within the context of the funds appropriated, the switch to FAR, which program officials stressed they would apply as lightly as possible, was an ominous sign. At the very least, the administrative costs for the contractors would skyrocket, with an inevitable effect on overall cost and timing. Equally significant, the switch to FAR effectively put the program administration on Boeing's home turf, where its long history and layers of management gave the aerospace and defense contractor a level of familiarity and expertise with the intricacies of the federal bureaucracy, dubious attributes which SpaceX neither possessed nor desired.

Heading into the late summer of 2011 with specifics of the new large booster project foisted upon NASA by Congress still under wraps, the Senate finally got its wish, when on September 11, NASA Administrator Charles Bolden, looking like a recalcitrant child made to apologize for some public transgression, stepped to the podium for a news conference in the Dirksen Senate Office Building, and flanked by a beaming Kay Bailey Hutchison of Texas and Bill Nelson of Florida, officially introduced the Space Launch System mandated by that same body. Bolden began his presentation not by discussing what the rocket might do for space exploration beyond LEO, but instead by observing that the new booster, which Nelson would famously refer to as a "monster rocket" only moments later, would "create jobs." There probably has never been a more honest assessment uttered about the real purpose of the SLS.

For all purposes, the SLS was little more than an alternate version of the just abandoned Ares-V, now topped by the Orion crew vessel, and dictated by the Senate. After months of debate over the configuration, the initial 70 ton version of the new booster would consist of a first stage based on a modified Space Shuttle External Tank, powered by four of the remaining stock of Space Shuttle Main Engines, an upper stage powered by the same J-2X originally slotted for the Ares program, and flanked by a pair of 5 segment solid rocket boosters, also adapted from the Shuttle program. The subsequent, 130 ton version also demanded by the Senate would be an almost entirely different booster, powered by yet to be built expendable copies of the SSME's, a new upper stage and new strap-on boosters which could be either liquid or solid, to be determined at a later date. Whereas Project Constellation had called for two launch vehicles to carry out deep space missions, with the crew carried to orbit atop the Ares-1, and everything else lofted by the heavy lift Ares-V, the Space Launch System, like the Saturn V before it, would carry both crew and cargo on the same flight. In case anybody missed the point, NASA released a rendering of the new booster which even sported the same black and white checkered paint job as the Saturn V.

With both the SLS and the Commercial Crew program falling under NASA's Human Exploration and Operations Mission Directorate and therefore lumped into the same budget category, the stage was set for an ongoing food fight over funding in which the SLS system, literally composed of the Space Shuttle's far flung network of supporters, clearly had the upper hand. Even worse, with Lockheed Martin still responsible for Orion, and Boeing for the SLS first and second stages, the two aerospace giants were on both sides of the fence, but with the SLS combination taking the form of sole source, non-competed cost plus contracts budgeted above $2 billion per year in a program expected to extend over the next three decades, and Commercial Crew a competitive acquisition now looking at budgets less than half that amount, and for only a few years at best, there was no mistaking where their interests actually lay. Though both regrettable and unnecessary, a vacuum of leadership coming from the Obama Administration and its allies in Congress allowed

Republicans in both chambers of Congress to try any means necessary to continue to hamper Commercial Crew, or at the very least, clumsily direct it to their preferred contractor, which certainly appeared to be Boeing and the CST-100.

It was hardly a surprise then, when Congressman Frank Wolf attempted to do just that, as his House sub-committee approved language directing NASA to "down select" to a single contractor.[92] With the SLS program now underway however, a subtle change was taking place in the Senate, where Florida Senator Bill Nelson, while still vigorously supporting SLS, but undoubtedly contemplating the possibilities of multiple providers launching from the vicinity of Cape Canaveral, and Boeing's announcement that it would build the CST-100 at KSC, began to release statements generally supportive of the Commercial Crew program.

Ironically, in cutting the budget request by more than half, from $850 million to the $397 million it finally received, opponents of Commercial Crew in Congress overplayed their hand, giving NASA the justification it needed to switch course once again and revert back to Space Act Agreements for the next phase by maintaining that the agency was being denied the requisite funding to proceed under FAR. The announcement, which came on December 15, 2011, was bolstered by a GAO report issued the same day generally praising the agency's approach to Commercial Crew but warning that without the budgets it requested, the program would be in serious trouble.

The reversion to Space Act Agreements was critical. Had the agency begun CCiCap under FAR contracting, the increased demands of unlimited NASA oversight, no matter how gently applied, and the absence of flexibility to do anything about it on the part of contractors, would have almost certainly resulted in a slowed pace and significantly increased cost for the program, no matter which proposal was eventually selected. If forced to down select to a single contractor, which was what the House legislation demanded, but the Senate's did not, the removal of any element of competition re-opened the door for cost spirals typical of large NASA acquisitions, while completely undercutting the rationale of supporting expanding space commerce. Under this scenario, and absent a budget increase, which was not coming, a few well-placed Congressional Republicans would be in a position to finally get their wish and see the program killed as increasing costs and slowing pace brought it into direct conflict with SLS and Orion.

For reasons which seemed to defy explanation, it was somehow either lost on, or dismissed by those members seeking to undercut the program, that regardless of what happened with Commercial Crew, the Space Launch System now under development would not see a first crewed launch until the end of the decade, or beyond. Now much more than a gap, NASA was looking at an ever widening chasm into which its existing skills would inevitably be sucked, further draining the agency of operational expertise. Equally significant, a series of unofficial cost estimates clearly indicated that NASA would only be able to afford to fly SLS once a year at best, and quite likely less frequently than that. Finally, just as the Augustine Commission had warned during Project Constellation, there was still no meaningful funding foreseeable to develop the type of payloads the giant booster was capable of carrying.

Ignoring the consequences, Congress signaled in early 2012 that it would slash the next year's Administration request for Commercial Crew once again. Equally unhelpful, for the third year in a row, the U.S. Senate failed to produce any budget at all, even though required by law to do so. The net result was that the U.S. government would now be run under a Continuing Resolution, and complicated by both sequestration and rescission. Though modestly improved from the previous year, the FY 2013 level of $525 million was still clearly inadequate, ensuring that the date of any first flight would slide further towards the end of a narrowing window.

For SpaceX, and perhaps for the Commercial Crew program as whole, the single most significant moment came on May 22nd 2012 with the Dragon's historic first flight to ISS. Although it was only a test flight, as almost everyone involved repeatedly cautioned, the implications for both SpaceX and the C3PO office were clear. If a significant problem had occurred, either with the Falcon 9 booster or the Dragon spacecraft, its opponents would have undoubtedly seized on the event as evidence that commercial industry was not up to the task. As it turned out however, the almost flawless execution of the mission appeared to sway a number of elected representatives who were on the fence, leaving the proponents of the old guard to seem as though they were fighting the tide of history. Momentarily defanged, Congressman Frank Wolf relented ever so slightly and signed an "agreement" with NASA Administrator Charles Bolden permitting the agency to proceed with a blended approach to the next phase, Commercial Crew Integrated Capacity (CCiCap), in which the agency would fund 2.5 winners under Space Act Agreements, but commence with entirely separate Certification Contracts which would bind the winners to defined reporting requirements. The final phase, in which one or more crew transport systems will be selected for development was scheduled to begin in April 2014, to be conducted under FAR requirements, but the three winning entrants would now have sufficient time to refine their proposals and remove much of the development risk. The question was, who would the 2.5 winners be?

Following weeks of speculation, and leaks late into the evening prior to the announcement, the official word came at 9:00 AM on August 3, 2012. The winners in the third round of Commercial Crew, the Integrated Capability Contract were Boeing, SpaceX and Sierra Nevada, with award amounts of $460, $440 and $212.5 million respectively. Although following its triumphant COTS 2/3 mission, SpaceX would have seemed to be an automatic choice as one of two primary winners, a late entry and concerted marketing campaign by Alliant Techsystems with its Liberty Launch vehicle raised real concerns that the company's considerable lobbying apparatus, and possible lingering resentment within NASA over the cancellation of Project Constellation, might have allowed it to secure one of the slots and put SpaceX in a vulnerable position versus proven aerospace giant Boeing and the Atlas V launch vehicle.

Within the context laid out in source selection statements, the distribution of the two primary awards to Boeing and SpaceX, as well as the half award, to Sierra Nevada Corporation to continue developing the Dream Chaser space plane made sense. With Boeing's larger award paving the way for a successful human rating of the Atlas V for its own CST-100 capsule, Sierra Nevada could use its smaller award for developing and flight testing the Dream Chaser at a more manageable pace, buoyed by the idea that with its strong affinity for winged spacecraft, fostered during 30 years of the Shuttle program, as well as the unique capabilities of the Dream Chaser itself, NASA would do everything possible to ultimately secure an opportunity for the spaceplane in the future.

The purpose of the third round of the awards was to advance a complete, integrated, final proposal to the point of Critical Design Review so that NASA could make a final decision regarding which system or systems would proceed into full scale development. For SpaceX, locked into a competition with politically connected aerospace giant Boeing, the largest single risk going into the CCiCap phase was that if the next NASA budget did not match the Administration's request, and it probably would not, the final round could be reduced to a single winner. Despite all its success, and regardless of any technological or financial advantages, the combined influence of Boeing, as well as Lockheed Martin through the Atlas V launch vehicle could spell trouble. In the environment of a looming presidential election which had seen Republican candidate New Gingrich lambasted for a proposal to build a colony on the Moon by eventual nominee Mitt Romney, who stated that he would "fire anyone" who even suggested as much, there was a very real risk that depending on the outcome of the 2012 Presidential Election, the political winds could blow hard against SpaceX, which was perceived as being solidly in the Obama camp. Later, as Romney began regularly bashing Elon Musk's Tesla Motors by name at campaign rallies and even in the Presidential Debates, it was not hard to imagine that a Republican victory could mean disaster for SpaceX in an era when facts rarely got in the way of a useful narrative.

With the general election still months away, putting itself in the best possible position to receive a final award regardless of the outcome meant selecting CCiCap milestones which would place the company well ahead of its competitors by the end of the upcoming round, giving it a commanding lead going into the fourth and final phase of the program where the prospect of a down select loomed ever more likely. As the specific milestones emerged, it became clear that SpaceX had done just that.

1	CCiCap Kickoff Meeting	08/2012	$60M
2	Financial and Business Review	08/2012	$20M
3	Integrated System Requirements Review	10/2012	$50M
4	Ground Systems and Ascent PDR	12/2012	$35M
5	Pad Abort Test Review	03/2013	$20M
6	Human Certification Plan Review	05/2013	$50M
7	On-Orbit and Entry PDR	07/2013	$35M
8	In-flight Abort Test Review	09/2013	$10M
9	Safety Review	10/2013	$50M
10	Flight Review of Upgraded Falcon 9	11/2013	Unfunded
11	Pad Abort Test	12/2013	$30M
12	Dragon Primary Structure Qualification	01/2014	$30M
13	Integrated Critical Design Review	03/2014	$30M
14	In-Flight Abort Test	04/2014	$30M
(15)	Dragon Parachute Tests	11/2013	$20M

(moved up from optional milestone 1 to Base 15 by Amendment)

Reflecting the increasingly intense competition going into CCiCap, SpaceX's optional milestones were completely redacted from public view with the exception of two; an orbital test flight with crew proposed for 2015, and a flight test to ISS with non-NASA crew to come after that. [93]

Setting the pace for the next phase, the SpaceX milestones included both a pad abort test, with a Dragon making an emergency ocean landing, right off the launch pad, and very significantly, an actual in-flight abort test conducted at the most aerodynamically stressful moment of the flight. As if that were not enough, Elon Musk soon announced that SpaceX would also demonstrate an achievement not required by the official milestones, but thrilling nonetheless, a powered touchdown of the Dragon spacecraft. Somewhere Robert Heinlein was smiling.

Dragon test article for Commercial Crew parachute test. Credit: NASA

Boeing by contrast, despite its history, inclusion of heritage systems, larger overall award and far deeper pockets, proposed a much more conservative agenda of design reviews and subsystem tests, deferring both a pad abort as well as an in-flight abort test to the next and final phase as part of a development award. Part of the reason could be attributed to the much higher cost of the Atlas V launch vehicle, which would now require modification to include both a second RL10 upper stage engine to provide the necessary thrust, as well as a launch adapter to connect the CST-100 to the Atlas V. Basically, Boeing would have to fund much more of the program out of pocket than it was inclined to do. The decision did not go unnoticed. In its selection statement, NASA "dinged" the technologically sound Boeing proposal for an unwillingness to make a substantial financial investment. If ever someone wanted a head to head comparison of the differences between old and new space, the self-selected CCiCap milestones could not set a clearer example.

The differences between the two companies were equally clear in the superficially similar capsule designs put forth. The Dragon and CST-100 are both relatively small capsules, each capable of carrying from 4 to 7 tightly packed astronauts in almost identical configurations. Both also use a forward mounted docking hatch, rear heat shield, an attached non pressurized assembly which is jettisoned prior to re-entry, and both are ultimately designed to land on ground as opposed to splash down in water. Finally, both employ a pusher escape system, and both are designed to be re-usable.

Unlike the Dragon, however, the Boeing capsule was designed to fly solely to low Earth orbit, its wings deliberately clipped by the absence of solar panels, swapped instead for limited duration batteries. For LEO flights, the Dragon as well would launch without its trunk and solar panels, but having already been tested in space, were available for extended flights when the need arose. Furthermore, despite using a pusher escape system, the CST-100 does not include the option for propulsive landings, and instead is designed for a conventional parachute landing, albeit on dry land, softened by airbags. The escape system, along with other crucial components, is contained in the expendable service module meaning that although the crew pressure vessel itself is potentially re-usable, many of the operational components are not, and must be replaced with every flight. One might wonder why, with all its history and experience, Boeing chose to introduce a serviceable, but barebones entry with a greatly reduced expansion path. It was a curious choice, and one almost certain to place the company behind SpaceX in many respects, particularly given the fact that from the outset, SpaceX had made no effort to conceal its design ambitions for the Dragon. The answer could be a matter of simple cost control. With its much higher overhead, and far more expensive launch vehicle in the Atlas V, it seems quite possible that the aerospace giant might not have been able to come anywhere near the lower pricing of its smaller rival with a similarly capable craft. As such, the alternative would be to offer a less capable craft for something on the order of a competitive price, and hope the name on the side of the capsule was enough to sway a close decision.

There is another, purely speculative angle as well. Boeing's partner in both United Launch Alliance, and fellow team member on the Space Launch System, Lockheed Martin had already spent nearly $5 billion dollars on its way to an estimated $10 billion dollar program cost to get the modestly larger Orion capsule through its first scheduled crewed flight in 2021, a figure exclusive of the SLS booster. [94] Moreover, Boeing itself is the contractor on the first two stages of SLS. While decision makers were repeatedly able to ignore the fact that SpaceX had developed a spacecraft in many ways comparable to Orion at a fraction of the price, had Boeing accomplished a similar goal, albeit at a lesser price difference for a vehicle which was also billed as capable of travel beyond low Earth orbit, it might very well have been too much to ignore. As it was, Boeing even highlighted its limitation in the spacecraft's name, with the "100" designation referring to the 100 kilometer Karman line as the boundary of low Earth orbit and the extent of its ambitions. Whatever the actual reason, even though the Dragon and CST-100 have similar features, one look under the hood reveals a world of difference, and sadly, what appears to be somewhat of an intentional hobbling of what might have been a more capable spacecraft.

Boeing's seemingly dispassionate and at times almost indifferent approach to the CST-100 project, underscored by statements that the company was not considering beyond LEO applications could not have been more different from SpaceX's clear enthusiasm for DragonRider, which included forward looking animations portraying the spacecraft returning to Earth for a powered landing, followed by a final shot of the Dragon touching down on a dry, reddish planetary surface which could only be Mars. And that perhaps was the difference. From the very beginning, SpaceX focused its efforts on building systems and subsystems in a classic spiral development which would eventually lead to a Mars transportation system. Boeing on the other hand, much like its fellow aerospace contractors, was always willing to talk a good game about leading the way in space exploration, often placing billboards around Los Angeles, Huntsville and Washington D. C. advertising itself as being in the forefront of space exploration, but in the end the company did not appear to be willing to go anywhere without NASA or another party reducing the risk and paying the bills, even if that meant going nowhere at all.

Artist's concept of Dragon arriving at an early Mars base. Credit: SpaceX

As it entered CCiCap, adversaries continued to focus criticism on NASA's plans to establish standards to insure crew safety and certify how they were met. With SpaceX having captured world-wide attention with the COTS 2/3 trip to the Station, the issue of safety became a sort of last ditch rallying point in efforts to derail a program which seemed to be well on the way to success. The ostensible line of reasoning was that without direct NASA oversight and authority throughout the entire process, systems developed under Commercial Crew could de facto not be certified as safe. It was a curious argument, considering the fact that NASA's own record consisted of 30 years of flying a system which was demonstrably unsafe at multiple points in its flight regime, and the agency had lost two orbiters and 14 souls in the process. Also overlooked, or more likely ignored, was the fact that both fatal accidents, though taking place on physically flawed systems, were directly attributable to human error, and NASA certification was certainly no guarantee against a repeat of the same problem.

Going beyond Congress's demonstrated antipathy, or at the very least indifference to NASA's Commercial Crew program evidenced by underfunding it, the legislative body's own general ineptitude and increasing intransigence led to programmatic confusion with the advent of government wide sequestration and a mandatory reduction of 5% of the discretionary budget after the House and Senate failed to agree on a common plan by a March 1st, 2013 deadline. Already partway through the 2013 fiscal year, the required cutbacks were intended to take place in the latter half of the year, resulting in an overall effective decrease of 9% of what was a budget still defined by continuing resolutions from the 2011 fiscal year. As a result, the funds available for NASA to actually administer the later milestones under CCiCap came under threat, raising the prospects of yet more artificially driven delays in the program. With the International Space Station program's future only secured through 2020, a further slip to 2018 for the first crew transfers would render the effort nearly pointless.

As an interim measure, both Boeing and SpaceX indicated as part of a Commercial Crew Program update in February 2013 that despite the Congressional obstructions, they were proposing to keep the program closer to schedule by launching test flights to LEO manned by non-NASA crews, with SpaceX looking at late 2015 and Boeing following shortly thereafter in 2016. Although NASA considered the plan acceptable, leaving it up to the contractors to assess their own levels of risk, the Aerospace Safety Advisory Panel weighed in once again, condemning the proposed private test flights as possibly creating a false sense of confidence in still "unproven" systems which although perhaps operating successfully on an initial test flight, could nonetheless fail when it came time to fly for NASA. While technically accurate, the warning ignored the fact that NASA's own plan for the Space Launch System calls for the first crew to launch on only the booster's second flight, and the third of the Orion spacecraft overall, whereas either the Atlas or Falcon 9 boosters would have conducted numerous flights before a crew ever went aboard.

Going in to the spring of 2013 and the long stretch which would see the progression of CCiCap milestones leading to very mature designs, the reality of sequestration, with cuts made in annual budgets which were already partially spent, and disagreements over how much, if any funds could be shifted around within different agency accounts, created a confusing situation for NASA, and for Commercial Crew (as well as every other discretionary federal agency affected by the cuts.) After initial concerns that the program might suffer immediate setbacks, Congress provided a small measure of relief by agreeing to add back in a limited amount of FY 2013 funding into the Exploration account, primarily to support SLS, but with a portion allocated to Commercial Crew as well, leaving the agency to worry about how to fund the last, and most significant set of milestones scheduled to take place at the end of the year.

Despite the temporary relief, Congress, and in particular Republicans in the House of Representatives indicated once again that when the Obama Administration belatedly released the FY 2014 budget request of $821 million for the Commercial Crew program, NASA could expect another significant cut. Reaching the limits of his frustration, NASA Administrator Charles Bolden took the occasion of a press conference immediately following the release of the budget request on April 10th, to strongly advise against any further cuts, stating that if his agency did not receive full funding, it would not make a 2017 goal for first flight to ISS.

With little choice but to move ahead despite the uncertainty, on July 19th, NASA released a draft Request for Proposal for the next phase of the Commercial Crew competition, and with it came new acronyms. Following Commercial Crew Integrated Capacity (CCiCap), the fourth and final phase would be called Commercial Crew Transportation Capability (CCtCap) and is intended to "develop a U.S. commercial crew space transportation capability to achieve safe, reliable and cost-effective access to and from the International Space Station (ISS) with a goal of no later than 2017."

CCtCap is divided into two phases; Design, Development, Test and Evaluation (DDTE) to procure the crew systems and conduct test flights, followed by Post Certification Missions (PCM) to conduct the operational flights. The two segments are covered by four Contract Line items (CLIN). CLIN 001 and 002 are for the testing and operational phases respectively. CLIN 003 is labeled "Special Studies" and enables NASA to pursue risk reduction activities over and above those required for the other three categories. The fourth item, CLIN 004 "Cargo in Excess of Requirements" offered something of a surprise in that it allowed the Contractor to propose providing additional cargo beyond the limited amount specified for Commercial Crew missions and that already being contracted under the Commercial Resupply program. Depending on the responses, which would include pricing, this item could introduce another variable into NASA's decision-making process, potentially benefitting the competitor with the most available margin.

Perhaps the most intriguing item, and one which offered further evidence of the change in NASA's approach to commercial operations was an item from the draft solicitation which suggested the agency is open to the prospect of commercial passenger or cargo service being performed as part of crew transport flights. Clause H.23 of the solicitation enables the Contractor to "propose to manifest Commercial Passengers, cargo or payloads on PCMs for contract price adjustment(s) or other contract consideration."

Four months later, with the funding still undecided, NASA released its final CCtCap proposal on November 19th, with a final award to be announced in September of 2014. For better or worse, without ever having provided a clear justification of claims that Commercial Crew could not be accomplished under Space Act Agreements, and in spite of the success demonstrated by both SpaceX and Orbital Science in bringing the spectacularly successful SAA driven COTS program to a close, NASA was now irrevocably committed to Federal Acquisition Regulations for Commercial Crew. Seemingly true to its word however, the final solicitation, while eliminating the CLIN 04 "Cargo in excess of requirements," clearly reflected an effort to minimize some of the impacts of the burdensome procures. [95]

In a sign of just how much the window of opportunity had narrowed based on the life of the Station, the CCtCAP solicitation indicated that in addition to a single crewed test flight to be conducted under CLIN 01, the phase of operational flights would consist of a minimum of two and a maximum of six flights. Presumably however, if the lifespan of Station was extended beyond 2020, then the opportunities for other flights would expand as well.

As it prepared to enter CCtCap, SpaceX faced two primary challenges. The first was maintaining its aggressive CCiCap milestone schedule and in particular achieving a successful pad abort demonstration of the new DragonRider crew capsule, and then following suit with an in-flight abort test in 2014 prior to the beginning of CCtCap. If it succeeded in both, then the one remaining obstacle was a trouble free introduction and early flight campaign of the new Falcon 9 v1.1 booster. This was the challenge which would bring SpaceX into competition not just with itself, but for the first time with the international commercial launch market as a whole.

Chapter XXV: International Competition

In planning to enter the commercial market with the Falcon 9 Block II, later updated to the Falcon 9 v1.1, SpaceX was no longer competing against other prospective rockets within the context of a relatively sheltered development program as it had done in winning a share of the COTS competition opposite first the Rocketplane Kistler K-1, and then the Orbital Sciences Taurus 2/Antares. Instead, it would have to meet well established and long proven governmentally backed competitors, often on their home turf, and literally take the business away from them. It would not be an easy task.

With the single largest market in the world, U.S. defense launches tightly wrapped up by the ULA monopoly, and basically off limits until SpaceX established its credentials elsewhere, there were two other markets open to the Falcon 9 in which to prove its reliability. Expressed in terms of the location to which the Falcon 9 could be expected to deliver, as opposed to the customers being served, or the payloads themselves, the overall opportunities are defined by the FAA as either to Non-Geostationary Earth Orbit (NGSO) or Geostationary Earth Orbit (GSO).

Within the first category (NGSO) SpaceX was already a significant participant through the NASA space station resupply contract, which falls under that designation. So too, do any future Commercial Crew flights. In fact, taken together, the combination of all projected NASA CRS and Commercial Crew flights through 2021, represent nearly 50% of the total world market for openly bid NGSO launches. Of the remaining launch opportunities, SpaceX was also already involved with both Orbcomm and Iridium. Consequently, the opportunity for new markets lay in the remaining category, the geostationary communications satellite launch business, and whatever portion of the depleted science mission portfolio a budget strapped NASA could get to the launch pad.

Establishing a foothold in the Comsat market meant demonstrating both reliability as well as the physical capacity to launch increasingly larger and heavier satellites to Geostationary Transfer Orbit (GTO). In terms of the performance required, it is a significant step up from the 200-300 mile orbit occupied by the International Space Station, demanding longer burn times, in-space engine restarts, plane changes and precise placement. More daunting than the mechanics required for achieving such demanding orbits however, are the barriers to entry presented by the established launch providers as well as new emerging competitors.

When the space age began under the shadow of the Cold War, virtually every aspect of the emerging field of space based technology was a part of a government program. It did not take long though, to realize that some applications, namely satellite telecommunications, were ideally suited for a purely commercial market, and in fact AT&T filed the first ever notice of intent to launch a commercial communications satellite, "Telstar" in 1960, before the U.S. government had even developed a policy for such an event.

As the United States established itself as the world leader in a steadily growing Comsat launch market, American launch vehicles remained tied to the military procurement programs from which they derived. Europe however, was intent on developing a launch vehicle of its own, one developed not to deliver the weapons of war, or even to support national programs of exploration, but instead to facilitate the rapidly emerging field of telecommunications.

Although it got off to a later start than the two Cold War rivals, the European effort, led by France, was optimized from the outset to pursue the geostationary launch market, an approach which began with the selection of its primary launch site. Because orbital mechanics give a significant benefit to launches conducted in an easterly direction due to the additional velocity imparted by the Earth's rotation, Europe needed a launch location with either open land or sea to the east to allow discarded stages to fall safely back to Earth. Unfortunately, it had neither.

What it did have however, left over from its days as a colonial power was French Guiana, located on the Atlantic coast of South America, slightly north of the Equator. Though difficult to access, the Guiana location offered a perfect launch site for France, one which gave a decided advantage in rotational speed compared to the 28.5 degree location of Cape Canaveral, allowing a rocket launching near the equator to offer more payload capacity for a given capability, or alternatively a lower overall capability in terms of booster performance required to deliver an equivalent payload. Furthermore, when launching to geostationary orbits above the equator, the preferred location for most large comsats, a low latitude launch site eliminates the necessity for a fuel intensive plane change, or "dogleg" maneuver required by rockets lifting off from higher latitudes headed for the same destination.

With the jungle cleared and flame trenches carved out of the granite below, Arianespace conducted its first launch on December 24, 1979. At the time, the U.S. dominated the embryonic international commercial launch market, while the Soviet Union was the undisputed leader in total annual launches. Thirty six years later, by the time SpaceX announced the Falcon 9, the U.S. segment of what was now a much larger market had plummeted to 2%, while Europe's share had risen to 24%, with Russia holding the lead at 43%. [96]

The development of the four stage Ariane 1 launch vehicle began in 1973, and after the initial launch six years later, proceeded with a steady program of product improvement leading to new, more capable boosters. Like many other first generation national launch vehicles, the Ariane 1 and its successors were based on low tech, storable hypergolic propellants for the first and second stages, but it did include a more sophisticated hydrogen/oxygen third stage, as well as a fourth stage which was more properly a part of the satellite delivery system itself. From early in its history Ariane employed a dual launcher strategy in which two satellites were lofted to their respective orbits on a single launch. The program hit a peak with the highly successful Ariane 44L, which based on an improved and stretched version of the previous series, offered a number of liquid or solid strap-on boosters to meet different launch requirements. Altogether it conducted a total of 116 launches with three failures.

As comsats steadily grew in size and capacity, the need for a larger, more capable launch vehicle than the Ariane 44L led to the design and introduction in 1996 of a completely new rocket, the Ariane V. Unlike its forerunners, the current Ariane V is powered by a single core LOX/hydrogen first stage flanked by two large solid rocket boosters which provide 92% of the liftoff thrust. Following booster drop off and the completion of the first stage burn, a cryogenic second stage engine propels the payload to GTO. The initial Ariane V flight took place on June 4, 1996 and resulted in a failure 37 seconds into the flight. The cause was traced to a software problem which ensued from adopting part of the software utilized in the Ariane IV, a problem not unlike that which was to doom the first Boeing Delta III flight several years later. The second flight also ended in a partial failure when a roll control problem on the Vulcain main engine led to a premature shutdown and a lower than expected final orbit. The program suffered two more failures during its first 14 flights, one partial and one complete, before going on to record a remarkable string of more than 50 straight successful launches, during the course of which it cemented Europe's domination of the large satellite commercial launch market.

Two versions of the Ariane V are currently in use. The commercial model is the Ariane V ECA, which uses a cryogenic upper stage, and is capable of placing a 22,000 lb. dual payload, or 23,000 lb. single payload into geostationary transfer orbit. A typical dual manifested mission consists of one larger and one smaller satellite, weighing 6 and 4 tons respectively, taking advantage of the booster's impressive capabilities to lower the cost for each. Duel manifesting however presents a unique challenge in finding and scheduling two generally unrelated commercial satellites to fly on a single mission to complimentary orbits, in the process effectively increasing the chance that problems in delivering one satellite could affect the launch time of the other. The alternate version is the Ariane V ES, which stands for evolution storable, and is a formidable launcher, capable of placing 46,000 lbs. into LEO where it is used to launch the Automated Transfer Vehicle to International Space Station.

Although nominally a commercial booster, the Ariane V is managed by an unusual and uniquely European arrangement. The operating company, Arianespace, is partially owned by various European governments, including the French National Space Agency, CNES, which is the Ariane program's prime contractor. At the same time, it is also in partnership with privately held European aerospace interests, led by defense giant EADS (now Airbus Defense and Space) Astrium business unit, which is the vehicle's prime contractor. Surprisingly, Arianespace itself maintains an employment level of less than 400 people worldwide. The picture is a little deceiving however because most of the costs of maintaining the Guiana Space Port, are paid by CNES, although it was developed by the European Space Agency. Adding a bit of dark romance to the whole affair, security is provided by no less than the French Foreign Legion.

The Ariane V averages between 5 and 6 commercial launches per year, and with most being dual manifested, regularly achieves a 50% plus market share for geostationary satellites, even though Russia generally conducts more individual launches in the same market. Arianespace also benefits from a unique support arrangement in which minimal annual profits are guaranteed through a yearly subsidy paid to the operating company by the European Space Agency to reimburse any losses in a down year.

In addition to the Ariane V, the Guiana facility now also hosts a modernized, commercial version of the Russian Soyuz launch vehicle, the Soyuz 2, which Arianespace acquires through a Russian company, Starsem, to serve the medium lift market. Rounding out the launch capacity at "Europe's spaceport" is a new booster, the smaller, 4 stage solid fueled Vega rocket built by Italy and first flown in 2012. Taken together the three launch vehicles, which are all controlled by the same integrated team, represent a full range of payload launch capability unmatched anywhere in the world. As of February 2014, Arianespace had conducted over 250 launches, and placed over half of the world's commercial satellites into orbit. Holding a steady market share at 50%, the Ariane V is by any measure is the gold standard in the large satellite launch business, and with typical French flair, Arianespace is not hesitant about flouting its accomplishments.

Throughout the 1990's as it consolidated its position as the world's leading commercial launch provider, Arianespace would look to the east for its major competitor, Russia, and the still considerable space assets of the disintegrating Soviet Union. When the U.S.S.R. broke apart, it left behind a sprawling and highly successful space infrastructure consisting of a broad range of launch vehicles spanning from the small Dnepr missile which Elon Musk originally attempted to procure for the Mars Oasis project, all the way to the massive Proton booster which launched all of the major Russian segments of the International Space Station. In between, lay Soyuz, the world's most commonly flown rocket, now with over 1800 successful launches to its credit, as well as the newer and more powerful Ukrainian built Zenit booster.

Far from a monolithic authoritarian enterprise, the history of the Soviet space program was marked by the rise to prominence of several different and competing state design bureaus, each led by legendary rocket designers who left their own mark on the systems which endure today. The one feature they held in common was the enormous Baikonur Cosmodrome, a launch complex 1700 miles south and east of Moscow in what is now the nation of Kazakhstan.

In the earliest days of the Cold War, both the U.S. and U.S.S.R. lacked the sophistication to accurately control intercontinental ballistic missiles with on-board computer systems. Instead, both the original American Atlas, and its counterpart the Russian R-7, were controlled by radio signals. In the case of the latter, the R-7 required at least three transmitting stations spread out over flat ground, but located hundreds of miles apart along the missile's flight path. To meet these demands, the Soviet Union engaged in the one of the largest and most expensive infrastructure projects in its history. Intended to remain a secret test facility hidden deep within friendly borders, the West was alerted to its existence when what was then known as the Tyuratam missile test range was photographed by a U-2 spy plane in 1957. The origin of the current name "Baikonur," which came into popular use after Yuri Gagarin's historic flight in 1961 is still a matter of some debate, as that name appeared to correspond with a remote mining village located several hundred miles away. By any name it represents a massive investment of resources. Launch vehicles, built in Russia are loaded onto railway cars and transported to Baikonur where they are prepared for flight and then transferred to a special wide gauge railroad which leads out to the widely spaced launch pads.

As technology progressed, after only a few years of missile testing with radio navigation, the Soviets successfully developed inertial navigation for the R-7 and its heirs, partially eliminating the need for such a remote and sprawling complex. When informed that the massive expense was rendered unnecessary by the new technology, Soviet Premier Nikita Khrushchev was reportedly enraged, but he might have taken comfort in the fact that the Soviet boosters still needed a large swath of mostly uninhabited land both to safely discard booster stages, as well as conduct recovery operations for the manned space program.

Although the U.S.S.R. soon added a second, polar launch facility at Plesetsk, 800 km north of Moscow and nearly within the Arctic circle, Baikonur remained the primary launch site for the Soviet Union, and over the years has hosted many of history's most significant space accomplishments, including the launch of Sputnik, Yuri Gagarin's first flight, the first woman launched into space, Valentina Tereshkova, as well as all of Russia's manned flights, the launches of the Salyut space stations, Mir core modules, and the current core modules of the ISS. With the retirement of the Space Shuttle, it has also become for a time, the only launch site on Earth for American astronauts.

For all its history however, Baikonur is still cursed by one thing which cannot be changed, its latitude, which at 45.9 degrees north, is far from ideally situated for communication satellites bound for an equatorial orbit. Despite this limitation, it is still the only launch location for the Ariane V's main commercial contender, the International Launch Services Proton-M booster. The Proton itself has a long and strange history, dating back into the 1960's and what was once, and hopefully forever will be, the world's largest ICBM project. Built to carry death on an unprecedented scale as a "super ICBM" launching 100 megaton nuclear weapons, the Proton was never deployed in that capacity, promoted instead by the Chelomei design bureau as a heavy lift space booster, and an alternative to the Soviet's Union's ultimately unsuccessful N-1 moon rocket built by the rival Korolev design bureau. Despite its size, the Proton is a relatively primitive booster whose first three stages are powered by the storable, but extremely toxic hypergolic propellants unsymmetrical dimethyl hydrazine and nitrogen tetroxide. The first stage features an unusual arrangement in which six RD-275 engines, and fuel tanks, are arrayed around the circumference of the single, common oxidizer tank, giving the false impression of being disposable strap-on boosters like the more familiar Soyuz. In the current Proton-M version, the fourth stage, used for orbital insertion is called Briz, and is also hypergolic, although occasionally the booster launches with a separate keralox fourth stage, the Blok DM.

Like most other Russian launch vehicles, the Proton's name is derived from the name of the first group of payloads it launched, which in this case was a series of high energy particle research satellites named appropriately enough "Proton." After a long and troubled development period lasting a dozen years and marked by repeated, often spectacular failures, the Proton booster finally entered regular service in 1978, but remained concealed to the West, and to the Russian public until it unexpectedly appeared on television launching the core element of the Mir space station in 1986.

Beginning in 1996, the Proton has been marketed commercially by International Launch Services of Reston, Va., a consortium originally established with Lockheed Martin to offer both the American Atlas and the Russian Proton as an integrated solution for medium and heavy lift missions. Lockheed Martin left the project in 2006, but ILS has maintained a strong commercial presence for the booster, which also serves Russian federal missions in a slightly different form. Only launching out of Baikonur, the Proton makes up in sheer size what it lacks in launch location, specific impulse and thrust to weight ratio. The Proton M, equipped with a Briz upper stage can launch up to 7100 lbs. directly to geostationary orbit, or alternatively for satellites with their own propulsion systems, 12,000 lbs. to geostationary transfer orbit. By contrast, it can launch an impressive 49,000 lbs. to the International Space Station's much more friendly 51.6 degree low Earth orbit.

Contained within those numbers however, is one of its key limitations versus the Ariane V, which by virtue of its higher performing cryogenic fuels, solid rocket boosters and nearly ideal launch location, offers almost twice the capacity to GTO, opening the door for much more lucrative dual manifested missions. Nevertheless, in recent years the Proton has maintained an impressive overall rate of 8-12 flights per year, coming from a combination of Russian federal missions, quasi-commercial missions and anywhere from 6 to 8 commercial flights hosted by ILS. ILS has also recently introduced a dual satellite capacity for Proton, which enables the booster to handle two smaller spacecraft at a time.

Despite its relatively lower cost, ILS has held a number two position behind Arianespace for more than a decade, based primarily on the latter's sterling reliability. The same cannot be said for the Proton. Along with much of the Russian space infrastructure, the Proton has recently begun to show its age. Following a long run of successful launches, five flights have failed since 2008, including three in a period between 2010 and 2012, and a spectacular flaming wreck in July 2013. Oddly enough, each of the failures has been on Russian Federal as opposed to commercially marketed ILS launches. In the first four cases the failure was in the Briz upper stage, but no two problems were the same. As a result, Proton was forced to lower prices in order to maintain its market share, even as the Russian ruble was rising. Despite its issues however, as well as several long running programs seeking to develop a replacement for the Proton, the Russian government anticipates continued reliance on the booster until 2030 at the earliest, and it is clearly in a position to keep the booster commercially relevant despite its limitations as long as it is willing to underwrite the program. Changes however, are underway. Displeased with the necessity of launching out of what is now a leased facility in a foreign nation at Baikonur in Kazakhstan, Russia is investing heavily in a new spaceport, the Vostochny Cosmodrome in the nation's far eastern Amur region, but plans to continue launching the Proton only from its current location, which is leased through 2050.

The second major commercial launch competitor from the Soviet era is the considerably newer Zenit booster, which unlike either Soyuz or Proton is a product of a Ukrainian state owned company, the Yuzhnoye Design Bureau, and is not considered a "Russian" product. Originally conceived in the 1970's as an entire family of modular launch vehicles based around a single core, the Zenit was first employed as a liquid strap-on booster for the massive Energia heavy lift vehicle. Debuting as the Soviet Union began falling apart, the Energia only flew twice. The first flight was a military test payload, while the second lofted the Russian Buran space shuttle on its one and only voyage in 1998. In both cases, the Energia and its Zenit boosters performed flawlessly. Designed to be re-usable, although no recovery attempt has been recorded, the current Zenit is produced in both a two stage, and a three stage variant. The overall architecture is similar to the Falcon 9 in that it is an all kerosene/oxygen booster, but that is where the similarity ends. The Zenit is powered by a single Russian built RD-171, high pressure, staged combustion main engine featuring four separate combustion chambers, and holds the distinction of being the highest thrust liquid rocket engine ever built, exceeding even the legendary Saturn F-1, which still holds the record as the world's largest single combustion chamber engine. The second stage is powered by a single main engine, with steering assistance provided by a separate, much smaller, four chamber vernier engine. Launched from Baikonur, the two stage Zenit can place nearly 30,000 lbs. into low Earth orbit, but it suffers from the general Russian affliction of having poor performance for reaching anything but the sort of highly inclined orbits which serve high latitude nations. For many Russian applications, this is not necessarily a hindrance, but for commercial service, it is a major problem.

The solution is Sea Launch, a multinational consortium established in 1994, and originally led by Boeing, which found an innovative way around this limitation. Following construction in the Ukraine, Zenit boosters are shipped halfway around the world to the Port of Long Beach, California where still moored at an 18 acre facility, they are loaded aboard a specially built roll-on roll-off cargo container ship, the Sea Launch Commander, and integrated below decks with the third stage and payload, to form a complete launch vehicle assembly. Taking place as it does aboard a foreign flagged ship, the operation avoids cumbersome U.S. import regulations, even though it occurs within spitting distance of U.S. soil. The Sea Launch Commander, as the name implies, also serves as the command and control vessel for the other half of the two vessel fleet, the self-propelled floating launch platform, Odyssey. The Odyssey is a unique, twin boom catamaran style ship converted in Norway from a North Sea mobile oil drilling platform. With the fully integrated booster transferred to the Odyssey and stored in an environmentally controlled container on deck, the Odyssey departs Long Beach several days ahead of the faster Commander, and sails southwest towards the equator and a location at 154 degrees West longitude where the nearest landmass is the coral atoll of Kiritimati, or "Christmas Island," in the Republic of Kiribati several hundred miles away.

Once arriving on station, the Odyssey floods its twin booms, sinking as much as 70 feet into the Pacific, shifting water ballast as needed to provide necessary orientation and stability. With the Zenit rocket transferred to the launch platform at Odyssey's stern, and the crew evacuated to the Sea Launch Commander located a safe distance away, the final launch countdown begins, controlled by line of sight radio telemetry from the command ship. The particular version of the Zenit rocket used for Sea Launch is the Zenit-3SL, which includes the base two-stage Zenit, topped by a Russian built Blok M third stage, very similar to the one also used on some Proton missions. Although something of a logistical nightmare, the Sea Launch system offers an elegant solution for equatorial launch, and is capable of placing very heavy satellites weighing up to 13,400 lbs. into GTO, but the venture is only made possible by the initial low cost of the Zenit booster. Sea Launch has encountered more than its fair share of problems however, including three major failures and one significant anomaly out of a total of 35 launches. One failure, in 2007, occurred within seconds of main engine ignition, resulting in a spectacular explosion which badly damaged the Odyssey's launch platform. The ultimate result of that mishap was a trip to bankruptcy court, and Boeing's withdrawal from an ownership position, even though it continues to provide the payload fairing and integration. Following reorganization, Sea Launch is now majority owned by the Russian government though Energia Overseas Ltd., a division of RSC Energia, with corporate offices in Switzerland.

With global transportation and two ships to operate, Sea Launch carries a high overhead, which can only be offset by a steady flight rate which the venture has rarely achieved. Compared to rock solid Arianespace, Sea Launch has been at best a troubled business whose primary advantage has lain in the fact that even though Arianespace is the vendor of choice, the leading satellite communication companies are extremely reluctant to find themselves dependent on a single company, and have been willing to accept the additional risk of assigning a limited number of launch orders to other vendors in order to maintain some schedule flexibility, and to keep Arianespace pricing in check. As the third alternative behind Arianespace and ILS/Proton, Sea Launch has been a beneficiary of that policy, but one with the most to lose from the entry of a new player in the market. Following its most recent failure in January 2013, Sea Launch is struggling to rebuild a customer manifest, and its future remains questionable. However, just as has been the case with the Proton booster throughout its history, the Sea Launch partnership is now partially in the hands of the Russian government as well, a fact which may keep it viable for a long time to come.

A more recent addition to the Sea Launch portfolio is a land based alternative, named appropriately enough, Land Launch, which uses a slightly different version of the same three stage Zenit-3SL, the 3-SLB, to boost satellites weighing up to 3.5 metric tons to GTO by launching out of Baikonur. Despite offering even lower prices, Land Launch, in direct competition with Proton, but lacking the Russian booster's performance, has only recorded a handful of commercial launches.

The final significant international competitor is China. A comparative latecomer to commercial space launch, in recent years China has poured considerable resources into its space program, building a solid track record in the process. Like its U.S. and Russian counterparts, the Chinese space program has its roots in the development of an ICBM program which still remains shrouded in secrecy. Much like early U.S. and Soviet liquid fueled ICBM's, the military requirements for launch on demand capability led Chinese designers to select storable propellants for its Long March series of carriers, an early decision which still forms the basic architecture of the family of space boosters sharing the same name.

China originally entered the commercial market in 1990, encouraged by the backlog of international launch orders resulting from the stand down after the Challenger accident. Offering low launch prices and subsidized through the Chinese government and the state owned Great Wall Industries Corporation, China raised the ire of both European and American launch providers, which succeeded in tying the rising superpower to a series of five year deals in which it could not offer prices more than 15% below average without provoking close inspections and possible sanctions. Despite this limitation, and the limited flight history of its vehicles, China began making commercial inroads in the mid 1990's, but soon suffered a series of flight failures, one of which is still shaping the market today.

In the early morning hours of February 15th, 1996, a Long March 3B carrying the Intelsat 708 communications satellite lifted off from China's main launch center at Xichang, near the Burma and Vietnamese borders. Just seconds after clearing the pad, the rocket experienced a major steering malfunction, and veered off course, accelerating down a valley until it suddenly arced into the air and smashed into a hillside. Loaded with toxic fuel, the resulting explosion devastated a rural village in the blast zone. Within moments, the Chinese army rushed to secure the site, where hundreds of onlookers had stood at a security fence only moments before to watch the rocket as it rose into the air. The actual death toll, originally assessed as 6, has been a subject of debate ever since. [97]

In addition to the human carnage, the accident also created a delicate national security issue. The satellite, built by Loral Space Systems, contained sensitive encryption chips, designed to prevent foreign parties from interfering with operations in space. With the hardware strewn out over the crash site, Loral employees were prevented from entering the area until the Chinese officials carefully picked it over. According to the public record, the chips were never located.

It did not end there however. Back in the United States, charges were leveled that in the process of helping with one of several review boards examining the accident, Loral shared sensitive U.S. technology and review methodology with the Chinese government, and in effect to the Chinese army. The greatest area of concern was the potential that the company had inadvertently improved China's understanding of the processes used for guiding ICBMs as well as for conducing launch vehicle failure analysis. Complicating matters was the fact that Loral CEO Bernard Schwarz, who was at the time the single largest donor to the Democratic Party, and a major supporter of the Clinton Administration, received a waiver to launch another satellite with China, in the midst of the ongoing investigation. This set the stage for a politically charged conflict in which the licensing authority for foreign satellite launches would become a political turf war between the departments of Commerce and State. In 2002, Loral eventually settled by paying a $14 million dollar civil fine, and pledging an additional $6 million to improve compliance, but without admitting any guilt. Both Hughes, which was acquired by Boeing, as well as Lockheed Martin also subsequently paid fines to settle accusations that in the effort to secure lower satellite launch costs, that they too had assisted China with efforts to improve its launch vehicle reliability during the early 1990's. If only that was the end of the story.

Not satisfied with the Loral settlement, Congress used the advent of the incoming Bush administration to push through a relocation of licensing authority from the Department of Commerce to the State Department, and to add essentially all satellite and launch vehicle technology, regardless of how mundane, to a weapons and munitions category which brought it under International Traffic in Arms Regulations, or ITAR. American satellite and launch vehicle manufacturers would now be regarded on the same level as African arms merchants, South American drug cartels and rogue states with nuclear ambitions. Given that the easiest way to manage something you don't understand is to simply refuse to deal with it, or failing that, bury it under a blizzard of regulations and requirements, the results were both entirely predictable and nearly disastrous to the domestic commercial launch industry. With the stroke of a pen, it became a lot less trouble to launch your satellite from French Guiana with Arianespace or from Baikonur with Russia, than from the United States. The impact on the launch market was only one consequence of the new regulations. ITAR also led to the creation of a new class of commercial satellite, one which was certified as "100% U.S. content free." Rather than keeping American technology at the forefront and reasonably secure, the regime instead spurred the global advance of foreign space technology, making sure it was securely out of U.S. control. While ITAR succeeded for a brief interval in effectively restricting China from launching any spacecraft with U.S. content and thereby driving it temporarily from the global commercial launch market, it also proved a boon to Arianespace and Russia. Furthermore, it pointlessly annoyed traditional business partners and close strategic allies by restricting the flow of even the most innocuous technology if it was related to space. For instance, ITAR prevents Virgin Galactic founder Richard Branson from stepping foot in SpaceX's Hawthorne headquarters on the grounds that as a British national, he might pose a danger to U.S. national security. The fact that his own suborbital tourism company Virgin Galactic is basing nearly all its manufacturing and operations out of the U.S. is inconsequential.

In the end, as so frequently happens with poorly considered, overbearing regulations, the ultimate result was the opposite of what was originally intended. The availability of ITAR free satellites actually created a ready-made market opportunity for a newly confident and re-engineered Chinese aerospace industry whose launch vehicle capability continued to progress after the imposition of American restrictions. In a perfectly ironic twist, recent years have seen China's manned spaceflight program come in to its own, even as the U.S. counted down the days until it could no longer launch astronauts from its own soil. With a series of successful missions, including five crewed flights and the launch of its first small space station, Tiangong-1, China now looms as a major force on the international launch market, one backed by an energetic and wealthy government ready to ensure whatever is necessary to keep a place it fought hard to regain. Perhaps not surprisingly, it is also where SpaceX sees the greatest source of long term challenge. In 2013, and in response to its obvious economic consequences, the United States finally began to move some components of space technology off the Department of State munitions list, placing it, under the discretion of the President, back under the oversight of the Department of Commerce, whose bias is much more towards facilitating rather than restricting America's global industrial competiveness. The damage however, is already done, and what remains is a "wet blanket" effect which scares silent even the most seasoned aerospace veterans.

Besides the four dominant international competitors, two other national space programs bear consideration as potential candidates likely to provide a challenge for SpaceX in coming years.

One nation perpetually on the edge of becoming a significant competitor in the international launch market, but which has never been able to get its pricing in line, is Japan. The primary Japanese large launch vehicle is the H-IIB. Designed by the Japanese Space Agency NASDA, and built by Mitsubishi Heavy Industries, the H-IIB and the slightly smaller H-IIA are both sophisticated rockets, powered by a liquid oxygen/liquid hydrogen main core and second stage, with supplemental boost provided by either two or four solid strap-on boosters. The H-IIB is used to launch the 30,000 lb. HTV cargo ship to the International Space Station, as part of that nation's commitment to the project. Although the H-II family has performed a limited number of commercial flights, and has a strong reliability record, the vehicle's cost and complexity, along with the strength of the Japanese yen, has thus far kept it from gaining any meaningful market share. Japan however, indicating a desire to enter the commercial launch business in earnest, has recently begun an effort to reduce the vehicle's costs, turning most of the responsibility for the booster over to prime contractor Mitsubishi Heavy Industries, one of the world's largest and most respected corporate entities. Standing in the way of progress until only a few years ago was one of the most unusual restrictions in the global aerospace industry. Due to the objections of Japanese fishermen, who asserted that rocket launches disrupt local fishing grounds, NASDA was prevented from launching out of its primary site on the island of Tanegashima, for much of the calendar year. Easing of the restrictions came only after the Japanese government agreed to pay the fishermen's unions an annual compensation for the very questionable "impositions."

The other nation "on the bubble" as an international competitor is India, which harbors long term ambitions of both servicing the commercial market as well as human space flight capability.

India has established a solid reputation with a smaller, solid fueled rocket used primarily for research and imaging applications, the Polar Satellite Launch Vehicle, or PSLV, which as the name implies, conducts most of its launches into near 90 degree inclination polar orbits, heading south out of its launch pad on the Bay of Bengal, over the Indian Ocean and into orbit over Antarctica. In contrast to the successful PSLV, India has encountered significant difficulties with its much slower moving program to develop its own entry into the comsat delivery market, a heavy lift rocket called the Geosynchronous Satellite Launch Vehicle, or GSLV. As an emerging industrial and technological power however, should it prove willing, given time, there seems little doubt that India is capable of fielding a larger launch vehicle economically and technically competitive with almost everything else on the market.

In many ways, the commercial arena into which SpaceX planned to introduce the Falcon 9 is a curious one, with a variety of very different launch vehicles at different stages in their life cycles, each fielded for different reasons. In both the United States and Russia, the original motivation for building ballistic missiles had peaked long ago, with the deployment of modernized arsenals which now differ considerably from the liquid fueled boosters which formed the first efforts. For the U.S., the primary motivation in supporting the EELV program was not about developing new methods of delivering weapons, or even fostering a commercial launch business, but instead about maintaining the stability of an industrial base which was useful for supporting what it really cared about, a highly sophisticated network of satellites capable of supporting a global military deployment through active weapons guidance, battlefield communications and for intelligence gathering. Until spiraling launch costs collided with tightening budgets, neither the military intelligence agencies nor their almost inseparable contractors seemed overly concerned with the demise of its presence in the commercial launch market.

Europe on the other hand, partly covered by the American defense shield during the 1970's, was afforded the luxury of optimizing its launch infrastructure for the commercial market from the outset. To its credit, Arianespace made the most of the opportunity, steadily evolving its launchers from humble beginnings to the market dominating Ariane V. For Russia, as military budgets decreased, its motivation for offering commercial launches was a matter of generating cash, maintaining prestige and preserving the core of its industrial base, hoping for better days to come. It is a wait which may already be paying off. After years of stagnation, the Russian space industry is now a primary focus of Russian president for life Vladimir Putin, who has pledged an investment of nearly $50 billion by the end of 2020, and has already begun to make good on his promise of significantly restructuring the creaking Soviet era space infrastructure. With its long and rich history, and a technology base in propulsion which is in many ways superior to that of the United States, including SpaceX, Russia's capacity to advance the state of the art in the commercial launch business cannot be discounted, as neither can the fact that two "American" boosters, the Atlas V and Antares, are absolutely dependent on Russian engines to make it off the pad.

Unsurprisingly, China has proven to be a very mixed bag, with a stop and start commercial launch business used alternatively to support a military effort, as well as a burgeoning space program clearly aimed at garnering prestige and making up for what it perceives as lost time compared to the original two space superpowers. Despite remarkable progress in recent years, China's launch industry is still based on a relatively inefficient hypergolic fueled core launch vehicle, and it is encountering a variety of growing pains as it seeks to field both cryogenic and keralox boosters for heavy lift, including the largest class of geostationary commercial satellites. Tellingly, China made news in 2011 by complaining bitterly about the low launch pricing offered by SpaceX. As if that weren't bad enough, SpaceX's continuing corporate policy against filing patent claims on its inventions so as to deny other nations, specifically China, a "blueprint" for copying had to be another source of irritation, albeit one to which the rising superpower would never admit.

In introducing the commercial Falcon 9 Block II and then upgrading it to the Falcon 9 version 1.1 before it ever flew, SpaceX was fielding a brand new rocket based on a clustered engine architecture against a diverse field of heavily subsidized competitors, whose launch vehicles were backed by a half century of experience, hundreds of billions of dollars, rubles, francs and rupee, and in some cases had been flying longer than the average age of SpaceX employees. Furthermore, each came from military industrial enterprises which bore a lot more similarity to each other than to a truly private startup. In the U.S. it faced an entrenched establishment, woefully uncompetitive on the commercial market which maintains its position via a revolving door military/contractor network and well-funded political connections willing to employ every trick in the book to protect its remaining business base.

This David faced not one, but an army of Goliaths. It was armed however, with one very powerful weapon, the knowledge that despite their history, none of the existing commercial class launch vehicles were designed from the outset with the goal of radically reducing the costs of space launch as its primary objectives. Despite all of the protestations and all of the patronizing explanations, that building and launching rockets was a job which only national governments and their preferred contractors could undertake, SpaceX approached the development of the commercial Falcon 9 with the understanding that the high cost of space launch was not inherent in the engines, fuel tanks and flight software, but was instead tied up in the various compromises made along the way, and in the inefficient and sometimes corrupt bureaucracies which directed the efforts. SpaceX was in a strange manner, secure in the knowledge that regardless of whatever flag might be waving at its competitor's launch pad, it was in support of a launch vehicle development process which was structurally incapable of presenting the most cost efficient product, and at least one player, the United States, was not even making the effort.

Beyond this assessment however a clear and fundamental difference emerges. SpaceX was founded and driven with one purpose in mind, "supporting the cause" and lowering the cost of access. Its competitor's motivations were murkier and frequently conflicting. Even Arianespace, the one competitor most directly aimed at the commercial market, is hamstrung by an overriding policy of geographic return which requires each nation participating in the project to receive economic contracts roughly equivalent to their degree of participation. The result is a painfully slow, classically bureaucratic decision making process marked by seemingly endless debates over which member country will participate under what conditions and to which degree. It cannot make a quick decision on anything. And yet, despite these limitations, Arianespace is the undisputed market leader, and the standard by which SpaceX would have to measure itself when the new Falcon 9 v1.1 finally entered service.

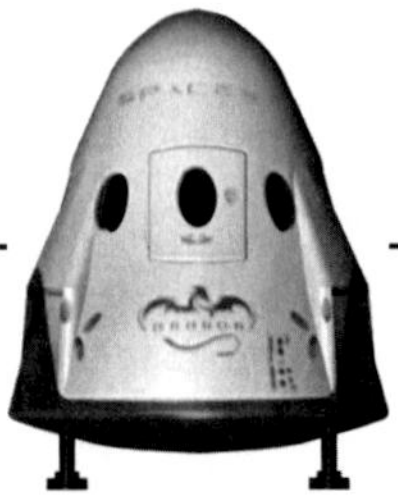

Chapter XXVI: The Falcon 9 v1.1

Over the course of its successful five flight history, the Falcon 9 first stage Merlin 1C engines had performed well, but they were not perfect. While SpaceX could legitimately claim a 100% launch success record, a premature engine shut down seconds before Main Engine Cutoff on flight two had raised some questions, as well as a lawsuit, but concerns faded with the next flight. They returned with a vengeance on flight four, when ground cameras captured the spectacle of a definitive chamber breach and engine failure, accompanied by images of the engine fairing flying off the rocket in flight. To SpaceX's credit, and in validation of the redundant engine approach, the Falcon 9 shook off the incident, and did exactly what it was designed to do, recalculate its flight trajectory using the remaining 8 engines and keep going. These were correctable issues, as demonstrated on the picture perfect fifth launch, but they were also representative of the fact that the first generation Falcon 9 was in many ways a precursor to the more capable launch vehicle intended to serve the three new markets SpaceX needed to enter, Commercial Crew, communications satellites, and national security launches.

The flight regimes differed for each of the markets. Commercial Crew, like Commercial Resupply, naturally focused on low Earth orbit, but unlike the CRS missions for which there was no flight "abort" option, carrying crew required a highly advanced fault detection system coupled with a launcher offering the greatest degree of flight redundant systems possible. Military missions on the other hand, would generally be to mid-range polar or elliptical orbits, demanding absolute reliability and precise orbital placement. Furthermore, they also required secure ground handling procedures necessary to protect some of the nation's most sensitive technology. Finally, the commercial telecommunications satellite business placed the greatest strain on the launch vehicle itself, requiring large payload fairings and multiple long burns of the second stage to place the heaviest of satellites in the highest of orbits.

The original Falcon 9 was a decent performer to the orbit occupied by ISS, but offered only limited performance for missions beyond LEO. There are a lot of benefits to selecting a kerosene/oxygen architecture for both the first and second stage of a launch vehicle, but efficiency to higher orbits isn't one of them. Consequently, in order to maintain its all keralox strategy which was a critical consideration for controlling costs, SpaceX planned from an early date to make the first flights of the Falcon 9 under the COTS program with an initial version of booster labeled "Block I", and then introduce a more capable "Block II" version for commercial service. It was this booster, powered by an improved version of the Merlin 1C which the company advertised on its website and in its Payload user's guide, even as it was performing its first five launches with the less powerful Block I, a discrepancy which led to occasional confusion as to what the vehicle's performance really was.

While the original Block I's almost 19,000 lb. capacity to LEO was sufficient to reach the International Space Station's 51.6 degree orbit with a modestly loaded Dragon capsule, coming in at roughly 9,200 lbs. dry weight and 7,300 lbs. split between pressurized and unpressurized cargo, SpaceX needed a more capable version to launch a fully loaded Dragon complete with occupied trunk assembly, as well as to provide ample margin for its DragonRider entry into the Commercial Crew program. As for its commercial ambitions, the 3.4 ton capacity to a 28.5 degree geostationary transfer orbit for the Block I simply wasn't powerful enough to accommodate most of the communications satellite market.

As originally proposed the Block II would offer a modest increase in performance to the Space Station, 20,502 lbs. vs. 18,739 lbs. as well as an increase in GTO capability from 7,495 lbs. to 8,597 lbs. The problem was that to adequately tackle the geostationary comsat market, SpaceX needed a booster with the capacity to place roughly 10,000 lbs. into a geostationary transfer orbit, and even at that level, it would still be missing out on launch opportunities classified by the FAA as "large" or "heavy" that saw satellites routinely massing 10,500–13,000 lbs. at liftoff.

The bottom line was that even though as of the end of 2010, it had already accomplished something on the order of a minor miracle in designing, building and conducting two successful flights of Falcon 9, including the launch and recovery of the first Dragon spacecraft, SpaceX still needed to quickly introduce a significant upgrade to its flagship product at a point when other new launch efforts would have just begun settling in to a long production run where the focus was minimizing change. In typical fashion, SpaceX went about it in a manner both innovative and almost mind numbingly ambitious, while at the same time long on promises and short on immediate results. As usual it thrilled supporters and infuriated critics. At the core of its new strategy was the fact that with a successful entry into the intermediate/medium lift category, SpaceX intended to not only introduce an upgraded vehicle for the three markets it wanted to enter, but also make another substantial step along the path to opening a new era of Mars exploration and settlement, the undiluted goal at the heart of the SpaceX experiment. In order to do so, the company needed three things; a heavy lift booster, a fully reusable ground to orbit space transportation system, and a dominating share of the global launch market to provide the revenue stream and experience to pursue both. Elon Musk's double barreled answer came in 2011, ironically enough, a year in which his company didn't conduct a single launch.

The first official word of a successor to the planned Falcon 9 Block II came from Elon Musk on April 5, 2011 during his speech at the National Press Club in which the SpaceX founder introduced the long rumored Falcon Heavy launch vehicle. In describing the triple core rocket, Musk noted that it would incorporate a new "stretched" version of the Falcon 9 as its central, sustainer core, with two Falcon 9 first stages serving as strap-on liquid fueled boosters in an arrangement much like that of the Delta Heavy. Significantly, a new version of the Merlin engine, designated the Merlin 1D would power the booster.

Following the initial introduction, details about the Merlin 1D emerged over the coming months in bits and pieces, through occasional interviews and through presentations at the annual AIAA conference on propulsion, a frequent venue for SpaceX announcements over the years.

The new booster finally began to come into view in early 2012 as SpaceX was preparing for its historic COTS 2/3 flight to the International Space Station, beginning with a short notice from NASA on May 14th 2012 that an "additional configuration of the Falcon 9" was being added to the National Launch Services II contract, calling it the Falcon 9 v1.1.

Behind the innocuous new designation was a major upgrade project which would bring the Falcon's performance deep into the heart of the EELV category with a capability of lofting 13.5 metric tons, or nearly 30,000 lbs. to LEO and 4.85 metric tons or 10,600 lbs. to GTO. It also produced a substantial increase in the Dragon capsule's cargo transport capability, nearly doubling it from 7,282 lbs. to 13,228 lbs. For commercial missions which required even greater capacity, the much larger triple core Falcon Heavy would outstrip every other booster in existence, but still fly at a price point which kept it under less capable competitors. At the same time, the new Falcon 9 v1.1 would also be the baseline operational vehicle for a renewed attempt at developing operational first stage reusability, a goal already well underway with the Grasshopper RLV test article announced in September 2011. If successful, it could lead to the single most important accomplishment in the space age, a rapidly reusable launch vehicle.

Taking the next step required among other things, a considerably more powerful first stage, one that generated roughly 1.5 million lbs. thrust, a nearly 50% increase over the capacity of the Block I. Its options were constrained by its goals. One obvious solution, adding more engines, would have meant widening the booster itself, and thus retooling the whole production line. Another option, attaching small strap-on solid boosters to the first stage like Atlas V and Delta IV, would drive up costs, tie SpaceX to one of two existing vendors, and pose a serious problem for the Commercial Crew program, whose heavier DragonRider spacecraft would require more takeoff thrust than the unmanned Dragon capsules used for cargo. That effectively left either repowering the booster completely, or significantly increasing the power of the Merlin engines as the only practical solutions. The solution selected, the Merlin 1D, was somewhere in the middle.

The Merlin 1D project began in 2010, with the adoption of an entirely new process for forming the thrust chamber and cooling jacket. Rather than electroplating, which as Musk observed, was "the slowest way you can make a metal thing, one molecule at a time," the new chamber is formed through the process of explosive deformation in which a complex metal structure is achieved by submerging the piece to be molded into a vat of fluid, generally water, which contains the hardened mold. A small charge is detonated and the stock piece is instantly deformed by hydraulic shock, plastered against the side of the mold. It is in essence, depth charged into shape. The process change yielded a stronger and lighter thrust chamber which is at the same time capable of operating at higher temperature and pressure.

Besides the change in manufacturing technique SpaceX introduced other proprietary improvements as well, creating a less costly and less labor intensive process, with increased use of robotic construction and decreased parts count. Combined with a new, internally designed and built turbopump capable of supplying dramatically higher flow of fuel and liquid oxygen, the results were impressive. Sea level thrust increased to 147,000 lbs., and the vacuum thrust to weight ratio improved to an astonishing 150 to 1, making it the highest level ever recorded for a liquid rocket engine. Marking a significant change from the Merlin 1C, the new first stage engines would be both restartable, and throttleable. The latter eliminated the need to shut down two engines during first stage boost to prevent over acceleration, and both were required to pursue eventual reusability and a much anticipated propulsive return and landing capability.

To handle the increased thrust load, and establish a basis for reusability, engineers changed the engine layout, going from three rows of three engines each in a grid, to an octagonal arrangement with eight engines arrayed around the base of the thrust structure, and a single, center mounted engine protruding slightly below the other eight. Termed an "octaweb," the new structure effectively nests each engine in a protective pod reinforced by the thrust structure itself, whereby even if it suffers a catastrophic failure, the surrounding engines are still protected.

Equally important, the new arrangement is designed to defend vital components against the heat of atmospheric re-entry and retrofiring necessary to bring it safely back to ground. This configuration also spreads the stress of the increased thrust to the skin of the first stage fuel tanks which carry the load. Significantly, the location of a single, center mounted main engine mirrors the design of the Grasshopper RLV test article, creating an optimum geometry for powered landings and allowing sufficient room for the engine to gimbal in any direction. Increasing the liftoff thrust also meant increasing the booster's overall fuel capacity, a goal met by "stretching" the first stage fuel tank from 29 to 42.6 meters, and the second stage tank from 10.1 to 12.6 meters, while maintaining the same 12' diameter for both. True to earlier predictions, this was where the SpaceX stage manufacturing technique fully paid off, with the higher capacity tanks achieved by simply adding additional barrel segments.

In addition to increasing the overall thrust for both first and second stages, and re-designing the Merlin 1D engine thrust chamber to resist fatigue, SpaceX took a major step in enhancing overall vehicle reliability with significant upgrades to the vehicle's avionics, going from single to triple string avionics, and three, all new, redundant flight computers using a voting architecture. Positional date comes from inertial guidance supplemented by a GPS overlay. The result was a system so robust that according to Musk, "You could put a bullet hole in any one of the avionics boxes and it would just keep flying." With each of the new Merlin 1D engines retaining three controllers of their own, and offering significantly more initial performance, depending on the payload, the new F9 could now lose two first stage engines at virtually any point, and continue on its way.

Engineers also made a significant change to the stage separation system, incorporating the pneumatic push rods with the clamps which hold the stage together, and in the process reducing the total number of connection points from 12 to just 3, further reducing the opportunity for a problem to occur.

As a result of the totality of changes being incorporated into the new upgraded Falcon 9, as SpaceX entered the final and definitive round of NASA's Commercial Crew Competition, it would now be offering what was empirically the most systemically reliable and fundamentally safe launch vehicle in the world, with a double redundancy at many levels. Boeing and Sierra Nevada Corporation by contrast, having selected the ULA Atlas V to boost their respective entries were constrained by a launch vehicle which while boasting a successful flight record, was designed to a lower threshold, and still contained a glaring single failure point in its most critical element, the first stage engine.

With its first NASA mission, CRS-3, not scheduled until the fourth flight of the new Falcon, and preceded by a demonstration flight and two commercial flights, SpaceX would at last launch its flagship booster with a working version of the two part 5.2 meter wide payload fairing which the company featured so prominently when the rocket was first assembled at Cape Canaveral in 2009. The addition of the fairing introduced two new elements to the launch procedure. The first is that at 17' in diameter, and considerably wider than the Dragon capsule, the "hammerhead" payload fairing presents a greater challenge for the vehicle's guidance system as it accelerates through the lower, denser atmosphere. Second, it introduced a new separation event which was the source of quite a bit of anxiety, and with good reason. SpaceX's competitor in the COTS program, Orbital Sciences Corporation, suffered two consecutive failures with its Taurus XL booster due to fairing separation issues in 2009 and 2011 respectively. Costing NASA the loss of its Orbiting Carbon Observatory and Glory satellites, and the American taxpayer nearly $750 million, the twin accidents brought new relevance to the old joke that the company's initials OSC really should stand for Ocean Submergence Corporation. Following the second highly embarrassing incident, OSC wisely changed the name of the booster it was developing for COTS from Taurus II, to Antares.

In light of the difficulties its competitor experienced, SpaceX undertook an extensive testing program for its new 5.2 meter fairing. The Falcon 9 had not flown with any payload fairing on the first five flights due to the presence of the Dragon capsule, but all of the earlier Falcon 1 flights had included a smaller metal payload fairing, and with the exception of the two vehicles lost prior to the designated point in flight, each deployed successfully when called upon. Extending its philosophy of vertical integration to this element as well, SpaceX designed and fabricated the ultra-lightweight carbon fiber structure in house. The Atlas V by contrast, uses a fairing supplied by Swiss company RUAG, which is derived from the same product it supplies for the Ariane V.

The new fairing, which is planned for use on both the Falcon 9 v1.1 and Falcon Heavy, at 17' diameter and 43' in height, is on par with the largest fairings in use anywhere in the world, but in keeping with a design philosophy first introduced on the Block I Falcon 9, avoids the use of explosive pyrotechnic bolts to blow the two sides of the fairing apart and away from the ascending second stage. Instead, the two halves of the fairing, which is attached to the upper stage by marmon clamps, and held together with 14 latch mechanisms, achieves separation with the aid of 4 pneumatic push rods similar to those used in stage separation.

After completing the initial fabrication, in March 2013, SpaceX shipped the massive new fairing to NASA's Plum Brook Facility in Sandusky, Ohio, part of the Glenn Research facility, for testing in the world's largest vacuum chamber. In addition to simulating the environment of near space in which the fairing would be jettisoned, the visit to Plum Brook also afforded the opportunity to test the fairing and payload deployment structure in one of the world's premier acoustic chambers, where both could be subjected to a high fidelity bone jarring simulation of the deafening noise and vibrations of an actual launch. Actually getting to the launch pad however, required completing stage qualification in Texas, and it did not come easy.

Following a series of new testing challenges which resulted in multiple aborts, engineers finally conducted a full length, mission firing of the new first stage on the Big Falcon Test Stand at McGregor early in July 2013. After years of anticipation, SpaceX was at last ready to begin the launch campaign for its oft delayed entry into the commercial market.

Big Falcon Test Stand, McGregor, Texas. Credit: SpaceX

Chapter XXVII: Securing Launch Contracts

By the time the first Falcon 9 v1.1 made it to Vandenberg, California for its maiden flight, SpaceX had compiled a backlog on its launch manifest of unprecedented proportions, coming in for more than a little criticism in the process. With nearly 50 missions on a list which already stretched out for five years, the company faced a major challenge in not only getting the new booster qualified and off to a successful start with a single planned test flight, it also needed to establish a rapid launch tempo which held its own risks. With some customers having already waited years, time was of the essence. The fact that so many customers were willing to wait as long as they did is a testament to either Elon Musk's salesmanship, the appeal of the Falcon 9's breakthrough pricing, or both.

After its brief commercial history with the Falcon 1, highlighted by the single launch of Malaysia's RazakSat 1 in 2009, the opportunity presented by NASA's COTS program redirected SpaceX from using its limited resources to aggressively pursue the commercial market with the Falcon 5, and instead focused the company on the development of the Falcon 9 and Dragon spacecraft. For customers at the time, that meant either sticking with SpaceX and waiting while the new launch vehicle was being qualified, or seeking a new launch provider. Though it created an uncomfortable situation for a new company attempting to establish long-term customer relationships, given the opportunity presented by NASA, it was the only reasonable choice.

NASA's COTS and CRS flights combined to create what amounted to a 15 launch initial buy for the Falcon 9, a near record amount only exceeded by the original 17 launch order won, and later lost, by Boeing in the opening bids of the EELV program. Effectively reduced to 14 by the combination of the COTS 2 and 3 mission objectives, NASA was by design, the undeniable anchor tenant for the Falcon 9, just as it would be for the OSC Antares a few years later. Nevertheless, SpaceX intended its new rocket to be first and foremost, a commercial booster, as that was the only means of establishing long term viability independent of the changing priorities and budgets of national space programs.

For the most part, those who would call themselves space enthusiasts are understandably much more interested in orbits involving the Moon, Mars and the outer planets than the diverse fleet of satellites circling the home planet. For the moment though, the economics of the existing satellite launch industry have a far greater influence over what will ultimately be launched out of Earth orbit than almost any other factor, and any meaningful reduction in the costs of launching satellites for Earthbound communications and observations is an equally important step in regularly traveling outside our home orbit. As the overview of international market indicates however, it has been a fairly stagnant market for many years, dominated by three primary vehicles, the Ariane V, Proton and Sea Launch Zenit. The newest of which, the Zenit-3SL, first flew in 1996.

One of the biggest challenges facing any new launch vehicle entering the commercial launch business is that the market is clearly limited, and is not particularly stable, except at the highest level, which of course, is the last place you want to start. The costs are enormous, the barriers to entry the highest, and the tolerance for failure is the lowest. Wisely, SpaceX started at the opposite end, with the Falcon 1, and by either fate or fortune, was in a position to move ahead when NASA assumed the role of anchor tenant in the medium and intermediate launch markets with the COTS, CRS and Commercial Crew programs.

There are other segments to the medium lift LEO market, such as the replacement of small satellite fleets for Iridium, Globalstar, and Orbcomm, along with a smattering of scientific research or technology demonstration satellites, but the former come in batches and the latter are very sporadic. The bottom line is that for anyone whose goal is to radically reduce the costs of spaceflight, there just simply aren't that many opportunities to do so outside of the market which already exists. This is the core of the almost impossible to crack chicken and egg problem which has bedeviled efforts to lower launch costs since the first commercial space missions.

After beginning operations with the Falcon 1, SpaceX initially focused its limited resources on developing the Falcon 5, and pursuing the more lucrative mid-sized satellite market, signing up early customers such as Orbcomm to provide services to low and intermediate orbits. Although the Falcon 5 would have been capable of delivering payloads to geostationary orbits, its limited capacity prevented it from addressing nearly 90% of that market. Consequently, had it not been for NASA and the COTS program, it is entirely possible that somewhere along the way, most likely in 2008, Elon Musk might have reached the conclusion, as Andrew Beale did before him, that between the EELV program in the U.S. and government underwriting in foreign markets, that his time and passion for changing the world was better applied to Tesla and the electric car industry.

Fortunately that was not the case, but even with NASA's participation and the switch to the Falcon 9, SpaceX still needed to hold on to both its existing customers as it transitioned to yet another launch vehicle, and then begin signing new ones in a hurry. One lucky break came when Orbcomm, which had originally signed a contract in September 2009 to have 18 Generation 2 (OG2) small satellites launched aboard the Falcon 1E, ran into schedule difficulty of its own when spacecraft manufacturer Sierra Nevada Corporation fell behind in its production schedule. With an eye towards its upcoming COTS missions, SpaceX suggested that it could instead launch the first Orbcomm satellites into orbit contained in the trunk of the Dragon. When both companies renegotiated the launch arrangement to reflect this opportunity, it sealed the fate of the Falcon 1E. As it turned out, NASA was not altogether happy with this arrangement for an initial ISS mission which proved challenging enough. In the end, the launch schedule was altered to place a single Orbcomm test satellite onto the second Falcon heading toward the station, which turned out to be the CRS-1 mission in October 2012.

The Orbcomm fleet, which was originally launched aboard Ariane IV, Pegasus and Cosmos launch vehicles, is part of the re-birth of the original generation of low orbit satellite fleets which drove the abortive reusable launch vehicle development era of the mid 1990's. Substantially downsized from earlier predictions, these constellations have emerged from bankruptcy, and under new ownership, begun to find new uses and niche markets for machine to machine communications in which only small amounts of data are transmitted. Typical users are fleet operators tracking trucking or heavy equipment such as earthmoving giant Caterpillar, or emergency and maritime assets where the satellites are able to keep fleet operators constantly informed on certain key status points wherever they may be around the world. One recent adoption is the U.S. Coast Guard's Automatic Identification System (AIS) which requires vessels above a certain size to be equipped with a transponder which relays the ship's present location, speed and course through the satellite fleet to all other similarly equipped ships. The system has proven invaluable in reducing the growing number of frequently fatal collisions between large container ships and smaller vessels in crowded sea lanes, particularly at night when even the most experienced mariners have difficulty distinguishing distance between themselves and massive, rapidly moving container ships which most assuredly cannot see them. As with many other technologies, once the ability to perform a certain service is developed, it is not too long before risk managers, insurance companies and governmental agencies determine that what was once optional is now necessary, and this tendency may be the saving grace for this segment of the market. In that light, the disappearance of Malaysia Flight 370 on March 8, 2014 and the surrounding mystery regarding what path the airliner ultimately took, may prove to be a compelling motivation for the widespread use of increased satellite tracking for commercial air traffic, for which small sat fleets such as Orbcomm and Iridium are ideally suited. In short, after an exceptionally turbulent start, the mid-range orbit satellite industry appears to be gaining a steady increase in customers and applications, which in turn bodes well for the launch market which serves, it but has been a dangerous place to try to make a living in the past.

No matter which rocket boosts a Comsat to geostationary transfer orbit, the satellite itself is going to be an extraordinarily complicated and very expensive spacecraft designed to work with unflinching precision in an extremely harsh environment. Furthermore, it is expected to do so for as many as 10 to 15 years, over which it may earn hundreds of millions of dollars of revenue. With an average cost in 2012 dollars of $250 million for a typical large Comsat, the price of the launch vehicle comprises a fraction of the anticipated revenue stream. Consequently, the reliability of the launch vehicle is of much greater concern than its price. There is a limit however, and if reliability were the only criteria, then neither the Zenit nor the Proton would regularly be receiving orders, and ULA would have maintained a steady business with the Atlas V and Delta IV. Because most launch providers except for SpaceX go to great lengths to conceal their prices, it is often difficult to determine with any precision what the going rate actually is, but Arianespace is generally estimated to be around $220 million combined between two payloads for the Ariane V, with second place Proton, gradually increasing from a rate of $80 million in the late 1990's up to $120 million per launch by 2012. Sea Launch is estimated to come in at around $90 million, and comparable Chinese boosters at $70 million. Based on these numbers alone, the substantial price differences between Ariane and its competitors indicates that cost does come in to play, and risk is mitigated through the purchase of launch insurance.

There are additional factors to take into account. The Comsat industry is dominated by a relatively small group of larger fleet operators with assets in the tens of billions of dollars. Companies such as SES, Intelsat and Eutelsat operate the largest and most expensive satellites, and are generally willing to pay a higher premium to insure a safe launch. At the same time, once a provider reaches a certain service level, they are also in a position to take on increased risk in the form of one-off launches of new providers. Besides these established operators however, there are a growing number of national or regional satellite owners whose spacecraft are generally smaller, and who are more price sensitive than major fleet owners. On the other hand, recent years have seen a growth in the instance of governmentally backed "package deals" in which a sponsoring government such as Russia or China may offer satellite construction, finance and launch as part of a comprehensive arrangement which effectively removes the launch from open competition. For SpaceX, the upgraded Falcon 9 v1.1 is still only capable of serving the small and medium category of geostationary comsats which comprise roughly half the market. With a gradual creep in mass towards ever larger spacecraft, the company must wait on the introduction of the Falcon Heavy to pursue the other half of the market.

In addition to the GEO Comsat market, one other general category tailor-made for the Falcon 9 v1.1 is that of Earth observing and scientific satellites launched into mid-range sun synchronous orbits. It was just such a customer, and one which was willing to wait for what seemed like an eternity for launch, that ultimately provided the first commercial payload for the Falcon 9. Canada's MDA Corporation, which in participation with the Canadian Space Agency built the multi-purpose CASSIOPE demonstration satellite, originally signed a launch order in 2005 with SpaceX to be carried out on a Falcon 1 from Vandenberg. After nearly an eight year span which saw the satellite wait through the Falcon 9's first five launches out of Cape Canaveral, with the approach of the first F9 v1.1 flight, the long suffering CASSIOPE would finally get its ride to orbit.

Not everything worked out so well however. SpaceX's very first arrangement for a geostationary Comsat launch was made with Avanti Communications Group and announced on September 14, 2007, calling for a launch between March and December of 2009. Had it occurred, the launch would have carried the HYLAS broadband and data satellite out of Cape Canaveral to a geostationary orbit at 33.5 West. When development of the Falcon 9 proceeded at a much slower pace, hindered by the difficulties with the Falcon 1 and the development of the Merlin 1C, Avanti eventually cancelled the contract, leading to a dispute over the down payment and inevitably, a lawsuit. The loss was particularly painful because the deal carried the potential of a total of three launches if all went well.

Following the Avanti deal, the next commercial launch to be booked for the Falcon 9 which still remained on the manifest as v1.1 entered service was a pair of dual-use Argentine remote sensing spacecraft for CONAE, the Argentine space agency, where they are scheduled to join an Argentine/Italian constellation, called SIASGE consisting of four satellites already in orbit. The launch of the SAOCOM 1A and 1B missions will be to Sun Synchronous orbits (SSO) where they will be used for communications, defense, disaster response, and environment management, including measuring soil moisture over the fertile Pampa Humeda grasslands.

What SpaceX needed, and had yet to receive as it was trying to enter the core of the commercial launch business, was an order from one of the blue blood players of the geostationary Comsat industry. That day came on March 15, 2010 when SpaceX announced an arrangement with Space Systems/Loral to launch one of their 1300 model satellites out of Cape Canaveral to geosynchronous transfer orbit, with the ultimate customer to be determined later. As the world leader in the construction of high power Comsats, the SS/L deal served notice that SpaceX was entering the big leagues at last. Less than three months later SS/L's apparent confidence in its new launch provider was vindicated when the first Falcon 9 roared into orbit.

SS/L was far from the only party with a vested interest in the outcome of that first launch. Following the Falcon 9's maiden flight, SpaceX signed what was the single largest deal in the history of space commerce, a $492 million dollar arrangement with the Iridium, LLC, to launch a new generation of satellites, Iridium Next, for the Satphone services provider. The Iridium announcement made it clear that although the deal had been long in the works, the determining factor was the success of the Falcon 9's first flight. The Iridium arrangement is unique, calling for a series of 7-8 flights, with some lofting as many as 10 of the 1,764 lb. spacecraft to its 485 mile orbit. Although it provided a major source of future launch revenue for the company, the Iridium deal also highlighted one of the risks of the small satellite industry. With Orbcomm on board as a SpaceX customer, and competitive Satphone provider Globalstar already replacing its own entire original fleet with launches aboard the commercial Soyuz, Iridium represented the last major opportunity for some time to come. The market tends to run in cycles, and 15 years is a long time to wait for a repeat order.

In terms of its actual development schedule, perhaps the most significant commercial launch agreement to date came on March 14, 2011 when SpaceX contracted to launch a geostationary satellite for longtime fleet provider SES of Luxembourg, owner of the world's second largest fleet of geostationary Comsats, numbering over 50 and covering 99% of the world's population. With SES already a reliable customer for both the Ariane V and the Proton, the addition of the SpaceX flight represented an important statement on the part of one of the largest traditional commercial satellite companies on the need to spark enhanced competition and better pricing on the international launch market, and a well-conceived plan to diversify its launch portfolio. At the same time, a statement by SES President and CEO Romain Bausch made clear that far from being given the keys to the kingdom, SpaceX was simply being let in the door.

"We feel comfortable in entrusting SpaceX with one of our satellites, thereby encouraging diversity in the launch vehicle sector and fostering entrepreneurial spirit in the space industry, Falcon 9 ideally compliments our roster of Ariane V and Proton boosters, and our framework launch understanding with Sea Launch."

SES, like most of the major fleet operators wanted diversity and price pressure on the market leader Ariane V, but they were not about to drop anyone wholesale just for a better bargain.

Although the SES announcement came more than a year after the similar announcements from Space Systems/Loral, it called for a comparatively near term launch in 2013, thereby making it the first geostationary launch the company would conduct. The challenge for SpaceX was that SES required the company to demonstrate at least one successful flight of the Falcon 9 v1.1 complete with a successful separation of the new payload fairing prior to July 2013, before it would allow its own bird to fly on an unproven provider. If SpaceX missed the deadline, SES could exercise its option to withdraw from the terms without penalty. Adding to the pressure, the SES arrangement held out the potential of a second order, provided the first effort proved successful. Although by early June it was clear that SpaceX would miss the deadline, launch preparations were far enough along that there was little chance the Luxembourg based operator would withdraw from the effort.

The first mission for the new booster, labeled as a "demonstration" launch and the one which would pave the way for commercial operations was that of the aforementioned CASSIOPE, a relatively small, 1,100 lb. spacecraft comprised of two elements, which taken together provided the satellite its name. First is the Cascade communications demonstrator payload, a "proof of concept" package to demonstrate data packaging, or "store and forward" services, and the second is ePOP, (enhanced Polar outflow probe) a suite of weather instruments to study the Earth's ionosphere. Combined on a new smallsat bus platform to maximize flexibility and contain costs, though sitting out the last 8 years in storage, CASSIOPE is still representative of a new breed of smaller, mixed use satellites designed to launch and operate at a fraction of the cost of larger platforms. In exchange for a deeply discounted launch price, the mission's sponsor, the Canadian Space Agency was willing to wait as long as it took to secure the ride.

After finally completing testing in McGregor in mid-summer, SpaceX shipped both stages of the Falcon 9 to its new home at Vandenberg, Ca. launch complex SLC-4 East, the very same pad which had hosted the last West Coast Titan IV mission and played a key role in forcing SpaceX to relocate to Kwajalein for its first Falcon 1 launch. Leased from the Air Force in 2010 following a long history of hosting Atlas-Agena, Titan III and Titan IV launches, nearly all of which were carried out to support National Reconnaissance Office missions, SpaceX demolished the remaining Titan IV infrastructure in 2011 and steadily went about the business of converting the pad to a similar arrangement like that employed as SLC-40 on Cape Canaveral, complete with a similarly spartan horizontal integration building. The one key difference was that with the company at that point planning to conduct the first Falcon Heavy launch out of Vandenberg, it constructed a much more robust version of the transporter/ erector assembly, both wider and taller than its East Coast counterpart.

Following a wet dress rehearsal, SpaceX conducted a 2 second static fire on September 12, but not unexpectedly, encountered several issues with the new ground support equipment, and elected to defer the launch to the end of the month in order to conduct a second static fire. The second attempt took place one week later, on September 19th, and with no issues to resolve, the launch team set the date for the next available opportunity, a two day window opening on Sunday, September 29th.

The CASSIOPE launch however, was much more than the debut of the upgraded Falcon 9. Coming in to the effort, SpaceX set audaciously ambitious goals, beginning with an attempt to perform a soft water landing of the vehicle's first stage in a two-part process. After stage separation, at roughly Mach 10, the Falcon 9 v1.1 would re-ignite three of its 9 Merlin 1D engines in an attempt to dramatically slow the stage before it re-entered the atmosphere and began the violent tumbling recorded by sensors on previous flights of the original Falcon 9, Block I. If the effort was successful, the next step would be to conduct a second re-ignition and burn, this time of the single, center mounted Merlin engine in an effort to further slow the booster to a near standstill right above the ocean, before it terminated thrust and splashed (gently) into the sea for rapid recovery by the same pre-positioned barge which the company hired to recover Dragon capsules.

With quite a lot which could go wrong with this particular flight, and to cover contingencies, SpaceX applied for and received a waiver for certain FAA risk thresholds, specifically the risk of blast damage from an explosion due to the anticipated development of a regional atmospheric inversion which would have the effect of increasing blast overpressure. The FAA regulation under question "prohibits the launch of an expendable launch vehicle if the total expected average number of casualties (Ec) for the launch exceeds 0.00003 for risk from far field blast overpressure."

Even if the initial first stage recovery effort proved to be a complete failure, the FAA waiver notice gave the hint of a possible silver lining. According to the waiver, after depositing the CASSIOPE satellite in its elliptical orbit, SpaceX planned to re-ignite the second stage and burn it to exhaustion. While it might have taken the form of a fuel intensive plane change maneuver, much like the company performed on the Falcon 1's first successful flight, some indications suggested the company intended to send the second stage out of Earth orbit altogether, and in doing so demonstrate not only the ability to perform a geostationary transfer orbit burn which would be needed for its upcoming SES launch out of Cape Canaveral, but also take a major step in marking out a future role for interplanetary missions.

In addition to the CASSIOPE satellite, the Falcon 9 carried several secondary payloads; the Polar Orbiting Passive Atmospheric Spheres, (POPACS) intended to study the effect of coronal mass ejections on the upper atmosphere, DANDE, the Drag & Atmospheric Neutral Density Explorer, a 50kg spherical spacecraft to measure the effects of drag in low Earth orbit, and finally, CUsat, a GPS calibration cubesat designed to test a new method of GPS measurements which could deliver accuracy to the astonishing distance of a mere millimeter.

With launch scheduled to take place in a window between 9:00 AM and 12:00 PM EDT, the day dawned clear and bright, and before fueling even began, it had already been a good day for commercial space. Hours earlier, the first Orbital Sciences Corporation Cygnus spacecraft, following a path blazed by SpaceX, had been successfully berthed with the International Space Station, effectively bringing NASA's COTS program to a triumphant conclusion. And although NASA had no official role in the debut flight for the new Falcon 9, with 10 more missions scheduled under the CRS program, the agency still had a vital stake in seeing the Falcon 9 v1.1 get off to a good start.

Following a pre-launch sequence only slightly altered from that of the first five Falcon 9 flights, fueling began just under four hours prior to liftoff, beginning with liquid oxygen and followed by RP-1. At T-10 minutes, the count switched to automatic control as the booster became fully alive. At T-2.5 minutes, Vandenberg's Launch Director gave the green light, and thirty seconds later the Air Force Range Officer verified the launch area was ready and the Flight Termination System was armed. With 60 seconds to go, the water suppression system, also named "Niagara" like its counterpart at Cape Canaveral, began to flood the pad with a solid torrent of water at 30,000 gallons per minute.

At T-3 seconds, the moment of truth arrived with ignition of 9 new Merlin 1D engines rapidly powering up to 1.3 million pounds of sea-level thrust. At T-0, with flight computers verifying everything was nominal and trending good, the hold down clamps released, and the slender booster rapidly climbed away from the pad. Seconds later, it adjusted course to head nearly due south and out over the Pacific on a pillar of brilliant orange flame. The Falcon 9 went supersonic 70 seconds into the flight, and 10 seconds later it entered Max-Q, and a key milepost for a booster potentially more susceptible to the bending forces imparted due to its increased length.

Once again, reality confirmed the results of computer simulations and ground level testing in Texas, and the rocket sailed effortlessly past the threshold.

At the 150 second mark, at an altitude of 56 miles, two engines shut down on cue, and seconds later, at MECO, the remaining 7 engines shut down. Next up came another moment of truth with an upgraded and simplified stage separation system utilizing only three collets and pneumatic push rods. Functioning perfectly, the first stage gently shoved the second stage away, and 5 seconds after separation, the new Merlin 1D Vacuum engine fired for the first time under actual flight conditions, beginning the steady climb to orbit. Two hundred seventeen seconds after liftoff was the most worrisome aspect of the entire mission, a first ever separation of the massive 5.2 meter composite two piece payload fairing.

Much to everyone's relief, the fairing separated without incident, and freed of its mass, the second stage continued to accelerate for another 337 seconds right up until the planned moment of SECO. Fourteen minutes and fourteen seconds after liftoff, the Falcon 9 released the CASSIOPE satellite exactly in its intended orbit. By any standard, SpaceX had an overwhelming success on its hands, and a huge load off its shoulders. The question was, could what was already a great day, turn into something more. The answer said as much about the difficulties of the launch industry as anything could.

Moments after it released the second stage to continue on its way to orbit, the now (mostly) expended first stage successfully re-ignited three of its nine Merlin engines, and for the first time in the history of the space age, began a measured burn to reduce its velocity and re-enter the upper atmosphere while under control and steady deceleration. Amazingly, it worked. With the returning stage now descending tail first, it started to enter the lower, thicker part of the atmosphere. As it did so, the stage began to rotate around its vertical axis, much like a rifled bullet exiting the barrel of a gun. Right on cue, the single, center mounted Merlin engine successfully re-ignited and in spite of the extraordinary circumstance of descent into its own rushing exhaust plume, the stage slowed even further. Unfortunately, even as the spinning stage began to slow, the rotation exceeded the capacity of the nitrogen gas reaction control system. In what appeared to be an exquisitely ironic twist on the corkscrewing motion which had doomed the nearly successful second flight of the Falcon 1, SpaceX initially concluded that the centrifuge effect threw fuel away from the pickup, causing an engine flame out. Subsequent analysis indicated that the rotation actually caused pieces of the tank's internal baffling to break free, which were then ingested into the engine, triggering a shutdown. As a result the stage slammed hard into the Pacific, leaving only debris for the awaiting barge American Islander to pick up.

Although not completely successful, the first stage recovery attempt had accomplished far more than nearly anyone expected, bringing the descending stage almost within range of the Grasshopper tests taking place in Texas, resulting in a remarkable photograph of the stage in the split second before impact, perfectly vertical and only 3 meters above the Pacific Ocean. As for the rotation, Elon Musk announced in a post launch telephone conference with reporters that based on the data collected, the addition of landing legs, operating like tail fins on an older rocket, should be enough to slow the spin and bring it within range of ACS's ability to compensate. In exactly one launch, SpaceX had nearly succeeded in achieving something many dismissed as virtually impossible, recovering a first stage through a powered burn.

It was all that much more surprising then, when word leaked out that after deploying the CASSIOPE satellite followed by an equally successful deployment of the secondary payloads, the Falcon 9 second stage had failed to relight as planned. Instead, and for reasons which were initially unclear, at the moment of second stage ignition, computers signaled an abort condition.

With an uncooperative stage on its hands, flight controllers elected to call it a day, issuing a command to begin venting liquid oxygen in order to avoid a potential explosion. Curiously, much like what took place after the very first launch of the Falcon 9, the dispersing oxygen created a halo around the second stage, leading to reports of a UFO as it passed over Africa and parts of the Indian Ocean. Furthermore, in yet another reprise of previous SpaceX incidents, reports began to surface suggesting that despite the venting, the second stage had in fact blown up in space. The basis for the report was the apparent presence of more objects being tracked in orbit by ground based radar than could be accounted for in the mission profile, which is precisely what happens when an upper stage or spacecraft "explodes" or more likely, ruptures. In this case, the company sought to quash the rumors quickly, issuing a statement that telemetry had confirmed the stage was intact, and the more likely explanation was that during the course of venting, several pieces of protective foam had broken away from the stage, and in the process created the unexplained objects radar was tracking.

The news was definitely more optimistic for first stage recovery efforts however. Based on the flight team's surprising progress in the first attempt to slow and hover the first stage, Elon Musk announced that although the company would not attempt recovery on the next two launches, it would try again on the next NASA flight, the CRS-3 mission scheduled for February 2014. Furthermore, with the upgraded Falcon 9 now easily exceeding the demands of NASA CRS launches to ISS, each of the remaining 10 flights would be marked by a recovery attempt. Increasingly confident of success, the boosters would begin to be outfitted with flight weight retractable landing gear, possibly as soon as the next mission, one which would also happen to include an upgraded Dragon capsule. Preparing for the possibility of touching down on solid ground, the company sought to establish a designated area on the extreme eastern tip of Cape Canaveral, possibly near Complex 46, a commercial facility operated by Space Florida.

In any event, and although it had been a very successful day for SpaceX, the fact that the second stage had not re-ignited on command raised a serious concern for the next two missions on the manifest, the SES-8 and Thaicom satellite launches both scheduled to take place as soon as practically possible out of Cape Canaveral.

Liftoff of the CRS-3 mission, the first Falcon 9 v1.1 to carry a Dragon capsule.

Credit: NASA/ Tony Gray and Tim Powers

In the weeks following the CASSIOPE launch, SpaceX identified the most likely source of the second stage relight difficulty as stemming from colder than expected temperatures in orbit, aggravated by liquid oxygen bleeding off from the upper engine as it went through a coast phase LOX chill, which partially froze the lines supplying the Merlin 1D vacuum engine with Tea-Tab ignition fluid. As a result, the engine failed to fully ignite, and the lower than expected initial combustion pressure triggered an automatic shutdown. Here again, ITAR regulations hindered the process. In order for the next launch to go forward, SES clearly needed to satisfy itself that the problem had been accurately diagnosed and corrected. ITAR however, prevented SpaceX from sharing its results with non-US citizens of the Luxembourg based company. The imperfect answer required working through a small team of American citizens employed by SES and embedded with SpaceX, which was limited to assuring upper management that the booster was now ready, but legally prevented from telling them the specific reasons for that conclusion. "Just trust me" is usually not sufficient when one is making multi hundred million dollar decisions, but in this case it would have to do. After addressing the issue by insulating the lines and relocating the LOX vent, SpaceX shared its results with both NASA and the Air Force, and set the launch date for November 19, which then slipped several days to Monday, November 25th three days before U.S. Thanksgiving holiday and the beginning of one of the busiest travel weeks of the year.

After an uneventful Wet Dress Rehearsal and static fire on the 21st, liftoff of the SES-8 satellite was scheduled during a window extending from 5:37 to 6:43 pm EST. As launch day dawned, the weather was overcast with a stiff breeze blowing in from offshore. Standing on its newly re-configured transporter/erector and at 224' feet, considerably taller than the previous five Falcon 9 boosters which had lifted off from SLC-40, the solid white booster blended in against the moody weather hanging over the Cape, a stark contrast to the gleaming Pacific setting and cobalt sky which had recently set the stage for the first launch two months earlier. It may have been an omen of things to come. Somewhat ironically, even though the Vandenberg launch preparations, the first ever for SpaceX in that particular setting, had gone off smoothly, breaking in a new booster, a new pad, and all new ground systems where any number of snags might have been expected, the SES launch, taking place in a much more familiar setting of SLC-40, proved to be the most challenging since the early years at Kwajalein.

With a first liftoff attempt less than two hours away, a clearly concerned Elon Musk emerged from the low concrete building immediately outside the gates of Cape Canaveral Air Force Station which serves as the Florida launch control center. After briefly stopping to talk to a small group of reporters, the SpaceX founder crossed the four lane entrance road at the south gate to walk along the shoreline, gazing at his booster several miles away across the Banana River. Although every launch thus far had been critical in its own way, this one stood out. After challenging, and some would say, taunting, the rest of the commercial launch industry over its high prices, and engaging in a long running war of words with market leader Arianespace, this would be a moment of truth. A successful launch could clear the way to start making good on 10 years' worth of promises, whereas a failure could be both profoundly embarrassing, as well as spell trouble on all three fronts which mattered to SpaceX, Commercial Crew and EELV, as well as the commercial customers it was time to satisfy. Having told reporters in a pre-launch press conference that "no stone hasn't been turned over at least twice," there was little left to do except wish it well, and then return to the launch control center and watch it all unfold.

At the pad, the countdown got underway on time, and appeared to be proceeding smoothly heading towards the final poll which is held at T-13 minutes, before the booster enters its auto sequence count at T-10 minutes. Going into the poll, engineers called a hold to troubleshoot an issue with a first stage liquid oxygen vent valve. While still troubleshooting the issue but hoping to clear it in time, the countdown clock was reset to T-17 minutes and a new liftoff time of 5:54. With high altitude balloons confirming that despite the strong ground level breeze the booster was clear for liftoff, the count ran down to T-6:11 when a new problem cropped up with the switchover from ground to on-board power supply, as computers detected a problem with narrowly set telemetry range limits. Resetting the limits, the countdown started back once more, only to be stopped by the call of "hold, hold, hold" at T-3:41. As events had unfolded, the launch control team had never been able to successfully resolve the original issue with the LOX vent valve. With the launch window not allowing room for more troubleshooting, the day's events were over. Adding insult to injury, as the launch assembly began the return to vertical, it snagged the line supplying fresh air to the payload, ripping it off.

Falcon 9 leaving the hangar aboard the transporter/erector. Credit: NASA

The Falcon 9 for the COTS 2/3 mission in the hangar at SLC-40. First stage and interstage sit on their respective carriages. Credit: NASA/Jack Pfaller

Following the scrubbed Monday attempt, SpaceX faced a scheduling challenge brought on by the idiosyncrasies of Thanksgiving week travel. For every launch out of Cape Canaveral, the FAA has to establish a zone of restricted air space immediately surrounding the launch area and extending off the Florida coast for obvious reasons. Ordinarily it is not a major problem as planes are re-routed as required. Tuesday and Wednesday being the two travel days before Thanksgiving however, and the East coast flight corridor full of aircraft carrying people home to friends and family, the FAA was not about to add to the congestion by further restricting airspace. That meant that the next available launch date was Thursday the 24th, Thanksgiving Day. Only one other US launch had taken place from Cape Canaveral on Thanksgiving Day, that occurred in 1959, and like many of the era, it had not gone well.

With a Thanksgiving Day launch now set for a window beginning one minute later than on the previous attempt, and the vent valve issue resolved, it was time to try again. Working no other issues, the countdown began on time and moved smoothly through its carefully scripted steps, all the way down to the moment of engine ignition which took place exactly on schedule, accompanied by a brief flair of green then orange light, and then nothing. Detecting a problem, launch computers had aborted the count at T plus 2 seconds, less than 1.5 seconds before release and liftoff. After going back into a hold to diagnose the problem, SpaceX announced that computers had detected a lower than expected ramp-up in engine thrust, which under normal conditions would have seen hold down clamps released as the engine reached full power at t plus 3.5 seconds.

As engineers continued to study the data, Elon Musk advised via Twitter that the launch team would try again by increasing the helium spin start pressure, but then went on to caution that the odds were less than 50% of passing all aborts. With the launch window closing once again, the countdown re-started, targeting a 6:44 pm EST liftoff. This time, it reached approximately t-60 seconds when it was halted for the final time, bringing the Thanksgiving Day launch effort to a close.

Confirming Musk's warning, SpaceX soon cited a lack of time to complete the necessary data review from the first abort. In a subsequent update via Twitter, Musk observed that it was "better to be paranoid and wrong," going on to say that the launch would probably be delayed for several days as engineers examined the nine Merlin 1D's via boroscope while the Falcon 9 remained on the pad. Pending analysis and resolution, a new launch date would have to wait until after the return component of the nation's holiday air travel, setting up the following Monday as the next available launch date.

By Friday it was clear that the latest launch abort had been brought about due to atmospheric oxygen degrading the potency of the pyrophoric ignition fluid kept in the ground supply. Detecting a slow buildup of thrust due to poor ignition, the computers had done their job and aborted the effort. Once again, as it had during the first COTS 2/3 launch attempt, and even though while still in the moment it was quite frustrating, the Falcon 9 had demonstrated that its control systems were tightly calibrated and extremely responsive. Although countdowns could be frustrating, as they sometimes were, once the new ground equipment which has caused the issue was properly broken in, it was a fair bet that having reached the split second of release and liftoff, the Falcon 9 would be in the best possible condition to ensure success. Working through the weekend to clean all nine gas generators, during which the number 9 center mounted generator was replaced "as a precautionary measure," SpaceX reset the next launch attempt from Monday to Tuesday, adding in an extra day to re-work all issues and assure itself of the results.

With clear weather and the booster once more fueled up, and this time hopefully ready for flight, SpaceX prepared for its third launch attempt in eight days. For the first time in what had been an exasperating campaign, events proceeded without incident, and the Falcon 9 roared to life on time at 5:41 pm EST, making its East Coast debut in a flight which still held the moment of greatest suspense not on liftoff, but on the all-important moment of second state re-ignition, presuming the booster made its preliminary parking orbit. Considering the difficulties thus far, the tension was if possible, even greater. Climbing quickly away from the pad and into the Florida evening, the Falcon 9 first stage performed flawlessly, leading up to a clean stage separation and second stage ignition. Although the company would not be making a recovery attempt on this flight, even though SES had indicated it was willing to sign off if asked, sharp eyed observers were clearly able to see the first stage cold gas thruster system actively aligning the spent segment after separation, presumably helping to gather more telemetry on the re-entry profile before the next attempt.

The SES-8 flight profile called for the Falcon 9 to establish a preliminary parking orbit where it would momentarily coast towards the equator, at which point the second stage would re-ignite at the T plus 27 minute mark for a five minute burn to place the 3,318 kg satellite build by Orbital Sciences into a 295 x 80,000 km geostationary transfer orbit. At the precise moment required, the Merlin 1D Vacuum engine flared to life once again, and performing nominally, burned smoothly, boosting the second stage to the furthest point from Earth yet achieved by the company that wanted to go to Mars, cleanly releasing its payload into its designated orbit. Moments later, the message came from SES that its satellite was safely deployed and in great condition. After more than a decade marked by periods of extreme frustration, near defeat and overwhelming triumph, all accompanied by seemingly endless weeks of long hours and hard work, SpaceX, now numbering almost 3,600 people, was entering a new era.

With the new Falcon 9 having proved itself, SpaceX could begin working down its enormous backlog of launches, a necessary pre-condition for addressing the one market SpaceX had been trying without success to penetrate since its early years, the USAF Evolved Expendable Launch Vehicle program.

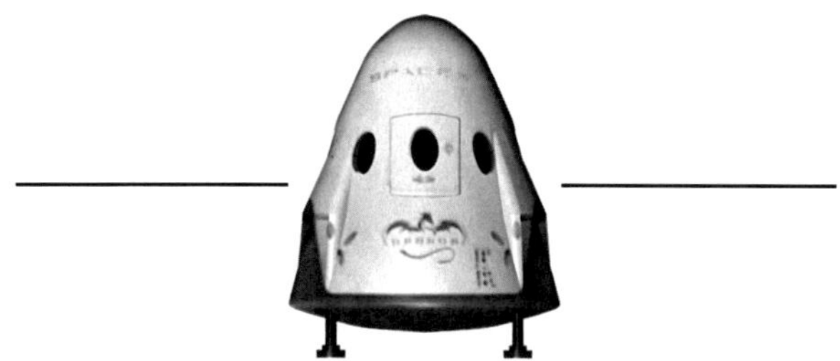

Chapter XXVIII: Challenging the EELV Monopoly

If there was one critical challenge which continued to elude SpaceX, even as it began putting together a remarkable string of commercial contracts for the Falcon 9, it was without a doubt breaking in to the largest single launch market potentially available to it, the United States military's Evolved Expendable Launch Vehicle program.

Much like Project Constellation and the Ares-1, the saga of the Evolved Expendable Launch Vehicle program is a critical part of the back-story to the history of SpaceX, and it is not possible to put SpaceX in the proper historical context, or consider its future without relating concurrent developments within the EELV program. Quite simply the magnitude of what SpaceX accomplished with the development of the Falcon 9 comes in to its sharpest focus when contrasted against the two launch vehicles which are its closest contemporaries, the EELV program's Delta IV and Atlas V. Taking place at the same time as the closeout of the Shuttle program and the rise and fall of Ares-1, the EELV program is the third component of a four way experiment in how the United States has managed access to space for the last 15 years.

The history of EELV, like so much else in America's space program for the last 30 years, is inextricably tied to the Shuttle. In order to get a reluctant Nixon administration to approve the Space Shuttle program as the successor to Apollo, its supporters needed to assemble as large a user base as possible. And because the program depended on building the case for reusability, it also needed to project the highest flight rates conceivable in order to minimize the per flight costs. The only way to do both was to bring in the Air Force as a major partner, and then capture as much of the substantial defense launch requirements as possible.

Although the Air Force ultimately, and with visible reluctance joined the Shuttle program, it brought along a number of requirements which drove up the cost and complexity of the system. Still, it was never fully comfortable in its partnership with NASA, maintaining separate and independent access to space, particularly for the large reconnaissance satellites it launched on the Titan III and Titan IV family of heavy lift boosters.

The risks of dependence on the Shuttle were brought to the forefront with the loss of Challenger and the grounding of the Shuttle fleet in 1986. Unfortunately for the nation's defense establishment, the Challenger accident was soon followed by loss of a Titan 34D on April 18, 1986 caused by a solid rocket motor burn through only 9 seconds into the launch. Coming as it did, on the heels of a previous Titan 34D failure on August 28th 1985, the Air Force faced nearly a perfect storm hampering its access to space.

The Challenger accident confirmed what was already a poorly kept secret. Although it offered remarkable capabilities, the Shuttle program was never going to come close to achieving the flight rate once predicted for it. In the aftermath of the disaster, it became clear that the policies put in place to drive those flight rates would have to change. The Shuttle would no longer be offered for commercial service, and it would cease being the preferred launch vehicle for the Armed Forces as well. The military was essentially being freed from the yoke of the Shuttle, and could pursue its own launch strategy, one built around its unique requirements, which prized reliability above all else. The problem was, its main launcher the Titan series, as demonstrated by two contemporary failures was not very reliable, and the production line of other boosters such as the General Dynamics Atlas series had begun to shut down as a result of the nation's failed Shuttle only policy.

In the short run, with a backlog of several years' worth of military payloads quickly developing, the Department of Defense had little choice but to order the restart of existing Atlas and Delta production lines. It was a temporary solution at best. As a natural result of the decision to basically force all payloads which could be carried by the Shuttle onto that platform, existing expendable launch vehicle capability languished, with rising costs, mounting delays and an unacceptable failure rate becoming commonplace. At the same time, the rest of the world, even if in awe of the American technology and the Shuttle's complexity, had not stood still. Europe was making steady progress with its own line of Ariane expendable launch vehicles, while Russia and China were preparing to enter the commercial market with much lower prices. The sudden stand down of the Shuttle following the Challenger accident accelerated the already emerging trend, causing a loss of access to space for the growing commercial satellite market and providing the perfect opportunity for Europe, Russia and China to make extensive inroads.

Although the status of the emerging international commercial launch market was not the most significant concern for the U.S. Department of Defense, both affordably and reliably launching its assets into space was, and the heritage launch vehicles activated after Challenger were falling behind on both counts. Following a continuing series of delays and launch failures, in 1994 the Department of Defense commissioned the DOD Space Launch Modernization plan, led by Air Force Lt. General Thomas Moorman, to review the issue and make recommendations.

The Moorman study, as it was called, looked at the challenge facing America's launch establishment within the context of a global market and falling launch prices. Roughly speaking, the U.S. providers, Delta, Atlas and Titan were conducting launch services between a range of $12,000 and $16,000 per lb. to GTO in 1994 dollars. Arianespace had been on the lower edge of the same threshold with the workhorse Ariane 44L, but was now developing the Ariane V with the goal of $8,000 per lb. and Russia and China were coming in at half that, $4,000 per lb.

With all three rapidly making gains in the commercial market at American expense, it was obvious that absent a major effort to reverse the trends, United States would soon find itself with no commercial launch business, and an even further escalation of military launch costs as the U.S. government became the sole customer for American rockets. The Moorman study examined four potential options, considering each on the basis of multiple criteria, one of which was the projected flight costs compared to international trends. The first option was to improve the existing expendable launch vehicle fleet to achieve a more reliable product. Although having the lowest development costs, it would result in the highest operating costs, and result in a system which was still hopelessly uncompetitive. The second option was to develop a new family of ELV's based on the foundation of existing systems, which it termed *Evolved Expendable Launch Vehicles.* This option was assessed as offering an attractive combination of modestly higher development cost and risk, resulting in lower long term operating costs. The two remaining options considered an all new, clean sheet ELV, and finally, an all-new reusable launch vehicle to be developed in concert with NASA. The final two options both offered significant improvements in per flight costs, but with a much greater up front development required. [98]

The results of the Moorman study were presented to the Clinton Administration, which responded with one of the major space policy decisions in recent history, and one which still shapes the overall environment into which SpaceX is still seeking to enter. Going forward, space launch development in the U.S. would be divided between NASA and the Air Force, with the former assigned responsibility for manned space flight systems and reusable launch vehicle development. Expendable launch vehicle development would be the province of the U.S. Air Force. One consequence of the new decision was that with an emerging split between DOD and NASA, the opportunity for pooled funding for any transformational new system as a successor to the Shuttle was gone, leaving it up to NASA alone.

Moving quickly the Air Force proceeded with the second option offered under the Moorman study and initiated the Evolved Expendable Launch Vehicle program, requesting bids from the nation's major aerospace contractors. Each of the contractor teams would submit proposals of their own design reflecting Air Force requirements, with the Air Force making the final selection.

Four teams submitted bids, led by Boeing, Lockheed Martin, McDonnell Douglas and ATK respectively. Of the four, Boeing put forward the most unusual proposal, one which ignored the premise of the request, and instead offered a two stage partially reusable launch vehicle based on the Space Shuttle Main Engine featuring parachute enabled water recovery. The idea was not terribly far-fetched. At one point NASA, interested in the potential of a recoverable thrust assembly for a cargo version of the shuttle, actually dunked an SSME in the Mississippi river, pulled it out, and hooked it up to the test stand for a successful firing.

Lockheed Martin offered a new version of the long lived Atlas booster family which Martin Marietta acquired in the purchase of General Dynamics Space Systems Division in 1993. The new booster, dubbed Atlas V, would be powered by a single Russian built first stage engine, which Lockheed Martin was already adapting for use on the short lived Atlas III commercial booster. McDonnell Douglas also offered a new booster based on its own legacy rocket, the Delta II. The new version, the Delta IV, was an all cryogenic two stage booster powered by a new main engine, the RS-68, an expendable engine loosely based on the SSME but with a reduced parts count and lower construction costs. The second stage, just like the Atlas V, would be powered by America's original cryogenic engine, the Pratt and Whitney RL-10. The final entry, by ATK was a forerunner of things to come, a two stage booster based on a Titan IV Castor 120 solid rocket motor which was a smaller cousin to the Space Shuttle Solid Rocket Motor.

Boeing and ATK were eliminated in the first round of the competition, leaving McDonnell Douglas and Lockheed Martin as the two remaining competitors. As part of the ongoing consolidation in the defense industry however, Boeing acquired McDonnell Douglas in August 1997, placing it right back in the ongoing competition, this time with the all new Delta IV.

Originally conceived as a winner take all contest which would result in a single, reliable and affordable launch provider with a mix and match set of common interfaces intended to eliminate "niche" capabilities, the program's basis was altered to a two vendor strategy in 1997. The changing rationale was promoted on the grounds of offering assured access to space through two distinctly different launch vehicle families, as well as the benefits of both price competition and competitively driven contractor innovation. The decision to adopt the two vendor strategy was premised, at the urging of final contestants, Boeing and Lockheed Martin, on the predicted boom in the commercial satellite launch market. At the same time, some of the Air Force requirements came into conflict with the trend towards larger commercial communications satellites, resulting in a launch capability not optimized for part of the market it was supposedly pursuing. Following signs which were already manifesting themselves during the product development phase from 1998 to 2002, the anticipated commercial market never materialized, but it was not the only significant event which triggered subsequent major changes to the EELV program.

The original EELV purchase order, Buy I, was placed in 1998, and Boeing was the clear winner with the Delta IV, awarded 19 launches compared to 9 for Lockheed Martin and the Atlas V. In both cases, the prices quoted appeared to confirm the success of the new acquisition strategy, with per unit costs initially coming in around $97 million for both boosters, 25% lower than the heritage vehicles being replaced. Equally encouraging, even though the commercial market had crashed as the program was being formulated in 1998, by 2002, the year both vehicles made their initial flights, the traditional comsat segment stabilized, and Lockheed Martin at least, was booking commercial orders. As it turned out, the Atlas-V conducted six commercial and two NASA launches prior to its first DOD launch in 2007. The Delta IV by contrast, hobbled by its more expensive cryogenic architecture and competing against an Atlas product powered by a deeply discounted Russian main engine, reportedly acquired at the bargain discount price of only $10 million per unit, only conducted a single commercial launch, and that was on its maiden flight, before it was pulled from the market due to its higher cost.

Behind the scenes, both Boeing and Lockheed Martin soon began to complain that because the commercial launch market was not meeting projections, they were not able to cover their overhead costs on the limited number of military orders received. Absent some sort of price support, those close to the program intimated one or both might not be able to stay in the launch business, with obvious implications for national security. Rather than downsize to a single provider, which wouldbe in keeping with the original intent of the program, the pleas found a sympathetic ear from the Air Force. Beginning in 2003, the Air Force began making supplemental payments to support both contractors. [99] One year into the operational phase of the EELV program, what had so recently looked like a bargain began to take on a very different appearance, corporate welfare.

In the first place, not everyone was convinced the supplemental payments were necessary or justified. While there is no doubt that the launch market collapsed, despite claims to the contrary, it was hardly "unforeseen." The signs a major crash was coming were clearly visible in 1998, even as the program was getting underway and only one year later, Iridium filed highly public bankruptcy proceedings in August of 1999, with others soon to follow. The bottom line was that both Boeing and Lockheed Martin had made remarkably poor business decisions, and in a move all too common for defense contractors, nonchalantly asked the taxpayer to pick up the tab.

Bad forecasting was only one of the critical factors which derailed the early cost gains of the EELV program. Although Boeing, with its all U.S. sourced Delta IV won the lion's share of the original launch order, the results were soon turned upside down when a wrongful termination lawsuit led to the disclosure that two Boeing employees were in possession of almost 25,000 pages of proprietary documents regarding the Atlas V, the rightful property of Lockheed Martin. An Air Force review concluded that included in that data was pricing information which had likely influenced the outcome of the initial bid. Even worse for Boeing, the allegations regarding the EELV program came in the midst of a major but separate bid-rigging scandal involving Boeing employee and former Air Force purchasing agent Darleen Druyun, which led to the convictions of both Druyun and Boeing CFO Michael Sears, as well as the resignation of the company's CEO.

The EELV documents allegation only served to make matters worse, prompting the Air Force to punish the company's malfeasance by rescinding the original Buy 1 launch purchase orders, and subsequently re-awarding seven of the nineteen total to Lockheed Martin, which then went on to receive an additional 3 launch orders during a second acquisition phase in which Boeing was forbidden to bid. The decision, made by Undersecretary of the Air Force Peter B. Teets, who happened to be a former president of Lockheed Martin, had far ranging effects, and set the stage for what happened next.

As Boeing emerged from its self-inflicted purgatory, it appeared that open and fair competition in the EELV program would emerge at last. Under the original two vendor strategy, the Air Force planned to conduct purchases of launch services in 1 to 3 year increments, thereby giving both vendors an incentive to provide the best pricing possible and the loser the option to sharpen their pencil in advance of the next round which was never too far away. Instead, as the next round of purchasing approached, Teets ruled that because of Boeing's unfair knowledge of Lockheed Martin pricing, fair bidding would be impossible. The curious solution was to allow the next round of purchasing to extend out at least five years, a decision which would effectively create a duopoly and exclude the possibility of any new entrants to the process until 2011. [100]

There was another issue as well. Boeing still faced a massive, possibly crippling lawsuit filed by Lockheed Martin over the document affair, which it brought in state and federal court under Racketeer Influenced and Corrupt Organizations(RICO) statutes, which contained the very real possibility of a potential trebling of punitive damages. Lockheed Martin's lawsuit was filed against the background of a diminished overall market and its own clear vulnerability, which included a reliance on a Russian main engine, as well as a decision not to build a three core heavy lift version of the Atlas V, as Boeing had done for the Delta IV. The lawsuit against its rival presented Lockheed Martin with a unique opportunity. Rather than continue to compete for limited business, it might be possible to ensure the survival of the Atlas product line by irrevocably tying it to Boeing, which had responded to the Lockheed Martin action with a countersuit. [101]

At this point, SpaceX, stepped into the fray with a lawsuit of its own, leveled against both parties and their attempt to create a joint monopoly. Although SpaceX had yet to field the Falcon 9, and in fact, had not even successfully launched the Falcon 1, the company alleged that a government backed monopoly armed with a five year exclusive contract would prevent SpaceX from offering the EELV class Falcon 9 which it was developing. Furthermore, the SpaceX lawsuit alleged that Boeing and Lockheed Martin were actively conspiring to corner the launch market, and jointly threatening to refuse to bid in the next round, Buy 3, unless they were granted the five year exclusion, subsidy payments and the authority to combine operations. Regrettably from SpaceX's viewpoint, its lawsuit was thrown out when the judge ruled that without a launch vehicle currently available in the EELV class, the company lacked standing to prove it was injured. For what it was worth, when SpaceX did in fact launch the Falcon 9 twice in 2010, it did so within the time frame of the original suit, so in one sense its legal complaint was ultimately justified. Based on the events that followed, there could be little doubt.

The SpaceX suit dismissed, Lockheed Martin pressed Boeing to combine their respective operations into a new joint venture shared 50/50 between the two aerospace giants. What was presented to the American taxpayer as two defense companies jointly deciding to join forces and do what was in the best interests of the country was belied by the little noticed fact that Lockheed Martin was holding open the threat of the RICO lawsuit until Boeing consented and regulatory agencies signed off. Far from a happy union, it was a shotgun marriage which gave birth to United Launch Alliance in 2006. From the outset, skeptics, which originally included the Federal Trade Commission, predicted that the only result would be still higher launch costs to the taxpayer as all the disagreeable results of a government backed monopoly began to take hold, as even the possibility of competition was effectively and conveniently eliminated.

The immediate outcome of the union was the elevation of the Atlas V to an on-par status with the Delta IV through formal enshrinement of a new assured access policy which distributed launches between the two lines on a 50/50 basis. What was originally supposed to be a competitive environment now took on the air of a small child's soccer game in which everybody got a trophy. In justifying the decision, Space and Missile Systems Commander Lt. General Brian Arnold told a skeptical press that the even distribution of launch contracts would only remain through the end of the decade, and "then we'll have open competition after that." [102] He apparently meant way after, considering the fact that the practical result was to secure a monopoly all the way to 2018.

Despite assurances to the contrary, and federal assistance in moving the Atlas V production line to Boeing's Delta IV facility in Decatur, Alabama, to no one's surprise, per units costs continued to climb, fueled by low launch rates and an ever increasing annual subsidy in the Launch Capabilities Contract. In yet another consequence of the decision, ULA soon decided to drop production of the very popular Delta II launch vehicle which Boeing acquired in the purchase of McDonnell Douglas, leaving NASA with little choice but to order launches on the more expensive and in some cases grossly overpowered Atlas V instead.

Even in the face of steadily rising launch costs, and the national humiliation of an almost complete withdrawal from the international commercial launch market because of those prices, Congress and the Department of Defense continued to tolerate the now corpulent monopoly due to the undeniable performance and sterling reputation of its two vehicles, Atlas V and Delta IV, which began putting together a string of successful launches, allowing ULA officials to make the argument that success must necessarily come at a cost. The flaw in that argument was that over the same time frame, Arianespace was putting together an even greater run of successful launches with the Ariane V, but at a price point which kept it in the forefront of the commercial market.

By early 2011 and nearing the end of the Buy 3 purchasing round, ULA's monopoly position finally came under threat, not from Arianespace, but from SpaceX which now had its own EELV class vehicle, the Falcon 9. With two successful flights under its belt, and a third high stakes flight imminent, the Falcon 9 offered a compelling alternative to the ULA products both in terms of pricing and its 100% U.S. production. SpaceX was only too happy to point out that the Falcon 9 presented an opportunity to rejuvenate the EELV program with the type of competition which was originally intended but never achieved when it implemented the two vendor strategy in 1998.

ULA had not been watching the rise of SpaceX and the Falcon passively however, and behind the scenes began pursuing a new program to secure its hold on the market by advocating a "block buy" of up to 50 launch cores, ostensibly to help it lower prices, but with the none too subtle effect of extending the monopoly for years to come. Initially it appeared that ULA might secure the block buy with little resistance, as Congress and the press focused on the final flights of the Space Shuttle program, and the ongoing fight with NASA over its intended replacement, the Space Launch System. The proposed changes to the EELV program, quietly underway for much of the previous year finally exploded in volume and intensity with the approach of fall. A clash years in the making, it came to a head in October of 2011, and was the first of two significant decisions, the second being the Commercial Crew program which would have a major impact on the sustainability of publicly funded spaceflight.

The immediate point of contention was whether the Air Force would proceed with the proposed "block" purchase from ULA, and thus the perpetuation of its monopoly, or whether it would allow SpaceX and other new entrants such as Orbital Sciences to bid on medium and heavy lift launches conducted for the Air Force and National Reconnaissance Office, and if so, how quickly and to what extent. The decision was framed by the emergence of two distinctly different government reports which cast in stark contrast the differences between SpaceX and ULA.

The first report, released in August, was a presentation of the results of a comparative exercise undertaken by NASA using the NASA Air Force Costing Model (NAFCOM) to determine what it might have cost the space agency to develop the Falcon 9 using either a traditional NASA development culture, or one which more closely represented the commercial approach taken by SpaceX. In the initial round, both calculations utilized a cost plus fee format, and measured Design Development Test and Engineering for the Falcon 9 and one flight vehicle. The results were an estimate of $3.977 billion under a NASA model versus $1.695 billion following a commercial approach. [103]

Noting that SpaceX's actual costs were nowhere close to the model's output, NASA personnel visited SpaceX headquarters to understand the difference and refine the inputs. The second round, which by this time included numbers for developing the Falcon 9 plus two flight vehicles, resulted in an estimate of $1.3 billion for a NASA development model with a cost plus fee contract, versus $443 million for SpaceX using a fixed fee approach, a number which more closely matched the actual costs cited by SpaceX.

The numbers themselves were not actual budget estimates, but what stood out was that in either circumstance, the SpaceX approach was much cheaper, coming in at only 1/3 that of a traditional development method. What was perhaps more insightful was the explanation provided by the company for its exceptionally controlled costs, and what it said about the rest of the launch industry as a whole. According to the presentation, SpaceX ascribed its costs to three primary factors. First, the controlled expense of a lean, production oriented workforce. Essentially, the cost of the Falcon 9 was a product of the total workforce expense of the company, then about 1600. The second factor was the reduced organizational complexity of a vertically integrated operation, highlighted by the fact that SpaceX estimated every dollar sent out of the company represented 3-5 dollars in additional costs. Consequently, the SpaceX figure of 80% internal production value for the Falcon 9 translated into an enormous comparative advantage. The final factor, reduced infrastructure costs associated with the DDT&E effort, and a high utilization percentage closely correlated to the first two, and the inherent simplicity of a design approach which used common engines and tanks for the first and second stages. While it seemed painfully obvious that the factors cited in the presentation represented at least one successful strategy for controlling costs in an industrial environment, it was also certainly at odds with that taken by the traditional launch vehicle industry. ULA member companies Boeing and Lockheed Martin, as well as Orbital Sciences Corporation and ATK's proposed Liberty launch vehicle, all depend heavily upon outsourcing for major components.

What was perhaps most interesting about the results determined by the model is that they tended to correspond with the real launch cost disparity between SpaceX and ULA which was at the core of the EELV debate. In the closest apples to apples comparison based on vehicle performance, which was a Falcon 9 versus an Atlas V 401, the SpaceX vehicle was priced for 2013 on the company's website for all the world to see, at $54 - 59.5 million for a commercial flight, while a government flight would also include mission assurance fees of around 20% or more depending on the contracting agency. Pricing for the comparable Atlas is intentionally much more obscure, a habit which prompted Elon Musk to observe that it was similar to a Middle Eastern rug bizarre, where the price changed dramatically depending on the buyer's ability to pay. Although never officially revealed, the price is frequently cited in a range between $150 and $180 million. A full accounting however, would have to include a portion of the billion dollar a year launch capability subsidy which ULA still receives under a second contract, which varies depending on launch rate, but would add at a minimum another $120 million to the price, bringing the total to a staggering $270 - 300 million. Even under the most conservative estimates, the Falcon offered a three to one price differential.

For the moment, ULA's advantage lay in the fact that the Block I Falcon 9 did not possess the capability of launching the larger defense payloads. That advantage was only temporary however, as impending upgrades of the Falcon 9 to the stretched v1.1 with Merlin 1D engines, provided not only a significantly enhanced capability, but also moved the company closer to the first flight of its Falcon Heavy, which poses a mortal threat in terms of a launch capability greater than the largest ULA product, the Delta IV Heavy, at a price which is still less than the smallest Atlas V 401 configuration.

With the Falcon Heavy several years away from first flight, and even further removed from passing a multi flight threshold, purchasers and policymakers had little choice but to work with what was currently available, not what might be. Consequently, SpaceX's second attempted foray into the EELV market came against the backdrop of an intergovernmental agency effort to better coordinate launch service acquisitions as a means of grappling with rapidly escalating prices from ULA. In March 2011, as part of a draft Memorandum of Understanding between DOD, NRO and NASA regarding launch vehicle plans, DOD served notice that it intended to proceed with the previously reported plan to commit to the large, multi-year purchase of 40 to 50 launch cores from ULA, while continuing to work on criteria for eventually allowing new bidders to enter the program at a later date. Given the size of the purchase it was undertaking as well as the tepid enthusiasm for prioritizing new entrants, it was hardly a ringing endorsement for the values of competitive bidding. NASA for its part, politely declined an offer to participate, citing a different budgeting process for its decision to pursue launch providers through its own more flexible NLS II contract, which allows an annual "on ramp" for new participants as well as a defined risk based criteria for entry. With SpaceX one test flight away from making operational trips to ISS, the Falcon 9 was already in the NASA stable, certified as a "medium risk."

The proposed Air Force purchase was the culmination of an intense lobbying effort on the part of ULA, which clearly hoping to forestall any competition, had been heavily promoting the massive block buy as a means of running out the clock before any new system could go into effect. The proposal called for the Air Force to purchase 5 launch cores per year, with NRO responsible for three. With an estimated budget of $15 billion over 5 years, the program's basic numbers worked out to an average of $375 million per core. The plan finally came under scrutiny when the GAO released a report highly critical of the proposed block buy, questioning the USAF's ability to make an informed decision based on the limited information available.

Among the findings, the rationale for the block buy was based on a poorly substantiated claim by ULA that the launch industrial base would be endangered if was not allowed to proceed. This claim however, relied on a survey ULA submitted to its own vendors, complete with a cover letter suggesting how they should respond in order to support the block buy and "enhance their collective business." GAO characterized the survey as "neither designed nor administered in a manner consistent with sound survey methodology procedures." Also noteworthy was part of the questionnaire which was intended to assess short and long terms risks, in which ULA actually supplied the vendor responses because as one company officially bluntly stated "we wanted certain answers," a tactic which in the common discourse is sometimes referred to as cheating. In addition to questioning the validity of the survey, GAO also noted the company was at a loss to provide details as to why its costs were drastically increasing for many key components, including engines. Equally troubling was an inability to explain what actually constituted "mission assurance charges" despite the fact that they can make up a substantial portion of a given launch's costs, leaving the buyer without any means of determining the relative value of the charge.

In questioning the number of boosters to be ordered under the plan, the GAO report found that ULA was requesting nearly twice the number of cores that DOD had actually been using during the previous five years, and that it would be committed to purchasing five new cores for each successive year, regardless of the actual usage rate. Despite a current backlog of launches due to payload program delays, this approach would in all probability result in storage and refurbishment charges for unused cores. The result conveniently enough, would be to limit launch opportunities for new entrants (read SpaceX) possibly well beyond the 5 year time frame. GAO also warned that with much of the cost increase blamed on uncertainty in the SLS program, DOD was running a risk of locking in unnecessarily high prices even as that program began to take shape. [104]

The GAO report was quickly picked up by Senators John McCain and Carl Levin of the Senate Armed Services Committee, who issued a joint letter on October 21, 2011 advising the U.S. Air Force to hold up on its purchase decision, at least for the moment. With Senator McCain's history of scrutinizing large defense contracts, this was hardly a welcome development. For its part the Air Force appeared to distance itself somewhat from the original block buy proposal and pointed to an October 14, 2011 announcement of a New Entrant Certification Strategy, as evidence of its sincerity in opening up EELV procurement to new vendors. The strategy would be similar to that developed by NASA, seeking to develop a risk based matrix for integrating new entrants while maintaining assurance for its most valuable payloads. On November 7, the USAF released its detailed plans for allowing new bidders, officially titled the United States Air Force New Entrant Certification Guide, and announced plans for an industry day to discuss the details.

These developments, though welcomed by SpaceX, and appearing to open the door to competition, did not clearly resolve the issue of the block buy, leaving it still under consideration. Three days after it received the Senate letter, the Air Force responded by requesting that ULA provide alternate pricing scenarios based on a smaller buy beginning with a three rather than five year minimum, and with annual purchases ranging between 6 and 10 cores per year. Given the close ties between the Air Force and its major defense contractors, as well as its apparent willingness to proceed with the block buy despite the obvious problems highlighted in the GAO report, it was difficult to judge to what extent the competition were really being opened up until specific contracts were announced.

ULA was by no means the sole reason the costs of the EELV program, and U.S. access to space had soared out of the bounds of rationality, it was readily enabled by the Air Force. The EELV program was the subject of a GAO report in 2008 which pointed out that even after being granted the right to form ULA in 2006, the program was for the second time, (the first being 2004) facing cost increases which would almost certainly trigger a breach of the Nunn McCurdy act, which was put in place to prevent defense procurement programs from spiraling out of control. Such a breach would essentially put the program on probation, and subject to routine outside scrutiny. Rather than address the real problem, that lack of competition and comfortable subsidies had drained ULA of any ambition it might have harbored, the Air Force's solution to the problem was to simply re-classify the program as in the sustainment rather than development phase and thereby avoid the light of examination which comes with the latter category. Not to worry though, the taxpayer would be assured of getting proper value by the use of "should cost" analysis in which Air Force personnel would verify that the program charged no more than what it reasonably should cost. Among the many flaws with such an approach, is the simple fact that assigning hundreds of personnel to review costs is itself an expensive proposition, one which the Air Force promptly neglected by not assigning the necessary staff. The net result, as now reported in the 2011 GAO audit, was that when ULA began its campaign to secure a block buy based on protecting the industrial base, and then to justify much higher prices based on escalating vendor pricing, the Air Force lacked the necessary information to make an informed decision as to the accuracy of the claims. So too apparently did ULA, which could not offer GAO a specific reason for the selection of 40 cores as the preferred block buy.

The year 2011 ended for the EELV program with President Obama signing a FY 2012 defense authorization bill which notably included amendments from McCain which gave the Secretary of the Air Force the option of essentially reversing the 2007 decision and re-classifying the EELV program as a defense development program, or alternatively, subjecting it to similarly equivalent scrutiny as if it had been anyway. It was a New Year's resolution McCain intended to keep. Following a January 19, 2012 announcement of a ULA "bridge" procurement of 9 launch vehicle cores for 2013, the Senator issued a letter questioning the Air Force's next step, a matrix provided by ULA with different prices for different purchase levels. Specifically, McCain questioned both the reliability and apparent conflict of interest on the part of ULA in providing the information from which it stood to benefit.

With the EELV program effectively re-classified as "in development" even though it was not regarded as a weapons system, it was only a matter of months before it once again triggered a breach, the third in its history, under the Nunn-McCurdy statute even as the per unit costs for FY 2013 rose to a staggering $421 million per core. At the same time, SpaceX which had not performed a single launch in 2011, was finally cleared for the COTS 2/3 flight which took place in May. If there was any lingering doubt that SpaceX was here to stay as a major player in the launch industry, and a legitimate contender for future defense launches, it was wiped away with the stunning success of the COTS 2/3 mission, and in particular the expertise demonstrated by the company in conducting orbital operations. Furthermore, the attention surrounding the historic mission finally brought the company into the general public eye, also garnering respect from some previously skeptical members of Congress. Following the successful mission, two influential members of the House Defense committee sent a letter to Secretary of Defense Leon Panetta urging the Department of Defense to take more aggressive action in opening up launch opportunities to new entrants. The letter also pointed out the fact that the current procurement arrangement made it all too easy to conceal the true cost of the program by utilizing two separate contracts for EELV, one which procured the launches themselves, and the second more dubious Launch Capabilities Contract, or ELC, which was little more than a barely concealed subsidy.

The Air Force finally responded with a mixed and somewhat disappointing message with the public release of its New Entrant Certification Strategy. In the first place it would still be going ahead with the requested block buy, albeit at the slightly lowered level of 36 cores, which were still under negotiation with ULA. After the term of the block buy expired at the end of 2017, the Air Force would fully open the program up for competition to any vendor which had passed its qualification threshold. But here was the rub, the Air Force deliberately rigged the terms of the strategy to continue to heavily favor United Launch Alliance. In the first place, any vehicle under consideration would need a LEO payload capacity in excess of 20,000 lbs. While the Falcon 9 easily exceeded this benchmark, the next most credible competitor, the Orbital Sciences Antares did not. Furthermore, the reason cited by the Air Force for the stipulation, that nearly all of its planned payloads exceeded that amount, completely ignored the fact that future trends in satellite development could very well lead to a requirement for mid-sized launch vehicles. In yet a further blow, the Air Force also demanded that any defense launch take place from an Air Force facility, even if the provider was already using a separate facility. Moreover, it also still maintained that launch integration had to take place vertically. Horizontal integration was not acceptable even if the contractor could demonstrate that it could be performed with no harm to the spacecraft, and would save the government money. Rubbing a little more salt in the wound, even as the Air Force was in discussion with potential new providers, it held ongoing payload interface meetings with ULA in which it essentially tightened its own specifications to match ULA's, instead of pursuing open standards. And of course, looming everywhere was the specter of the direct subsidy the Air Force was supplying to ULA in the form of the ELC, or launch capabilities contract.

However, the Air Force did announce that it would reserve as many as 14 launch opportunities of less critical payloads for which new entrants could bid, provided they achieved qualification. These flights would be open to a number of vendors and vehicles, including SpaceX, Orbital Sciences, ATK, and Lockheed Martin, which was hoping to re-introduce the cancelled Athena solid cored small launch vehicle. Besides these potential flights none of which were initially identified, the Air Force offered something else, a pair of awards to SpaceX to be conducted under the entirely separate Orbital/Suborbital (OSP-3) program, which carries a lower threshold of risk. The first was for a Falcon 9, and another, later flight, dubbed Space Test Program-2 (STP-2) for the Falcon Heavy. The first award, for Falcon 9 V1.1, to be conducted in 2014 was for the DSCOVR mission, a spacecraft with a long, strange history.

Originally proposed as the Triana mission by Vice President Al Gore in 1998, and named for Rodrigo de Triana, a lookout aboard the Pinta, and the first person to spot the New World from Christopher Columbus's expedition of 1492, Triana was meant to be placed in a deep orbit, with a camera trained on the Earth at all times, providing the only 24 hour per day image of the entire planet at once, with live streaming over the internet. Invariably, it was unofficially re-designated "GoreSat," in the process becoming an improbably politically charged spacecraft. Scheduled for launch aboard the Space Shuttle Columbia, the project was put on ice, reportedly at the insistence of Vice President Dick Cheney, who objected to its environmental overtones and ties to his predecessor. Instead of being launched into space, it was quietly put into storage to wait for a better day. That day finally arrived when the Air Force decided to grant SpaceX a trial launch, and unwilling to risk one of its own satellites, began looking for a suitable payload. The Air Force sponsored test flight of the Falcon 9 presented a new opportunity, a new mission and a new name. While it will still be used as an environmental monitoring spacecraft, the Air Force, taking advantage of its somewhat unique vantage point, intends to use DISCOVR to monitor from its remote orbit, the vast array of satellites and space debris occupying numerous Earth orbits "below." It is in effect, seizing the "high ground" for a unique observation post. The second launch opportunity, STP-2 and granted to the Falcon Heavy, was not initially defined.

Following the process announced in the New Entrant Certification Strategy, the U.S. Air Force Space and Missile Systems Center announced on June 11, 2013 that it had signed a Cooperative Research and Development Agreement (CRADA) with Space Exploration Technologies Corp. paving the way for the company to take the next steps along the pathway to entering the EELV business as outlined in the New Entrants Certification Guide which was introduced in October 2011. According to the press release the CRADA:

"Enables the Air Force to evaluate the Falcon 9 v1.1 launch system according to the Air Force's New Entrant Certification Guide (NECG). As part of the evaluation, SMC and SpaceX will look at the Falcon 9 v1.1's flight history, vehicle design, reliability, process maturity, safety systems, manufacturing and operations, systems engineering, risk management and launch facilities. SMC will monitor at least three certification flights to meet the flight history requirements outlined in the NECG. Once the evaluation process is complete, the SMC commander will make the final determination whether SpaceX has the capability to successfully launch NSS missions using the Falcon 9 v1.1."

Wasting no time, SpaceX indicated that beginning with the very first launch, it intended to submit information to the Air Force as outlined in the CRADA. At the same time, a series of government documents outlining future plans for the EELV program, including a planned extension all the way to 2030, suggested that the need for new entrants was greater than ever. Projecting numbers cited in the 2012 Selected Acquisition Report which was released to Congress in May 2013, the Department of Defense estimated the Air Force would need to spend $39.9 billion to conduct 2 additional launches in FY-2017, and 56 new launches between FY-2018 and FY-2030, a staggering average of $687 million per launch. The reality gap between what the Air Force was projecting, and what SpaceX was charging, as evidenced by a manifest of more than 40 launches, simply beggared belief.

Inexplicably, even as SpaceX pursued the Falcon 9 development program in earnest beginning in 2006, and backed by NASA's own expertise, which was then reinforced by five successful flights of the original Falcon 9 booster, the Department of Defense assumed that prices would go on rising, even in the face of incontrovertible evidence that SpaceX was effectively driving them lower and from a global perspective, no launch provider, including market leading Arianespace, whose own track record of success exceeded that of the EELV boosters, was anywhere near as expensive as the DOD program. Perhaps it stemmed in part from an understanding that between the unwieldy Delta series and a dependence on Russia for the Atlas V's engine, ULA had no real answer for the challenge posed by SpaceX, and dramatically lower pricing other than a further appeal for continued monopoly, which its benefactors in the Air Force were willing to accommodate just a little longer. In many ways, it mirrors a looming credibility crisis created by SpaceX on the civilian side of the space program, where the performance and pricing of the Falcon Heavy undercuts virtually the entire rationale for the Space Launch System.

The one critical distinction is the fact that while an aggressive program of human deep space exploration is a debatable luxury, secure and reliable access to space for national defense purposes is an absolute necessity, providing a partial explanation for the defense establishment's seeming reluctance to embrace the opportunity offered by the Falcon family of launch vehicles until fully proven. Still, with a large flight manifest and intensive vetting by highly sophisticated commercial customers, as well as NASA's own confidence, the risks posed to national security by a problem with the Falcon 9 are shrinking rapidly, even as the credibility gap created by dependence on the Russian RD-180 main engine continues to grow.

That dependence was dramatically reinforced when Orbital Sciences both filed a protest with the FTC, and then sued United Launch Alliance in June 2013 over alleged efforts to prevent it from acquiring access to the same RD-180 engine used on the Atlas V. Although OSC scored a successful demonstration launch with its Antares booster powered by reworked Soviet era NK-33 engines, and then followed it up with a successful second launch which placed the first Cygnus on course to the ISS, its future was limited by the finite stockpile of forty year old engines. According to court documents, OSC had originally approached ULA regarding access to the engines, which are imported into the U.S. by RD-Amross, a 50/50 partnership between the manufacturers N.P.O. Energomash, and Pratt and Whitney Rocketdyne. After the expiration of a 5 year exclusivity contract allowing Lockheed Martin to recoup its initial outlays, OSC presumed that the engines would now be available for other launch providers. Instead, ULA informed OSC that it was unilaterally extending the contract for an additional five years. When OSC noted that it would not need the engines prior to 2016 and the end of the second contract anyway, ULA responded that it was levying yet another extension. Exasperated, Orbital Sciences filed suit over what appeared to be a blatant and illegal attempt to restrict trade, asking for both a jury trial and a consideration of trebling of more than $500 million in damages from lost business under RICO statutes. The threat apparently worked, as OSC subsequently announced that it was suspending the suit and negotiating for the engines, but reserved the right to re-instate if need warranted. Whatever the ultimate result, the spectacle of two American companies fighting over a Russian main engine for their respective launch vehicles has underscored just how badly the EELV program has failed to protect the American industrial base, and just how important the introduction of new all U.S. sources in the Falcon series is to the same base. That it comes at a much lower launch price, and almost no development cost to the Defense budget is more than enough justification to re-consider the dubious benefits of a monopoly which has seen costs rise to unprecedented levels.

Throughout the process of rolling out its decidedly wary New Entrant Certification Strategy, the Air Force has remained somewhat guarded as to how many flights SpaceX will have to conduct before it can launch the most sensitive payloads, beginning with a refusal to discuss which payloads fall into which categories of risk, all the while hinting that nearly everything would be in the highest risk category. ULA for its part, has helpfully suggested that SpaceX should be required to complete at least 21 flights, while also implying that every last payload should be considered to be in the highest risk category, and thus off limits until that threshold is met. That its own Delta IV flew an operational mission after only one precursor flight is of course somehow irrelevant.

Ultimately, the outcome depends on SpaceX. While the "playing field" is hardly level, SpaceX now has the opportunity make a definitive statement with the early flights of the Falcon 9 v1.1. If they go well, continuing the streak of successful launches which began with the original Falcon 9, then the company will have passed any reasonable threshold, and most unreasonable ones too, well in advance of the first date that full and open competition takes place. On the other hand, an early stumble could provide a convenient excuse to perpetuate the ULA monopoly even longer.

To some, SpaceX's difficulties in breaking in to the EELV market could perhaps be dismissed as relatively inconsequential, particularly considering the fact that the company has already secured an enormous backlog of commercial launch orders which will carry it to nearly 2020 in any case. There are several ramifications worthy of further consideration however, beginning with the Commercial Crew program. With the two remaining funded contenders, Boeing and Sierra Nevada both using the Atlas V as its chosen launch vehicle, SpaceX is facing a competitor which is still receiving an outright subsidy of much of its fixed costs, creating an artificial narrowing of the price spread between the two vehicles. In the event Congress forces a single winner outcome, this artificial price support could well play a critical factor.

There is another angle as well, and one which boils down to a simple issue of rank hypocrisy. As detailed in the GAO audit of 2011, ULA based its justification for a block buy and continued high pricing on the premise that it was necessary to protect the shrinking U.S. industrial base. This might be a partially credible argument, if ULA, and Lockheed Martin in particular had not directly contributed to destruction of the same industrial base by continuing to import the Russian RD-180 main engine, while neglecting to exercise its option to license US production of the engine, a possibility the company periodically brought up whenever someone questioned the wisdom of continued dependence on Russia. Rather than doing so, it elected to keep supporting the Russian industrial base at the expense of the American, all while doing everything it could to prevent the entry of SpaceX and the one new hydrocarbon engine developed in the US in 40 years. Furthermore the Atlas V uses other significant foreign components in the form of both the interstage connector and the large payload fairing provided by RUAG. While there is nothing wrong with using foreign components for civilian items, it is something else entirely to demand an outright subsidy, as well as monopoly privilege, all of which places a burden on the taxpayer and reduces funds available for other defense procurement, while attempting to prevent the entry of a lower cost, all U.S. provider to a domestic industrial base, which you claim is so endangered it needs subsidizing in the first place.

For SpaceX, the real significance of securing EELV business lies not with the Falcon 9 as much as it does with Falcon Heavy, and the potential influence on its ultimate cost. In unveiling the Falcon Heavy, Elon Musk suggested that its final price would depend on the flight rate which it achieves, with a minimum target of four flights a year necessary to maintain the $125 million baseline announced in 2011, adjusting annually for inflation. While SpaceX would also be pursuing commercial orders for the Falcon Heavy, securing EELV business represents a means of almost guaranteeing the flight rate required to keep the booster at the breakthrough level of nearly $1000 per lb. making it the springboard for an entire series of exploration missions, many of which are privately funded, which were otherwise inconceivable.

It is also not difficult to see why it was so very important to ULA to categorically prevent the entry of the Falcon Heavy for as long as it possibly could. Although it goes to great lengths to conceal its prices, each EELV flight vehicle secured by NASA offers a rare bit of solid data. In the case of the Delta Heavy, Lockheed Martin successfully convinced NASA of the need to conduct a test flight of its Orion spacecraft aboard a Delta Heavy, before its first flight on the SLS. The Delta IV sold to NASA for the 2014 test flight was priced at $375 million, three times the price of the higher performing Falcon Heavy. Taking in to account the ELC subsidy however, the actual cost of a Delta IV to the Air Force, or to the NRO, rises to roughly $500 million.

While the Air Force does not owe SpaceX, or any other launch provider orders as a matter of industrial policy, it does owe the taxpayer and the nation it defends the best return on its defense dollar, and the best defense posture possible. Aside from other matters, there is no doubt that the ballooning costs of the EELV program have severely impacted overall space defense spending, and thus national security, as progressively more resources are required to merely get payloads to orbit. For instance, in August 2013, the Air Force permanently shut down the system of ground based radars, informally known as the "Space Fence" which monitor objects in orbit, due to lack of funding. Although the Air Force blamed sequestration for the decision, the fact remains that the amount in question was substantially less than the EELV subsidy.

Yet another Air Force program, this one aimed at partially bypassing the expense of dedicated defense satellites by placing military components onboard commercial satellites, was also cut short due to funding. The irony in this case was especially rich. Commercially hosted payloads not only offer the benefit of lower prices, they also make for a resilient, dispersed system which is more difficult for a hostile power to disrupt, further undercutting the entire rationale for overpriced EELV launchers. Nevertheless, if the budget amounts put forward in the DOD's long range strategy for EELV are any indication, the future holds quite a bit more of the same. Regrettably, there is absolutely no reason this should be the case.

For the moment however, the issue is very much unresolved, but that moment is passing. With a successful debut for the Falcon 9 v1.1 under its belt, and a long client list awaiting launches, it is patently unreasonable to assume that even in the event of an early mishap, the Falcon 9 will be anything but an extremely reliable booster readily available for DOD missions well in advance of the first available flight opportunity. But, as the history of the EELV program strongly suggests, the bond between the military and its closest contractors is difficult to break, no matter how much sense it makes.

As should be plainly obvious however, the issue of the EELV program and its effect on domestic launch vehicle pricing is one of the key contributing factors to the "malaise" in the U.S. space program into which Elon Musk and SpaceX stepped with the introduction of the Falcon 9. It is not just the effect on the military, after all, who among us is not used to apocryphal stories of $500 hammers and $5000 toilet seats? The public has come to accept and even expect a certain amount of inefficiency with the military industrial complex, even as it also expects technical brilliance. It would be a different matter however, if the price of hammers to the public was $300 dollars each, all due to a nearly unbreakable monopoly granted by the Department of Defense. In a way, this is just what has been the case with launch vehicles, particularly where NASA science launches are concerned. From the perspective of the space enthusiast, the great sin of the EELV program is not that it is wasting defense dollars, or even that over a five year time frame it nearly doubled NASA's launch costs for science payloads. Instead the greatest problem looking forward is that by establishing an apparent price point for space launch well above what by all rationale means it should have been, the unfailing result of any monopoly; the EELV program, its contractor United Launch Alliance, and the two companies which own it, Boeing and Lockheed Martin, substantially altered the perception of what the nation could afford, and what it could not, in re-setting its priorities in the post Shuttle era. In doing so, it may have opened one door for SpaceX through its New Entrant Certification program, but at the same time it set the stage for the return of the mega booster in the form of the Space Launch System, and a growing political controversy which may prove to be every bit as much of a challenge to SpaceX moving into its second decade, as simply achieving orbit was during its first.

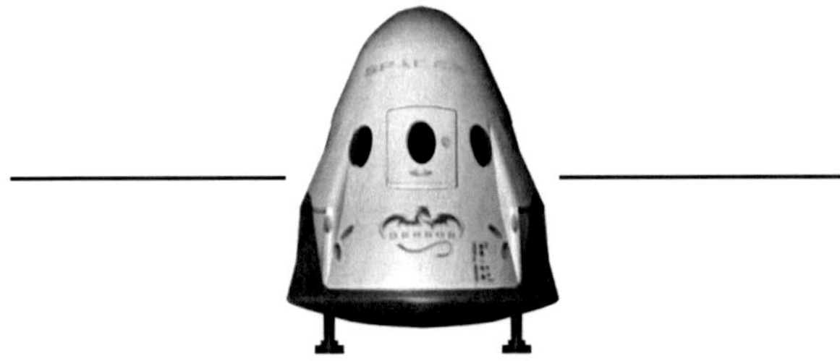

Chapter XXIX: Raising Questions

SpaceX in the second decade of the new millennium is the undeniable darling of the rapidly emerging NewSpace industry, and seemingly the answer to nearly every space enthusiast's prayer for a better and quicker route to a frontier which has been kept tantalizingly at bay for far too long. While the company appears poised to fulfill that promise with the inauguration of a transformational launch vehicle in the Falcon 9 v1.1, making the most of that opportunity, and fulfilling its own ambitions will depend on more than its technical accomplishments.

The rise of SpaceX would not have been possible without well-crafted and competently managed public/private partnerships at key points throughout its history, and that is one factor of the real rocket equation which is not likely to change. Despite the occasionally brash approach of its founder, and positively impudent cheerleading from space enthusiasts on the sidelines, from the outset SpaceX cultivated and employed the communications savvy to develop highly productive relationships with the people who mattered, the various government agencies who had the power to help advance or to stall the SpaceX agenda.

Early on, the Office of Secretary of Defense was crucial in providing the all-important first launch order for the Falcon 1, for a satellite which in fact never left the shelf, even as DARPA ultimately underwrote the booster's first two flight attempts. Despite the frustrations stemming from the Air Force's reassignment of the Vandenberg launch pad to Lockheed Martin, the Air Force provided SpaceX with invaluable assistance in multiple regards, both indirectly as a customer in its own right, beginning with Falconsat-2, which was lost aboard the company's maiden flight, and directly, when push finally did come to shove in the first Falcon 1 launch campaign, with the provision of a chartered C-17 transport aircraft. Its contribution to the Falcon 1 program pales in comparison to the crucial role the Air Force played in helping to secure the SLC-40 site at Cape Canaveral for the Falcon 9, an arrangement which no doubt irked its own main contractor United Launch Alliance. Furthermore, the USAF provided a vital, and perhaps last lifeline through the use of its high power satellite communications network as the CRS-2 Dragon spacecraft was running out of options. Going forward, SpaceX will still face an uphill, and in many ways blatantly unfair struggle in its ongoing efforts to secure EELV launch orders from the DoD against a competitor receiving a direct subsidy. Finding the right balance of aggressively pushing to enter that segment without alienating the decision makers who sign the orders will continue to be a challenge, one compounded by the pressure of maintaining a 100% mission success record.

If there is one critical inflection point in the company's history however, it is without a doubt the opportunity presented in 2005 when incoming NASA Administrator Michael Griffin elected to resurrect a nearly dead program aimed at partially re-supplying the logistical needs of the International Space Station by utilizing commercial providers in an entirely new way. In winning COTS, and then securing 12 more launches as part of the Commercial Resupply contract, SpaceX gained the flight opportunities necessary to secure the future for the Falcon 9 booster. For other companies, it might have been enough to settle into a comfortable period of guaranteed launches, while modestly seeking opportunities to enter into other markets, just as Orbital Sciences Corporation has done with its Antares booster. For SpaceX, it was only a start, and the finish line is 225 million kilometers away.

Perhaps fittingly, the two disparate systems, as well as the COTS program which supported them, reached critical milestones within hours of each other on September 29th 2013, a day the space media dubbed "Super Sunday." The maiden launch of the Falcon 9 v1.1 and OSC's successful berthing of its own Cygnus re-supply vessel at the International Space Station marked a red letter date in space history, with the former paving the way for the first commercial mission out of Cape Canaveral, one which serves as a thoroughly serviceable bookend to the first era of SpaceX, and of the NASA program which defined it.

Aerial view of SLC-40.

The success demonstrated by SpaceX, as well as by Orbital Sciences, has undeniably answered the one principle question raised by NASA in initiating the COTS program, "Are commercial companies capable of providing services to the International Space Station?" As so often happens with most intriguing questions, the answer, a resounding "yes," only serves to raise more questions, a lot more. Three however, stand out, and it is the answers to those three questions which provide a useful lens through which to consider the potential changes to the future of America's space program made manifest in the introduction of the first generation Falcon/Dragon system, and the almost dizzying potential in the second generation systems which are already entering into service.

The first relates to the Commercial Crew program. Here, the issue is clearly more political than technical, coming down to a simple question of whether or not Congress will adequately fund the program all the way through to completion, whether or not SpaceX will be one of the winners, and will NASA administer the program under FAR contracting with a "light hand?"

The answer is murkier than it should be. Had Congress indicated its willingness to support the program along the lines requested by the Obama Administration, and demonstrated a fidelity to the program's stated intention of producing at least two winners by fully funding it from the inception point beginning with Recovery Act funding in 2010, then the pace of the program and the selection of SpaceX for one of the slots on the basis of its merits would seem to be a foregone conclusion. Regrettably, this does not appear to be the case. With political opposition to the program unmistakably driven by the nation's long standing Shuttle constituency, reborn in the SLS program, and seeing an existential threat in an alternative approach to space acquisition and contracting, the opportunity for ongoing interference is ripe. Furthermore, two major contractors, Boeing and Lockheed Martin, are playing on both sides of the fence and the gaping disparity in the levels of funding involved should leave doubt as to where their loyalties lie, and where their political contributions are directed. Ironically however, the best hope for Commercial Crew may also reside in the fact that even though it is in some ways competing with SLS for funding, the International Space Station is like any large program, difficult to kill, particularly when taking in to account the fact that a major recipient of ongoing engineering contracts is Boeing itself.

Clearly, the future of the Commercial Crew program is inextricably intertwined with the fate of the International Space Station Program, and will rise or fall on that measure. Originally scheduled to be de-orbited shortly after 2015 in order to make room in the budget for Project Constellation, ISS is now assured of continued operations through 2020, and as of 2014 the Obama Administration has signaled its willingness to extend the Station to at least 2024, a critical but only partially settled decision which will still have to be supported by the next President, and funded by a future Congress. It will also require the consent of Russia, and likely the continued participation of other International Program Partners.

Given the schedule slips in the Commercial Crew program intentionally inflicted by Congress, even though SpaceX could potentially be in a position to fly a test crew aboard DragonRider as early as 2015, NASA does not envision beginning operational crew transfer flights until late 2017. Any further schedule slips, and there will be plenty of chances both technical and political for them to arise, could easily push the program so close to the 2020 timeframe that if Congress fails to extend the life of the Station to 2024, Commercial Crew could be killed with depressingly little fanfare. After all, by the earliest time a NASA chartered Commercial Crew flight could take place, American astronauts will have been launching exclusively aboard the Russian Soyuz for the previous seven years. What is three more, particularly if the only destination is a twenty year old space station scheduled for destruction?

Alternatively, an affirmative decision among the ISS partners to extend the Station's life to 2024 or 2028, would almost just as easily almost guarantee the future of the Commercial Crew program and not insignificantly, secure eight more years of support engineering contracts for Boeing. If recent experience is any guide, it seems much more likely that rather than making an affirmative decision one way or the other, Congress will delay the issue for as long as possible, essentially kicking the can a little further down the road, and in doing so create the window NASA needs to get at least one system operational.

Whether that system belongs to SpaceX, Boeing, or perhaps less likely Sierra

Nevada, will no doubt be a matter of intense debate (and GAO protests) but one thing seems clear, with NASA facing a Hobson's choice in the Congressionally driven funding priorities between SLS and Commercial Crew, costs are going to matter, and for whatever funding Commercial Crew does ultimately receive, the much higher price of the Atlas V is a major issue. While there is something perhaps a little unsettling about discussing NASA human spaceflight policy in terms of the cost per seat, that is precisely one of the measures by which the selection will be gauged, if not actually decided. Established by Russia, and funded by the American taxpayer, the cost per seat aboard Soyuz has steadily increased in the post Shuttle era, rising to $70.7 million for flights in early 2017, and it is difficult to see how the Commercial Crew program's costs will not be compared against that number. [105] How is it stacking up?

Currently, the Station can only host a compliment of 6 full time crew members due to the limitations of the two three person Soyuz spacecraft which remain docked throughout a given crew's mission to serve as rescue vehicles. Here is where it gets interesting, and for Boeing and Sierra Nevada, problematic. The current crew rotation calls for two flights a year, supporting a crew complement of three Russians, two Americans and generally either a single Canadian, European or Japanese astronaut. Although seldom noted even in the space press, the United States pays the launch costs for non-Russian partner astronauts. With the introduction of U.S. crew transfer vehicles, each of which can hold up to seven astronauts, the ISS partnership can finally reach its intended full time compliment of 7 members, an increase which will have a profound effect on the amount of man hours available for research each week.

However, once a new provider comes on line, NASA is likely to loft only four crew members per mission in order not to exceed the maximum nominal crew complement of seven, meaning that for comparative pricing purposes, four is the relevant number. Based on firm quotes provided by both SpaceX and Boeing which were released by Bigelow Aerospace in early 2013 as part of the opening round of solicitations for its own planned station, there is already a substantial price spread between the CST-100 launched aboard the Atlas V, and the Falcon/Dragon combination. The actual price difference, $26.25 million versus $36.75 million per seat marks a substantial spread between the two companies and begs the question of why anyone would pay more to book a flight on the Boeing system if SpaceX is an option. [106] In terms of NASA's Commercial Crew program however, it is also indicative of an overall per flight difference between the two proposals of $140 million, a number already cited by SpaceX, versus an imputed figure of $257 million for the Boeing/ULA combination. For a four passenger flight, the result is $64.5 million per seat to fly on the Atlas V, an amount perilously close to what NASA is already paying Russia.

SpaceX by comparison would come in well under the Russian Soyuz at $35 million per seat on a similar four astronaut rotation. Either way the $200 million plus differential between the two systems based on two crew rotations per year argues convincingly in favor of SpaceX and DragonRider, a difference which will be further magnified when considered over a multi-year contract. There is also likely to be a significant difference between the entirely separate costs of developing the respective systems during CCtCap, numbers based on confidential estimates submitted to NASA during the bid process which will prove to be the determining factor in the agency's decision to support two winners for development, or down select to a single provider. It is likely for this reason that Boeing has taken pains to emphasize the fact that it is still holding open the option of launching the CST-100 on SpaceX's Falcon 9 after it is qualified aboard the Atlas V, a somewhat ironic statement from a company with its own booster in the Delta IV, but made possible by SpaceX's significantly lower costs and its commitment to offer the Falcon 9 to all comers at the same internal price it charges itself to launch the Dragon.

Beyond the immediate benefits, there are also the intangibles. Elon Musk's company took a major step forward with the successful introduction of the new Falcon 9, and in its willingness to choose more aggressive goals under CCiCap, including both launch pad and in-flight abort tests, critical milestones which neither of its competitors was willing or able to take. In doing so, the company demonstrated a commitment to its system and its actual cost effectiveness which will not go unnoticed. Furthermore, in extending its plans to include powered landings of the Dragon capsule, as well as powered return of the Falcon 9 first stage, even if it is not employed on crewed flights, SpaceX is offering a national capability and a safety redundancy so far ahead of its competitors, that there is no question that in the case of a single funded winner, who that should be. Barring an unforeseen and uncorrectable critical flaw in the booster itself which appears before a decision is reached, a failure to select the Falcon/Dragon combination would not only raise serious questions about the nature of the process, it would also set the stage for a major, and quite possibly irrecoverable PR disaster for NASA itself, as SpaceX went its own way.

The second issue raised by SpaceX's success with the Falcon/Dragon is that of its potential to expand the range of LEO operations beyond ISS. For all that COTS has achieved, it is important to remember that as originally outlined, the program justification embraced the idea of government playing the role of anchor tenant in LEO, but with the clear expectation that in the long run, it will not be the only one. It is a policy explicitly meant to help break the chicken and egg dilemma which has plagued commercial space for decades.

While critics of the Commercial Crew program in Congress have repeatedly seized on this aspect of the program, and the lack of a pre-existing market as evidence that the assumptions are unfounded, NASA has remained steadfast. It appears that faith may be rewarded. The rise and perseverance of Bigelow Aerospace, culminating in NASA's decision to test the company's BEAM module at ISS in 2015 gives a reason for optimism. Whether it is as early as 2020 or as late as 2028, at whatever point ISS reaches the end of its operational life, in signing the BEAM agreement, NASA has already established a working relationship with the company which could readily lead to a gradual transition to ongoing agency sponsored research aboard a Bigelow station. It is after all, unlikely that at whatever point ISS is de-commissioned, NASA will have learned all it ever needs to know about living and working in a zero-g environment, and if current plans are maintained, the agency will have no direct avenue for pursuing further research.

In the long run however, Bigelow's success, which is absolutely dependent on the introduction of Commercial Crew systems, will be determined by the value its prospective customers; sovereign governments and private corporations, place in having access to extended stays in LEO. Even here, the ISS program has an enormous role to play. Although it has yet to produce the type of breakthrough advancements one might hope for, research aboard ISS, hampered for so long by the slow pace of construction and limited transportation options, is indisputably beginning to produce results, particularly in the burgeoning field of protein crystallography for mapping disease pathways. The number of biomedical firms pursuing this research is steadily rising, in no small part due to the contributions from both SpaceX and Orbital Sciences in terms of transportation, entrepreneurial NewSpace companies such as NanoRacks in logistics, and the wider range of research sponsored by CASIS. Both the research space and the costs of working aboard ISS are limiting factors however, and ultimately the opportunity to conduct additional research, at a more rapid pace and at significantly lower costs than can be performed aboard ISS would seem to be an early commercial driver for Bigelow, complementing its initial focus on non-traditional space faring governments. As for the skeptics, who will continue to be correct until suddenly they aren't; if a failure to achieve established goals within a specified timeframe was a reason to throw in the towel, then the entire aerospace industry, both old and new, would be obligated to give up and stop trying, because with the rarest of exceptions, it almost never happens. Fortunately, that is not the case, and if there is any one lesson to be gained from the history of space commerce thus far, new projects always cost more and take longer than forecast to come to fruition, and new markets emerge slowly and with great trepidation. After more than 30 years of effort which began with the Space Shuttle and Spacelab, privately funded research in Earth orbit is at last beginning to find its way.

There is of course one largely untapped market for services to low Earth orbit which is not dependent on any scientific or manufacturing breakthrough whatsoever. All space tourism requires is substantially more affordable access to orbit. If and when that develops, the permanent expansion of human spaceflight in LEO, the unspoken subtext to COTS, is certain to occur, and this is one area where SpaceX is by no means the only company actively at work.

Thus far, precisely seven people have made privately funded trips to orbit, and each was conducted aboard the Russian Soyuz. It must be a worthwhile experience, as former Microsoft executive Charles Simonyi made the trip twice. Only one other person, international soprano singer Sarah Brightman is currently scheduled to make a "spaceflight participant" trip to ISS, a limitation driven both by the exorbitant costs, and more importantly the lack of available seat space aboard Soyuz. The logistics will change with the introduction of Commercial Crew, as additional seats aboard both Soyuz and the U.S. provider become available. Furthermore, in stark contrast to the Shuttle era, the Commercial Crew program explicitly recognizes that its providers may seek to bring additional non-NASA personnel aboard. Nevertheless, it will still be an expensive ride.

Virgin Galactic, XCOR Aerospace and Blue Origin are each pursuing the quest for affordable space transportation but working from the opposite end of the problem, beginning with reusable suborbital systems established explicitly for space tourism, and hoping to ultimately extend their reach to orbit. In contrast to the seven orbital trips, more than 900 people have already paid substantial deposits for suborbital journeys and the long delayed inauguration of service is expected to begin in 2014, with the always flamboyant Richard Branson making the maiden flight on his namesake spaceline accompanied by his members of his own family. Furthermore XCOR, with its two seat rocket powered spaceplane is set to begin its own operations shortly thereafter. While both are committed to ultimately achieving orbital operations, it would seem that route chosen by SpaceX, beginning with expendable boosters designed for existing markets and then transitioning through partially reusable solutions to Elon Musk's goal of a rapidly reusable space transportation system is the one likely to bridge the gap between the Russian Soyuz and the new U.S. companies, and in the process benefit all parties. Undeniably, it still has a long way to go.

The established "seat price" of $26.25 million per person aboard a Dragon bound for a Bigelow station establishes a price point affordable to most national governments and even some corporations, but it is also so far out of reach as to be meaningless to all but the smallest cadre of prospective space tourists, likely comprised of only a few dozen individuals worldwide. On the other hand, the introduction of even partial re-usability for the Falcon 9 offers the potential for measurably increasing the size of the overall market, while also helping to give it a destination other than the insides of a somewhat cramped capsule. In the process, it is likely to secure a healthier business climate for Bigelow, providing the bootstraps for pulling up an entire industry.

Based on the surprising success of SpaceX's first attempt to recover the Falcon 9 first stage during the CASSIOPE flight, the advent of partial reusability appears to be much more a matter of "when" than "if." Once that moment arrives, the economics of space launch begin to accelerate a long waited shift already begun with the baseline Falcon 9, creating what Elon Musk has described as a "forcing function" in which other launch providers have no choice but to follow suit or face extinction. Though progress will doubtlessly still be slower than what many would desire, it will also be relentless. As prices come down, the market for space tourism will increase.

The final question raised by the success of SpaceX in the context of NASA's COTS program is perhaps the most profound, and it is one which goes directly to the core of the single biggest challenge facing the U.S. space program and a clash in the making since the moment the Falcon Heavy was introduced. Having demonstrated the ability to deliver cargo to LEO, and with the near certainty that on a technical basis, crew transport services may soon be secured as well, it would seem to follow that despite the arbitrary line drawn by NASA in delineating the boundary between LEO and the space beyond, that the COTS format can be extended to support missions beyond low Earth orbit. The question is, will it?

In order to consider the expanded role SpaceX, Orbital Sciences, and Bigelow as well as any number of other commercial suppliers including "Old Space" companies could play supporting NASA in the coming years, it is important to first identify what is called for in existing plans and policy, what NASA calls the "program of record" as a point of departure. It is a painful exercise.

As of early 2014, NASA has only one planned crewed flight of the Space Launch System, notionally slated for 2021, but considered likely to slip at least a year. The purpose of that mission is either a trip around the Moon, or much less likely, a rendezvous with a small asteroid which would be robotically captured and delivered to the vicinity of the Moon. The latter option, known as the ARM, or "Asteroid Redirect Mission" was proposed in early 2013, and is the subject of a spirited debate. Although an interesting and undeniably creative proposal, it seems likely that NASA has neither the time, nor the budget to first identify a target asteroid, mount a mission to reach it, and then perform an automated rendezvous and "capture" using untried technology and techniques, only to then commence a slow cruise to the vicinity of the Moon powered by a solar electric transfer vehicle. With the ARM facing stiff Congressional opposition in addition to its myriad technical and logistical challenges, the first mission for SLS will almost certainly be a trip around, or in the vicinity of the Moon.

As a "shake-down cruise" of SLS, a lunar circumnavigation seems a reasonable proposition, although it has been pointed out that with only one unmanned test flight of a partially completed system, (the Orion will *still* not be fully outfitted) slated for 2017, NASA is accepting a higher level of risk than it did with the sequential flights of the Apollo program. In terms of actual accomplishments, the first mission for SLS may be somewhat less than inspiring, repeating what the United States previously accomplished during the flight of Apollo 8 in 1968. Fifty years after the fact, it is also likely to be perceived as somewhat of a hollow achievement, rendered vacuous by the striking absence of a comparable next step, the triumphant lunar landing which came seven months after the first time we saw this movie. Instead, NASA will have the unenviable task of consoling the taxpayer that half a century after American astronauts last orbited the Moon, and despite the enormous expenditure required to stage the sequel, the nation has no plans, and more importantly no capability, to land there once again. Instead, it might journey to an asteroid sometime in the coming decade, or perhaps the one after that. With no sense of a greater purpose, and no credible context, it is the cost of the mission which is likely to make the greatest impression. By the time SLS launches its first crewed flight in 2021 or shortly thereafter, the United States will have spent somewhere in excess of $30 billion dollars to achieve it. [107] Even if there were no plausible alternative, the perception of cost to value is not likely to be a favorable one.

The problem is there are alternatives. On its first commercial launch, the Falcon 9 v1.1 second stage propelled the SES-8 satellite into a highly elliptical orbit which took it nearly a quarter of the way to the Moon. Though a quirk of the orbital insertion plan dictated by SES, it was also a harbinger of things to come. Unless SpaceX falters very badly in the coming years, and in particular the Falcon Heavy fails outright, prompting the company to completely abandon its pursuit of one half of the entire commercial satellite market, by the time SLS finally lifts off with a crew aboard, it will be on a mission easily duplicated by SpaceX hardware at a fraction of the cost, a fact the press is not likely to overlook. Then what? NASA will be faced with attempting to defend a program it did not want, and an agenda it cannot afford, against a newer, smarter and far more affordable alternative which the agency itself helped to create. Moreover, it will be happening not in the environment of 2014 in which only Russia and China are launching people to low Earth orbit, but instead in a very different scenario in which both private citizens and NASA astronauts are launching aboard one or more commercial vehicles, and hundreds or even thousands are making suborbital flights. In short, the optics are going to be awful.

For its part, SpaceX is understandably reticent to say anything about the looming conundrum, and indeed it would be both highly improper and foolhardy to do so in anything but the most carefully parsed manner. Fortunately, the company does not have to say anything at all, in part because so many other parties are addressing the issue and also because in a very real sense, it is not SpaceX's problem. SpaceX and NASA may or may not work together in deep space exploration in the future, but either way, it is abundantly clear that the company will proceed in some fashion independently of NASA in any event, so why bruise what has been a remarkable relationship?

As every year goes by, an overall expansion of space commerce brought about by falling launch prices and increased access is likely to make the relative value proposition of follow on flights of SLS and Orion increasingly tenuous. That however, is not the reason there is a distinct lack of defined missions for SLS beyond 2021. The reason instead stems from the unacknowledged (officially at least) reality that due to its extreme development and flight costs, there are no foreseeable funds to develop payloads or conduct meaningful missions. The simple fact is that while NASA can very well oversee the successful design, development and initial launches of SLS/Orion, reveling in the brilliant flame and ear splitting roar which may momentarily silence even the most jaded of critics, (really, who doesn't love a good rocket launch) the traditional sole source, cost plus fixed fee format which is the absolute opposite of the successful approach pursued in COTS/CRS, is threatening every other budget account in the agency, including that for SLS sized payloads, and it is still not enough. Within that context, and with no defined missions beyond simply launching, for the time being, there is literally nothing to which commercially provided support missions could be attached and no program for developing them.

Given the fact that SLS marks the second consecutive time NASA has undertaken what multiple independent bodies have assessed as an unaffordable, or at least highly problematic massive booster project, and the current effort is a reprise of the first, it is worth examining how we got to where we are. It is the only context in which to understand why the Congress, much of NASA and almost the entire traditional aerospace establishment is so committed to ignoring the revolution in launch costs brought on by SpaceX which has brought the barbarians to the gates of the established order.

After the Obama Administration cancelled Project Constellation, and America's plan to return to the Moon as a step on the way to Mars as unaffordable, one of the key findings of the Augustine Committee, its official position was to adopt one of the five alternatives put forward from that same group. As the most recent major, independent assessment of America's manned space program, its findings are still relevant, even as SpaceX in just four years has demonstrated a number of its assumptions to be highly debatable, if not flat wrong.

The committee reviewed the existing "program of record" as well as alternative plans, under two budget scenarios. In addition to looking at the issue financially, it also examined possible goals in terms of the primary destination or area of interest. While coming to the conclusion that Mars has been, is, and will be the primary focus of the U.S. manned space exploration, it also found that there was no practical way to adopt a "Mars first" program, as some interests advocated. With Mars out, at least for now, that left either a return to the Moon, or a series of missions into deeper space, including the LaGrange points, a near Earth asteroid, Martian orbit or the diminutive Martian satellites of Phobos and Deimos as target destinations. The critical distinction between the two alternatives is that the latter, what the committee deemed the "Flexible Path" to Mars, simply eliminated the necessity to design and build hardware to land on the Moon. Although they would not take place for many years, trips around Mars or to its tiny, low gravity moons did not require the hardware expense of a lunar landing program.

In addition to the overall budget, and choice of destination, the other major area of distinction involved which vehicle would serve the need for heavy lift; some form of the Ares-V as envisioned by Constellation, a less capable Shuttle based booster, or possibly a new heavy lift booster based on existing EELV's. Other considerations were extending the ISS program to 2020, and whether or not to endorse a Commercial Crew program.

The selected option, which was clearly favored by the Committee as well, was the "Flexible path to Mars" which would see the agency forego the Moon's gravity well entirely, and focus instead on a gradual series of deeper forays into Cis-lunar space. This was not because the Moon was no longer an interesting destination. In fact based on research published after Project Constellation was conceived, the discovery of water ice at both poles and in certain mid latitude regions as well has arguably made the Moon a far more compelling destination than it ever has been. Not to worry, we were going to an asteroid instead.

Initially this was taken to mean as the President suggested in 2010, a near Earth asteroid mission sometime in the mid 2020's, to be conducted after the nation updated its technological toolbox and worked to perfect its understanding of the microgravity environment aboard ISS with an extension of its operating life through at least 2020. As the committee determined, even if NASA had gone through with its Constellation driven decision to terminate the ISS program in 2016, and even taken together with the budgetary savings gained from retiring the Shuttle in 2011, the costs of developing and operating the Ares-1 and Ares-V vehicles were so high that the commission concluded "there are insufficient funds to develop the lunar lander (Altair) and lunar surface systems until well into the 2030's, if ever." [108]

Instead, the committee found that under the then current FY-2010 budget scenario, the only plausible alternative was to extend the Station to 2020, cancel the Ares-1 and Ares-V, and instead produce an "Ares-V Lite," a 130 ton capable version of the booster which could perform lunar missions under a two launch scenario. Even then, the United States would see the next two decades slip by with little advancement because as the report went on to glumly assess, "this option does not deliver heavy-lift capability until the late 2020's and does not have the funds to develop the systems needed to land on or explore the Moon in the next two decades." [109]

The remaining three options presented by the Augustine committee; an "executable" version of Project Constellation's Ares-1/Ares-V "program of record," an Ares Lite "Moon First" scenario and the "Flexible Path to Mars," all depended on a permanent $3 billion per year expansion of NASA's budget, and regular (2.4%) cost of living adjustments going forward. It never happened. In fact, by 2012 NASA's budget had remained flat or decreased since 2010, falling to its lowest since 1960 when expressed as a percentage of the total Federal budget.

Though there was much to criticize in the committee findings, beginning with an endorsement of the Orion crew vessel as acceptable, and ending with a failure to take into consideration the type of reduction in costs to orbit already underway at SpaceX, which would have made the fifth option, a mission architecture based on heavy lift EELV's achievable, the "flexible path" at least attempted to match resources to expenditures.

What ultimately resulted from the controversial rollout of the President's new policy in 2010 was not the flexible option put forward by Augustine, or a modification of it proposed by the Administration, but something else entirely. New main engine propulsion research, which the Administration had targeted at $3 billion over several years was dropped, and funding for Commercial Crew was slashed as Congress, driven by space state senators, Bill Nelson, Richard Shelby and Kay Bailey Hutchison completely disregarded the stark and clearly defined budgetary shortcomings identified by the Committee, and directed NASA to proceed immediately with building the Space Launch System, a sort of alternative version of the Ares-V lite, built out of Shuttle components. Viewed from the perspective of Augustine's recommendations, there are two major problems with new "plan of record." First, in its determination to mandate a booster which would fly as soon as possible, the Senate ordered NASA to begin with the 70 ton capacity rocket the agency is now developing. Unfortunately, while it will be the world's largest rocket, it will still offer only half the capacity to perform any of the baseline lunar missions NASA previously envisioned under Project Constellation or the Augustine Commission put forward in its own recommendations without conducting multiple launches. It is a limitation created in part by the massive weight of the 73,000 lb. Orion capsule which has been problematic since the beginning of the program, the unusual heft a curious specification which caused some to wonder if it were chosen for the specific purpose of excluding other potential launch vehicles. Whatever the ultimate reason, in tying NASA to Orion, which the Obama administration initially sought to cancel, and the 70 ton SLS, Congress ensured that the agency could do almost nothing of meaning until the debut of its larger offspring nearly two decades later.

The second problem is equally glaring. Ignoring the unambiguous financial constraints which led the Commission to conclude that any program of exploration beyond LEO was not sustainable without a lasting increase to NASA's budget, not only did Congress fail to provide the increase in funding to support the booster it mandated, it then reduced the overall existing funding levels for NASA, including the Exploration account, first by enacting lower budgets, and then by imposing sequestration. As a result, even though as of 2014, NASA is now committed to building the 70 ton capacity SLS, ostensibly to be followed by the 130 ton version as well, despite glowing press releases and NASA channel promos delivered by the voice of Optimus Prime, the agency is at a loss as to what to do with it. Believing otherwise is the real science fiction.

Beyond the initial crewed flight aboard SLS, the makeup of future missions and whether or not they can be supported or enabled by commercial providers is an open question which depends very much on the policy direction provided to NASA by Congress and the White House. Unfortunately, Congress appears to be interested in nothing other than preserving existing space state jobs, a stance many have come to regard as pure corporate welfare, and in its second term, the Obama administration is both disinterested, and distracted by much bigger problems.

The harsh reality underlining the SLS/Orion program is that without the ability to enter and exit gravity wells, i.e., land on the Moon, there is very little which can be accomplished with SLS which seems worthy of the expense. That is not to say there are no other proposals for utilizing SLS, in fact there are quite a few; such as placing a smaller "deep space station" at various points in the Earth Moon system, or actually travelling to a real Near Earth Asteroid instead of hauling what amounts to little more than a large rock to the vicinity of the Moon. Each may hold appeal to a certain constituency, and be justifiable provided one is prepared to accept the inevitability of SLS, but none seems sufficiently energizing to bring together a coalition to push it through, much less secure the funding for the hardware it entails. Should that change however, the involvement of commercial providers offers the opportunity of lowering costs and increasing the opportunities for mission success. Provided of course, everyone is willing to continue overlooking the gaping cost disparity between SLS and commercial providers which will be obvious.

Still, momentarily leaving aside alternative exploration scenarios such as those based on in-space fuel depots and reusable launch vehicles, for a nation which professes a belief in the value of human space exploration, the Space Launch System is a (barely) plausible construct in early 2014 because there is no existing and demonstrated large booster alternative. That is almost certainly not going to be the case well before 2021, and whether policymakers like it or not, it is the filter through which subsequent plans will be considered. Until that moment comes however, officially at least, and adhering to the mantra repeated over and over at each SpaceX and OSC launch; COTS, CRS and Commercial Crew were created to allow NASA to "do what it does best, and go exploring."

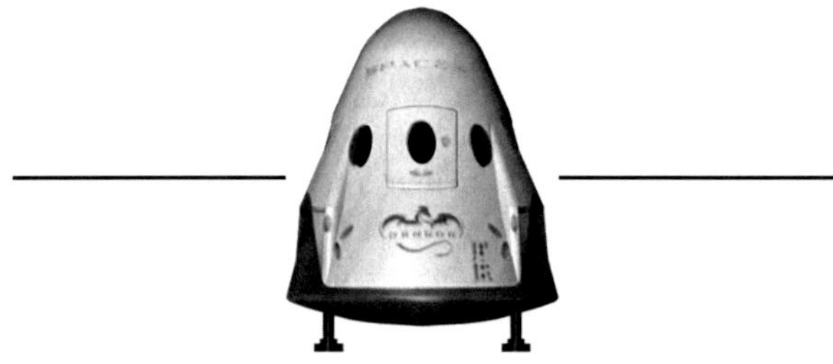

Chapter XXX: Private Plans

It should perhaps come as no surprise that based on extremely compelling cost/benefit projections, existing or very near term SpaceX technology is at the core of a number of private concepts for human space exploration beyond low Earth orbit. They include three independent proposals focused on human trips to Mars and at least one very achievable plan for a return to the Moon. In terms of ease of execution, the first of the Mars centered plans was that put forward by the Inspiration Mars Foundation, the brainchild of lifelong space enthusiast and the world's first self-funded space tourist, Dennis Tito. Introduced on February 27th, 2013, as SpaceX was preparing for its third trip to ISS and the CRS-2 mission, the Inspiration Mars Concept seeks to take advantage of a repeating cycle in orbital mechanics which presents the opportunity for a "free" circumnavigation of the Red Planet which occurs roughly every 14 years. Essentially, a spacecraft launched out of Earth orbit en route to Mars can circle the planet's far side and return to Earth without ever having to conduct an additional propellant burn beyond the one which places it en route along the journey in the first place. That is the "free" part.

Tito's idea, formed in collaboration with veterans of the Biosphere Two project, was to inspire the U.S. towards a new era of space exploration by sending two people, a married couple, on just such a flight in January of 2018, on a journey which would last 501 days, and after an all too brief swing around the Red Planet, arrive back on Earth for a hair raising re-entry and landing in early 2020. It seemed at the same time both a very risky and a very achievable plan, but one which absolutely hinged on making use of existing resources for both the launch vehicle and recovery capsule in order to meet a hard and fast launch date. As originally introduced, the plan was base-lined on using a Falcon Heavy booster and a DragonRider spacecraft as the most obvious, although not the only prospect for the recovery capsule. Once launched to orbit, the two spacefarers' home for the next year and a half would consist of the capsule itself, coupled to an inflatable expansion module similar to the Bigelow BEAM experiment scheduled to be attached to ISS in 2015.

After an enthusiastic and generally positive reception, the Inspiration Mars team initially stated it was settling down into perhaps the most demanding aspect of the entire endeavor, designing and testing a mostly closed loop life support system which would have to function reliably for more than 500 days. What might be the most significant concern, radiation exposure, was also a major challenge, but one which the Inspiration Mars medical team proposed addressing in an interesting manner. Rather than attempting to absolutely shield the prospective voyagers from a level which will almost certainly exceed NASA's maximum allotted lifetime exposure for its own astronauts, Inspiration Mars suggested skewing its selection criteria in favor of a middle aged couple who are in a word, statistically less likely to have the radiation exposure catch up with them during their expected remaining life spans. Once arriving safely back on Earth, the history making pair would receive state of the art genetically tailored medical care premised on the basis that they bear an elevated risk of radiation induced cancers. Though some would consider it a controversial approach, the Inspiration Mars team is anything but dismissive of the risks of spaceflight. Its lead physician, former NASA doctor Jonathon Clark, lost his wife, astronaut Laurel Clark, aboard the Space Shuttle Columbia.

With the first two years of development work personally funded by Dennis Tito, the Inspiration Mars team appeared to be in the enviable position of being able to thoroughly vet the mission concept prior to fully committing to a course of action, and a decision which might be made somewhat easier by a successful introduction for the Falcon Heavy, and a win for SpaceX in the Commercial Crew competition, both of which were likely to occur before Inspiration Mars reached its hard timing deadline.

It came as somewhat of an unwelcome surprise then, when Inspiration Mars appeared before a Congressional committee in November 2013 to announce the findings of its preliminary study, and in short, beg NASA to take over the mission as the first flight of the Space Launch System. Under the revised proposal, on its first flight, the Block I SLS would boost to low Earth orbit a Mars departure craft consisting of an Orbital Sciences Cygnus Cargo craft modified into a hab structure, and a partially completed Orion capsule to serve as a "re-entry pod." In order to break Earth's gravity, NASA would also need to complete a proposed alternate upper stage for its new rocket. As for the two person crew, they would launch into LEO aboard a commercial crew vessel, now presented as a Boeing CST-100. [110] SpaceX was nowhere to be found. Neither was a scintilla of support for the now wildly improbably proposal. In the way of explanation for the change, Inspiration Mars, which had been working with NASA under an unfunded Space Act Agreement, offered that it had from the outset assigned a NASA centered mission as its primary area of study, with a "backup" team looking at commercial hardware. The ostensible reason was the difficulty in coordinating a three launch campaign to form a privately centered mission. The reason for the objection did not make much sense, and even if the real reason was SpaceX was not interested, a credible (but more expensive) plan could have been put forward using the ULA Delta IV Heavy and Atlas V boosters. Furthermore, if Tito and company had been willing to drop its "Mission to Inspire America" angle and strictly focus on the simplest solutions available worldwide, two massive and less expensive rockets, the Russian Proton and European Ariane V were potential booster candidates as well.

Clearly aware that the NASA approach was dead on arrival, at least as far as a 2018 launch window was concerned, Tito used the occasion of his Congressional testimony to point out that another, somewhat less advantageous window opened in 2021, and that if the United States did not seize the opportunity, somebody else, and that meant China, just might. [111]

With time running out, the Inspiration Mars proposal is going nowhere, and taking with it what might have been a major turning point in mankind's pursuit of the Red Planet. Although a six hour swing around the night side of Mars might seem a very small payoff for a dangerous and expensive 500 day journey, it also held the potential to change the perception that Mars is always 20 years away, and requires a massive mega booster based strategy to reach. This may have been its undoing. Anticipating a 20 to 30 year gravy train of cost plus contracts for SLS, it defies imagination that ULA's parent companies, Boeing and Lockheed Martin would have ever allowed it to provide the launch vehicles which could have wrecked the SLS program before its first crewed launch. As for SpaceX, the company has no comment regarding its degree of interaction with the Inspiration Mars team, other than a statement by President Gwynne Shotwell that her company was never contacted as Inspiration Mars was putting together its original proposal. And, while a long list of Boeing, Lockheed Martin, and NASA officials are listed in the credits for the revised proposal, only one SpaceX name is on the list. Whatever the real reason for its sudden change, even though the 2018 Inspiration Mars mission is not going to happen, Denis Tito, Taber McCallum and the rest of team performed an admirable service in putting forward a credible proposal for reaching the vicinity of Mars in very short order, and in the process reducing much of the risks associated with subsequent missions. Viewed from another perspective, in shifting to an SLS based proposal, Inspiration Mars at least came up with a highly innovative use for NASA's mega booster which one might have expected the agency to arrive at independently, were it more focused on exploration than saving jobs.

One person unconstrained by conventional thinking, and determined to get to Mars anyway possible, is Dr. Robert Zubrin, president of the Mars Society and author of the book, The Case for Mars. Consequently it should come as no surprise that the first proposal for a Mars landing and return mission using SpaceX hardware came from Zubrin. Published in a May 14th 2011 article in the Wall Street Journal, only a few weeks after Elon Musk introduced the Falcon Heavy, Zubrin's concept demonstrates just how achievable such a mission could be provided one is willing to approach the idea from a different direction.

Zubrin's proposal is a very much a minimalistic concept, and one which carries a risk profile significantly greater than that typical of NASA missions, beginning with the fact that like Inspiration Mars, it is comprised by a crew of only two. The Zubrin proposal consists of just three launches of a Falcon Heavy, each equipped with a Dragon capsule, and each placed on course to Mars by a cryogenic engine which would either have to be acquired externally, or developed by SpaceX. The first mission would send the Earth Return Vehicle (ERV), a Dragon capsule with a kerosene/oxygen return stage into Mars orbit where it would linger until needed. The second launch would send an unmanned Dragon capsule equipped with the Mars Ascent Vehicle to the planet's surface. Upon landing, a small chemical plant aboard the Dragon would begin converting Mar's atmosphere, which is comprised of 95% carbon dioxide, into 9 tons of liquid oxygen, which combined with 2.6 tons of methane shipped from Earth, would provide the necessary fuel for the astronaut's ultimate ascent to orbit and rendezvous with the awaiting ERV. Once the necessary infrastructure for a safe return was in place and verified as operational, the two person crew would board the third Dragon capsule for the flight to Mars. As with other plans, once safely on the path to the Red Planet, the crew would detach the Dragon from its trunk and rotate the spacecraft to dock with an inflatable habitation module packed inside during launch. After expansion, the inflatable hab module would expand the available living space from a very confined 400 cubic feet, to something approaching a manageable level. In further evidence of a "whatever it takes" approach, Zubrin went on to suggest that if available space were still a problem, rather than seeking a bigger booster, why not just choose smaller astronauts? [112]

Though clearly minimalistic in more ways than one, one thing which stands out, as Zubrin points out, is just how easily extra margin could be built in with the inclusion of additional Dragon flights sent in advance of crew departure. For example, sending just one additional support flight could significantly increase the mission's scientific potential by adding additional research equipment, an upgraded rover, and multiple spares for mission critical equipment, as well as providing additional supplies to increase mission margins and astronaut comfort for the lengthy stay on the surface.

Furthermore, an additional precursor mission to test the onboard chemical plant and Mars Ascent Vehicle could remove a great deal of risk, while still pre-placing additional assets to aid the first mission.

Amazingly, one of the two riskiest elements of the mission, Mars Entry, Descent and Landing is already estimated to be within the capacity of the first generation DragonRider capsule. It is just that capability which prompted NASA's Ames Research Center to promote its "Red Dragon" mission concept, which would see an unmanned Dragon capsule tricked out into a scientific research platform and equipped with nearly 1,000 lbs. of instruments, launched to Mars in 2018. Arriving at Mars on a direct entry trajectory, the Dragon would slam into the atmosphere at 6.0 km/s using only its heat shield for initial deceleration, guided by software developed for the Mars Science Laboratory's famous "7 minutes of terror." Slowing from hypersonic to a supersonic velocity at Mach 2.24, the Dragon would then fire its SuperDraco engines for a powered touchdown at 2.4 m/s. [113]

Conceived as a constrained cost, Discovery class mission, any decision regarding Red Dragon proposal has been delayed from 2013 to 2015 due to budget cuts within the agency's Science Mission Directorate, yet another consequence of the James Webb Space Telescope's cost overruns. With an additional two years in which to prove the Dragon's capabilities, as well as to conduct the first launch of Falcon Heavy, the occasion of a 2015 decision on the next Discovery class mission may offer an early opportunity for SpaceX, or provide yet another referendum on just how resistant NASA will continue to be regarding its game changing capabilities. Either way, the Red Dragon proposal is an impressive testament to the work already performed by SpaceX, underscored by the fact that members of the same NASA team which has successfully landed hardware on Mars believes the upgraded Dragon spacecraft is fundamentally capable of landing on the planet as well. As such, and even though the prospect may strike terror into certain areas of the agency, the ability to soft-land large and varied payloads across the Red Planet for less than the cost of a single Shuttle mission may create another small but vocal constituency in the Mars faction of the planetary science community, while building a bridge in the often fractured relationship between it and NASA's human space flight programs. Although there has been little in the way of a formal proposal to date, the obvious synergy between a Mars sample return mission, the "Holy Grail" of Mars science, conducted for its own sake, and a precursor demonstration flight prior to any crewed mission is unmistakable, and is not likely to be missed.

While Zubrin and the Mars Society lack the resources to put together anything approaching an actual mission, both have a valuable role to play in advocating loudly for a re-assessment of NASA's plans and priorities, as well as putting forward ever more complete proposals which take advantage of emerging infrastructure such as the Falcon Heavy and DragonRider as they move from theoretical to proven. As a result, with every passing year, the potential for a privately funded first mission becomes greater, and it is not unreasonable to believe that just as space tourism is now entering the popular consciousness, the perceptions of what can be achieved in space by non-governmental organizations may undergo an overall shift as well. If the right coalition for a privately sponsored first mission to Mars does emerge, it seems likely that the Zubrin model could serve as the template. After all, the only critical piece of hardware not in the current path of development is the ascent stage, which, taken with the need to demonstrate in-situ oxygen production poses the largest overall area of risk and likely expense, even as it constrains the overall challenge.

It is the desire to avoid precisely the issue of developing an ascent stage which provides the rationale for the most far reaching concept utilizing existing SpaceX hardware put forward so far. It is that of Mars One, a not-for-profit foundation based out of the Netherlands, which proposes to send four person groups of colonists on one way trips to the Red Planet beginning in 2024.

As with the original Inspiration Mars and Zubrin concepts, the Mars One plan is base lined around the use of the Falcon Heavy to boost key payload segments into low Earth orbit. According to the organization's roadmap, colonization would begin with a 2018 launch of a proof of concept lander and a communications satellite to provide a link from Mars back to Earth. Two years later, operations would begin in earnest with the delivery of a remote controlled rover and trailer, which would be used to first locate a suitable site for the colony, described as a level plain marked by readily available water ice, as well as adequate sun exposure for deploying rolls of thin film solar panels which are to be the outpost's principle power source, a necessary compromise to avoid the near political impossibility of obtaining a nuclear reactor. Following an initial mission to deploy the rover, the next window, opening in July 2022, would see the launch of no less than six unmanned cargo modules, each based on a variation of the SpaceX Dragon capsule. Two of the modules would be fitted out as life support units, equipped with on-board plants to convert the thin Martian atmosphere into breathable air by adding inert nitrogen and argon, which would be combined with oxygen harvested from water in the soil, gathered by the rover and heated to release its contents. Using this process, Mars One estimates each life support unit could collect and store 1500 liters of water and 120 kg of oxygen over 500 days.

With the basics of life support provided by two of the landers, another pair would function as the "wet rooms" for living quarters, with the balance of available space provided by inflatable habitat segments covered with soil to provide protection from radiation. Furniture and other amenities would arrive as if shipped from Ikea, tightly packed and awaiting an excruciating assembly process with no prospects for outside help. The remaining two landers from the initial six spacecraft fleet would serve as the primary cargo containers for the first colonists.

Following a homing beacon and touching down within a 10 kilometer radius of the selected site, the workhorse rover would locate each of the landers, hook it up and tow it back to the site to be linked together to form the basic infrastructure of the colony. Once the habitation and support modules are linked and confirmed to be operational and stocked with locally acquired water and oxygen, the next phase begins with the launch of the first four person crew of colonists. Once again, existing SpaceX hardware is presumed to serve mission critical roles. Like Inspiration Mars, the crew carrying vessel consists of a Dragon capsule connected to a habitation module, but in this case powered by a propulsion module comprised of two separately launched fuel tanks. The Mars ship is joined together in low Earth orbit by a two person "assembly crew" launched aboard the mission's Dragon capsule/lander. Once their task is completed, and the ship is verified as ready to go, the first colonists are launched into LEO aboard a conventional Dragon spacecraft, and swap places with the assembly crew, who having completed their assignment, return to Earth aboard the just arriving capsule. After making a 210 day transit, the crew enters the landing capsule for the descent to the surface, abandoning the habitat and twin propulsion stages to solar orbit. With the cargo units for the second wave of settlers arriving later in the same month, after recovering from their time in space and acclimating to Mars' 38% Earth gravity, the initial crew begins the hard work of assembling the modular living quarters and preparing the infrastructure for the next crew due in two years. After the initial manned landing, new four person crews arrive every two years, adding redundancy and new infrastructure to the tiny settlement which is expected to be growing much of its own food, expanding its reach and rather fervently experimenting with techniques for constructing additional living space out of the limited suite of local materials. [114]

Besides the controversial one way format, reported by some news sources as a being a "suicide mission," the principal problem with the Mars One plan is not surprisingly the financing plan, which is not nearly as defined as the technology roadmap. The overall concept calls for a mix of revenue streams, most of which would come from media sources, focused around a reality television show format which follows the course of the entire process; beginning with the competitive selection from the more than 200,000 people who submitted applications in the initial round, all the way from training through the mission itself, and the travails of life on a new planet. It is a plan based on the funding model which has driven the Olympic movement for the last 30 years. Beginning with the Los Angeles Games in 1984, and reaching stride with the Atlanta Games in 1996, the Olympic movement has become a massive global advertising and sponsorship event with revenues from all sources exceeding $8 billion in the four year cycle leading up the London Games of 2012. In terms of a much more compressed time scale, the 2013 Super Bowl, an almost exclusively North American phenomenon, generated nearly $1 billion in advertising for an event which lasted just over 3 hours. For its part, the 2010 World Cup generated $3.7 billion in overall sales, with $2.4 billion coming from television rights.

It is not unreasonable to assume then, that the first manned mission to Mars, presented to the entire global audience, could be expected to generate a level of revenue well into the multiple billion dollar range, punctuated by the building narrative of its four penultimate moments of Launch, Departure, Mars Entry, Descent and Landing, and most of all, the First Step. Given all the various media through which the event could be marketed, as well as what would likely be an enormous groundswell of direct public support as the mission crossed the threshold from a flight of fancy to the flight of a lifetime, the financing plan may not be as much a question of "could the money be raised," but could it be raised in the time frames necessary to allow for a development profile with most of the expense coming well in advance of the pathfinder launch.

There is no way to know in advance how corporate sponsors would or would not respond to an event such as the first human landing on Mars, but several points are worth considering. First, the Mars One project is conceived as truly global enterprise, a fact borne out by the amazing diversity of more than 200,000 people from over 140 countries who filled out applications to compete to join the initial four person crew. Although met with considerable skepticism in the space press, to the point of being called fraudulent, Mars One surprised some detractors in late 2013 by issuing a study contract to Lockheed Martin into outline a small, proof of concept automated probe based on the Phoenix lander which successfully touched down on the Red Planet in 2008, confirming the presence of water ice shortly thereafter. It is a modified version of the Phoenix Lander, equipped with both water collection and thin solar film demonstrator experiments which Mars One expects to launch in 2018, along with a small communications satellite built by Surrey Satellite Technology Ltd. While a study contract is quite a long way away from a construction and launch contract, it does mark a meaningful development and a necessary first step, and as the first privately funded robotic mission to Mars, would be a pathfinder event in more ways than one.

While the Mars One plan has yet to be taken seriously by many, there are several reasons which suggest it may not be as improbable as it sounds. In the first place, it has the virtue of requiring a minimal amount of new technology development, provided one takes for granted the actual development of the commercial hardware being sourced. With preliminary pricing cited by SpaceX, as well as other critical vendors, it is also possible that funding could fall within the profile of the Olympic Games it is modeled after.

Second, in terms of the laundry list of hazards of going to Mars which NASA invariably cites as requiring further study before a mission could be mounted, the one-way format adopted by Mars One effectively halves the risk of in-space radiation exposure inherent in two way missions. Another in-space element presumed necessary for Mars missions, advanced closed loop life support is bypassed as well. Mars One colonists would simply depart in spacecraft adequately outfitted for the 6 month transit through conventional means, and once arriving on Mars, derive water and air from the planet itself. Furthermore, with an Earth Return Vehicle removed from the equation, the single largest area of immediate risk becomes that of Entry, Descent and Landing. In following the roadmap however, it is a process which would have been demonstrated at least eight times prior to happening with a crew aboard. By comparison, Neil Armstrong and Buzz Aldrin successfully piloted the Eagle module to a lunar landing without the benefit of even a single automated trial run. Presumably, in addition to the eight pre-cursor missions, by the time the first Mars One crew departed in 2023, the DragonRider spacecraft would have dozens, if not hundreds of powered landings on Earth. Viewed in that context, even though enormous long term risks remain, permeating the project to an extent that is inconceivable for a national space agency such as NASA to even consider, it might very well fall within the bounds of what sponsors consider acceptable. One need only consider NASCAR or Formula One racing to understand that advertisers not only accept extreme risks, they embrace it.

Also, from the vantage point of early 2014, the idea of landing Dragon capsules on Mars appears farfetched due to the simple fact that SpaceX has yet to either officially unveil the DragonRider capsule, or demonstrate its propulsive landing capability. If the company's history tells us anything, it is very clear on the fact that sooner or later, Elon Musk will catch nearly everyone by surprise by tweeting out a video of a Dragon taking off, hovering and landing, just as he has done with the Grasshopper test article. Once released, and then demonstrated as the pad abort test for the Commercial Crew program, perspectives will change.

There are a few other points worthy of consideration as a well. First and perhaps foremost is the fact that unlike Inspiration Mars which is facing a hard and fast launch window in late 2017, one which already appears to have derailed the project; Mars One has the luxury of a recurring launch window which opens every two years. Provided progress is being made, the entire schedule can slip two years at a time into the future without creating a significant hardship. Moreover the greatest progress is likely to come from SpaceX itself, as the rapid advances the company is making in recoverable booster technology presents the very real possibility that the overall costs of mounting the mission may in fact plummet.

Other elements, such as deploying and operating the rover, already demonstrated by NASA and JPL with the Mars Science Lab and Curiosity rover, become a matter of comparatively straight forward analog testing on Earth. If for instance a team can remotely hook up, tow and configure simulated habitats somewhere in the Arizona or New Mexican deserts then presumable it can do so on Mars as well. As for other components, the life support system and "wet rooms," not being required to work in Zero-G, are actually much more closely related to gravity based systems on Earth than they are to the more exotic and complicated space based counterparts.

The one piece of major hardware not built by SpaceX included in the initial concept, the transit habitat is based on collaboration with Thales Alenia, and is basically a Shuttle derived Multi-Purpose Logistic Module outfitted for 6 months habitation. There is no rational reason why it should not be both available and affordable, and if it is not, Bigelow Aerospace may very well offer a preferable and far more spacious alternative in any event.

Though still confined to be sure, its occupants would actually face a less demanding environment than likely to have been encountered by Inspiration Mars, excepting of course the non-trivial fact they have left the home planet, presumably never to return. Even here though, "never" is a very long time. While the economics of settlement require more people arriving at Mars than departing from it, the same advance of technology and circumstances which makes the trip conceivable, also argue that a return option will eventually become available. Consequently, from the perspective of each individual strapping into the seat for launch, it would not necessarily mean abandoning all hope of ever seeing Earth again.

Finally, Mars One is first and foremost conceived of as a one way venture designed to enable the first human settlement on another planet. In order to have a chance at succeeding, it simply sidesteps the costs and challenges of including elements of in-situ fuel processing, ascent, and Earth return vehicles which are not a requirement of mission success. That is why it is an intriguing and legitimate concept for a private organization, but wholly inappropriate for NASA, whose charter no more extends to the political act of forming new settlements than it does to conducting war. For SpaceX however, whose employees routinely wear "Occupy Mars" shirts, and led by Elon Musk, a man who has repeatedly asserted that his purpose is to provide the means for establishing a second branch of human civilization, the Mars One concept has the distinct advantage of being most directly aligned with that long term goal.

While Mars is the primary focus of interest for SpaceX and the long term U.S. human space exploration repeatedly espoused by NASA, the Moon is still considered by many to be a preferred destination in its own right, and by some (although certainly not Zubrin) as a logical and even necessary pre-cursor to Mars missions. Not surprisingly, just as is the case with Mars, the potential use of SpaceX hardware is increasing the prospects of a privately funded return to the Moon. The most prominent group is the Golden Spike Company, a corporation founded by former NASA Science mission head Dr. Alan Stern to conduct privately funded excursions to the lunar surface. Much like Bigelow Aerospace, which also harbors lunar ambitions, the Golden Spike business plan is based on offering commercial services to non-space faring nations, enabling them to conduct missions for science or prestige in a manner which is simply unachievable in any other way.

The Golden Spike plan would see the use of small two-person landers designed for brief forays to the lunar surface, very reminiscent of the first Apollo flights. It has been established as an open architecture system with proof of concept proposals based on both the ULA Atlas V and SpaceX Falcon 9 and Falcon Heavy boosters, but it is the latter which clearly offers the greatest potential in terms of both performance and pricing. In a typical Golden Spike mission scenario, a lunar lander and ascent vehicle is launched into low lunar orbit (LLO) by a Falcon Heavy carrying both the lunar craft and a ULA provided Centaur departure stage. Having reached LLO, it remains until it is joined by a two person crew launched in a DragonRider capsule which also serves as the Earth Return Vehicle. After a lunar orbit rendezvous with the awaiting lander, the crew transfers to the sortie vehicle and descends to the Moon for a few days stay, after which they ascend to lunar orbit for a second rendezvous and the trip back home.

Golden Spike envisions a number of different scenarios based on the specific combination of vehicles chosen, and anticipates that it could actually begin operations with a lunar circumnavigation taking place aboard a Dragon launched by a standard Falcon 9, in which case a smaller chemical engine contained in the Dragon trunk would provide the necessary velocity change to escape Earth orbit, swing around the Moon and fly back to Earth. A likely next step would see a Dragon capsule launched into a lunar orbiting mission by a Falcon Heavy in a partial test run of the complete system. With the addition of the separately launched lunar lander as it completed development, Golden Spike would be ready to begin a historic return to the Moon, followed up by an average of two missions per year. Shortly after its announcement in late 2012, Golden Spike issued a study contract for a lander proposal to Northrop Grumman Corporation, the company responsible for the original Apollo landers. It came back with the "Pumpkin," a minimalistic two person craft designed to fit within the size and weight constraints of existing launch vehicles.

Beside participation by SpaceX, one item which stands out about the Golden Spike plan of operations is the potential role played by the United Launch Alliance Centaur upper stage. With a history dating back nearly 40 years, the RL-10 powered Centaur is one of the most frequently flown and dependable space hardware elements in the world, and in some ways is the only American counterpart to the series of long established Russian hardware elements such as Soyuz, Proton and Progress which has seen continual use for decades. Despite its own role as the Atlas V upper stage in the pitched battle between SpaceX and ULA, the Centaur is both rock solid reliable yet versatile at the same time. In fact ULA has independently produced a number of study concepts over the years featuring the use of the Centaur for deep space propulsion, and even as the backbone of a standalone lunar lander and ascent vehicle. There is absolutely no question that if called upon, the Centaur could perform in a number of new roles, including that identified by Golden Spike. While SpaceX is focused on developing its own high energy upper stage, one which would presumably advance the state of the art well beyond Centaur, the fact remains that until it is available, the combination of launch provided by the Falcon and Falcon Heavy, and in-space propulsion powered by the Centaur provides an unmatched combination of strengths which could readily open up the Earth Moon system if put to the challenge.

As for the capsules included in the Golden Spike plan, although the original proposal suggested the missions could potentially be performed with either the Dragon or Boeing CST-100 capsules, the potential for the latter is constrained by Boeing's insistence that the CST-100 was designed for low Earth orbit only, and is not equipped with a heat shield sufficient for lunar returns. In fact the most futuristic element of the CST-100 is probably the cool blue cabin lighting the company chose to highlight in media publications. The Dragon however was designed to provide up to 45 man-days of life support, and is more than capable of supporting a crew of two, including up to a seven day loiter in lunar orbit as the mission unfolds on the surface. With its heat shield already designed for lunar re-entry velocities, Dragon's only deficiency in supporting lunar transfer, orbit and return is the lack of adequate reserves in its in-space propulsion system, for which there are a number of options. [115]

Using conservative assumptions, Golden Spike projects an overall development cost of just over $6.4 billion, with a recurring lunar landing mission cost of roughly $1.0 billion, or $500 million per seat, which it plans to partially defray with television and marketing rights. It is a funding profile which places the overall concept in a difficult position. On one hand, absent governmental participation, it seems somewhat unlikely that the company will be able to find the financial wherewithal to absorb the substantial development costs, even as the mission costs would appear to fall within the discretionary spending capabilities of quite a few countries, should any prove willing. On the other hand, unlike Mars One, the Golden Spike roadmap offers human flight participation from nearly the moment Falcon Heavy and DragonRider are available, and at an escalating level of involvement which could begin with pre-cursor flights to LEO, followed shortly by the first circumnavigation flight.

The real question perhaps should be why the United States itself is not one of those countries? After all, a return to the Moon via commercial providers at a fraction of the price anticipated under Project Constellation would appear to be a win/win for the U.S., simultaneously achieving the oft stated goal of resuming deep space exploration, while at the same time trumping other nations such as China by doing so in a way which could not be easily matched, through the energy and imagination of unfettered capitalism. One might think that given Congressional opposition to the Obama Administration's Asteroid Redirect Mission, generally accompanied by periodic calls for a return to the Moon, NASA might be directed to participate heavily in Golden Spike, particularly given the fact that it offers ongoing business opportunities not just to SpaceX, but to aerospace stalwarts such as ULA, Lockheed Martin and likely Northrop Grumman as well. Perhaps regrettably, that does not seem to be the case. There are two answers, both plausible.

The first is like so much else accompanying SpaceX, until the Falcon Heavy flies and the Dragon is flown with crew, the entire concept is notional. Some might point out however, that it is no less notional than the equally undeveloped SLS/Orion, and unlike NASA, SpaceX has recent history to its credit, having introduced three different new launch vehicles, Falcon 1, Falcon 9 and Falcon 9 v1.1 all within the last decade. As the two new mission critical elements come into service however, assumptions may change.

The other reason is far more intractable, at least for now, and perhaps far more likely. Even more than any of the proposed Mars missions utilizing SpaceX hardware, Golden Spike's lunar ambitions pose a direct challenge to the entire, tottering rationale for the Space Launch System and the Orion spacecraft. Highly theoretical Mars missions are one thing, but in putting forward a perfectly plausible lunar transportation architecture built out of fully demonstrated and well understood commercial components such as the Atlas V, Falcon 9 and Centaur upper stage and Dragon spacecraft, the Golden Spike proposal has the power to unambiguously demonstrate that the Emperor has no clothes, provoking the press to ask loudly how it is that a $40 billion dollar monster rocket and capsule which cannot accomplish anything other than flying four people around in empty space for the next 20 years, is preferable to a two person lunar return architecture available with minimal development costs, a fraction of the operational costs, and on a time frame of now?

Whether or not Golden Spike is able to develop sufficient interest among prospective clients is an open question, and its initial reception has been somewhat cool to say the least. Nevertheless, even more than with Mars One, time may be on its side, and in late 2013 the team took an important step forward in announcing a partnership with Honeybee Robotics, a leading NASA contractor which has supplied tools for three different Martian landers, including the Curiosity Rover, to begin planning an initial series of small, automated lunar rovers capable of collecting samples from far field operations and delivering them to the site of anticipated crew landings, a joint human/robotics architecture which could multiply the scientific return of future Golden Spike missions by several times.

Although Golden Spike does not have the backing of major corporate donors or a team of angel investors, it does have a broad and diverse group of experienced space leaders in its Board Members including Inspiration Mars' Taber MacCallum, and Max Vozoff, SpaceX's former program manager for the Dragon. Golden Spike has also compiled an impressive lineup among its Spaceflight, Scientific, and Creative Council members and a Board of Advisors notably including Apollo 13 astronaut James A. Lovell, who as the only person to have flown to the Moon twice without making a landing is no doubt particularly motivated to see Golden Spike succeed. And perhaps it will, provided the company holds together through early lean years, it is very well possible that greater changes in national space policy, as well as working demonstrations of the Falcon Heavy booster and DragonRider spacecraft change the landscape of what is considered possible. It goes without saying that even more so than with Mars plans, the successful introduction of the Falcon 9-R would be a boon to Golden Spike. If it does survive until the introduction of new hardware elements, the Golden Spike business plan presents an interesting dilemma for NASA.

It is difficult to imagine that facing the prospect of Golden Spike actually beginning operations, NASA would not want to participate in some manner, both to reap the benefit of valuable data, and to be able to legitimately make the claim that it was involved with an American return to the Moon. To do otherwise would risk oblivion, as a successful mission would make the agency's nebulous plans look all the more ridiculous. On the other hand, taking a more visible role, and thereby lending credibility and expertise to get Golden Spike into an operational phase would be equally damaging to the SLS program, with NASA's own hand driving the stake through it.

While the proposals offered by Inspiration Mars, the Mars Society, Mars One and Golden Spike will no doubt be joined by others, there is actually no clear indication that SpaceX is, or would be willing to see its products put to use in high profile and extremely risky ventures over which it does not maintain full operational control. Clearly, there are several sound reasons why it might want to avoid just such a circumstance. Furthermore, hard pressed in its own more immediate development efforts and concentrating on winning NASA's Commercial Crew competition, it is also quite possible that at least for now, actively working on the refinement of any deep space proposal is simply beyond the company's reach. Nevertheless, merely by putting forward sophisticated, well thought out concepts which challenge NASA's basic assumptions regarding the size, expense and time frames for renewed exploration, each of these organizations is helping to re-define the terms of the debate which is sure to come, making it progressively harder for the politically driven components of American space policy to remain standing impassively on shaky ground.

In the end however, the path of future events may depend less on the will of Congress, or the President, than it does on what Elon Musk wants to do.

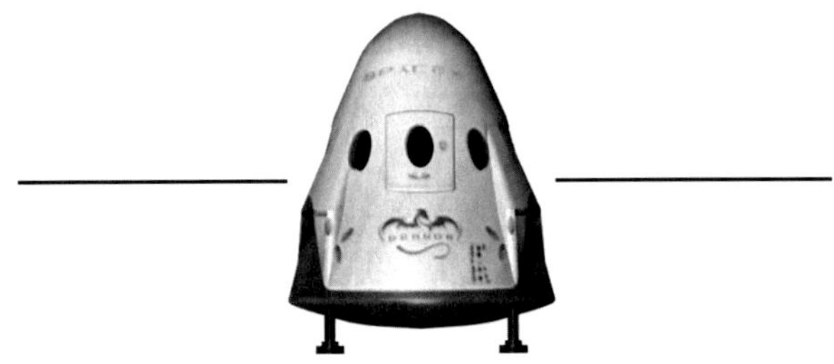

Chapter XXXI: Following SpaceX Down the Rabbit Hole

For anyone willing to listen, Elon Musk has clearly and consistently stated that his purpose in forming SpaceX was not to make money, or to validate a particular rocket design, but to enable the human exploration and settlement of Mars at the earliest possible opportunity. An avowed atheist, Musk is utterly practical about the reason as well, to create what amounts to a backup hard drive for the human race, and a separate, thriving second branch of humanity. And yet, relentlessly optimistic, Musk rarely if ever speaks in terms of the type of possible looming global calamities which might otherwise accompany an expressed need to prepare for the worst. Instead, at the center of each of his projects, SpaceX, Tesla, Solar City and Hyperloop, lies a belief that the future is supposed to be better than the past, that technology should progress, and each time he has stepped in, it is to "fix" what he perceives as an aberration in what should be a steady, if episodic march towards a better future. And even though the rest of his other projects are devoted to improving life on Earth, he is grounded in the belief that "a future where we are a space faring civilization is far more exciting and inspiring than a future where we are not." The particular vision associated with that belief, is a series of statements Musk has made in recent years regarding his desire to see "millions" of people emigrating to Mars.

At least for the moment, it should perhaps be considered as more of a concept, or even a wish, than an actual plan. The fulfillment of Musk's vision depends on a long list of developments falling into place, beginning with the steady march towards a rapidly reusable Earth to orbit launch system and ending with a fully reusable, two-way Mars transportation architecture requiring a new large booster and in-space assets on a scale greatly exceeding the current Falcon 9, Falcon Heavy and Dragon systems. Elon Musk refers to the as yet undefined system as the "Mars Colonial Transport," and in contrast to many earlier pronouncements for which the SpaceX founder confidently, if inaccurately attached time frames, he is notably vague on when his ultimate vision of Mars colonization, or even a first mission, might take place, stating "hopefully within the next 10-15 years." By any measure, it is still a strikingly bold prediction.

It is also one which would appear to signal a major, and historic point of departure from NASA's own meandering path towards Mars and the mutually beneficial partnership which has allowed SpaceX to reach heights in its first decade that even Musk might have doubted when he founded the company. If in fact the United States remains wedded to the Space Launch System and the Orion Multi-Purpose Crew Vehicle for the next 20 years, then the conflict is clear. SpaceX must eventually go its own way.

SpaceX was begun in an almost comically quixotic manner, as a result of Elon Musk's personal frustration with the stagnant state of space exploration, which gave rise to an unlikely plan to spur a new era of NASA led space exploration with the creation of a viral image, a flower growing on Mars. Over the course of its first decade, even as it came of age through a burgeoning relationship with NASA, the company maintained a weather eye on its long term goal, with each element of its system meant to contribute toward that goal. At the same time, it is also the goal espoused by NASA, the pot of dusty red gold at the end of the flexible path, and it is the difference between the two organizations and the paths they have chosen to pursue the Red Planet, that provides the most useful context to view the prospects for the future.

The fact that a company could go from the trials of launching a single diminutive rocket to low Earth orbit to fielding a system capable of reaching Mars in a little over a decade almost defies imagination, and to some extent, perhaps it does just that. Understandably confused by a steady stream of statements coming from NASA leadership and other sources that Mars is decades and untold billions of dollars away, and almost certainly a task only the world's space powers acting together could hope to achieve, it is difficult to absorb the reality that humanity is in fact standing on the brink of settling another planet, but less than eight years after SpaceX first launched the Falcon 1, that is precisely where we are.

With the debut of the Falcon 9 v1.1, SpaceX has already flown the critical piece of hardware in fielding the far more powerful and Mars capable, Falcon Heavy. While the company still needs to demonstrate its ability to successfully launch the triple body core booster, as well as the particulars of its performance enhancing cross valve fuel supply, both are so far within the body of what has already been achieved by other systems that its ultimate success should be taken as a foregone conclusion.

The Space Shuttle, the Russian Energia, the Ariane V and most directly the Boeing Delta IV have all successfully proven the art of conducting space launch with large scale strap-on boosters, and a central sustainer. Furthermore, what might appear to be the hardest challenge in developing the Falcon Heavy, cross plumbing, already has a history of sorts in the Atlas and Atlas II boosters, which launched with three first stage engines, but jettisoned the two outboard engines while the central sustainer kept burning on its way to becoming the most commonly launched American booster of all time. As for most basic areas of concern, firing 27 engines at a single time, it is worth repeating yet again that the world's all time most popular booster, the Russian Soyuz, with over 1800 launches to its credit and counting, lifts off with five main engines, and 32 separate thrust chambers all engaged. With the Falcon 9 v1.1 and Falcon Heavy sharing a common second stage already proven in flight, there are no other significant impediments remaining.

In short, after whatever amount of testing is required in Texas, the Falcon Heavy will fly, and it will fly true. Indeed the most significant indication that the writing really is on the wall will come not when the booster first lifts off, but as with its other projects, with the first successful, full scale mission test firing in McGregor. It is a rumble certain to temporarily deafen ears in central Texas, but one which may echo much louder at NASA's Marshall Space Flight Center, home of the Space Launch System, and even across the Atlantic at the headquarters of Arianespace in Evry-Courcouronnes, France, home of the Ariane V. After that moment, years of theoretical discussions become the prelude to something suddenly tangible, and the future quickly begins to reveal itself. With the advent of the Falcon Heavy, the United States will have at its disposal, as Elon Musk pointed out when introducing the booster in 2011, a heavy lift vehicle capable of launching 53 tons to LEO, putting it squarely within the range of capability of a notional "super heavy" EELV outlined as one of three possible architectures for exploring deep space by the Augustine committee. In the absence of a public/private joint venture, with the introduction of these elements, possessing both the will and the means, the principal question surrounding the SpaceX story will become when, and how, will the company first reach Mars?

As the original Inspiration Mars proposal underscored, there is little doubt that in terms of hardware at least, a flyby of the Red Planet comes within reach with the debut of the DragonRider capsule and the Falcon Heavy. Furthermore, as Zubrin's concepts suggest, with the addition of an ascent stage, and an in-situ LOX plant, in terms of the major hardware elements, a historic first landing becomes (just barely) within reach at nearly the same time. Would SpaceX seek to take advantage of the opportunity?

In order to do so, SpaceX, which has thus far maintained a sharp focus on its well defined business plan, would necessarily be required to step into the role of a mission operator and commit a substantial portion of its own resources. For a man who risked nearly everything on the fourth flight of the primitive Falcon 1 after first enduring three consecutive failures, it might come as more of a surprise if SpaceX pulled away from its ultimate goal, than if once again, and far wealthier, Elon Musk does not carefully weigh his opportunities and commit whenever he believes the moment is right.

Thus far, the SpaceX founder has had very little to say regarding the prospects for an all-important first Mars mission, other than to note that it would probably be conducted through a "mix" of public and private elements. Furthermore, Musk states that as far at least as far as SpaceX's closely held internal plans are concerned, the Dragon is not intended to haul colonists to Mars. It may be however, that the key word is "colonists," with explorers being a different matter. On the other hand, it is possible that wary of creating his own "flags and footprints" legacy, Elon Musk is waiting on both SpaceX's own introduction of even more robust hardware already under development, and the opportunity for a new public/private partnership to develop, hoping that given time and a steady pace of advancements, once a first mission begins to come together, it will not be a one-time only affair conducted with expensive and unsustainable hardware, but the beginning of a steadily building series of missions which lead directly to the first permanent settlement. The pre-cursor work may already be underway.

The key component in the SpaceX vision is the introduction of what Elon Musk frequently cites as a "rapidly reusable space transportation system." Assuming the company succeeds in recovering and re-launching a first stage sometime over the course of its ever growing commercial manifest for the Falcon 9 v1.1, an event which would perhaps be one of the most important developments in the history of rocketry since Robert Goddard's liquid fueled rocket flew 40 feet into the air above his aunt's Auburn Massachusetts farm in 1926, it would still only mark the beginning of a long development path. This is a point Musk himself frequently makes in emphasizing the word "rapidly." Keeping in mind the various challenges SpaceX has faced with its Merlin engine, notably on Falcon 9 flights 2 and 4, there is a lot of work to be done in fielding a rocket engine robust enough to be used for both launch and recovery, refueled and launched again, time after time, day after day, all without requiring a tear down and rebuild. Still, the process can only begin with the first successful recovery, and SpaceX is holding nothing back in pursuing that goal, even as its immediate competitors look the other way, apparently consigned to business plans consisting of two elements, maintaining government backed monopoly and hoping SpaceX fails. With literally no other player in the world actively pursuing RLV efforts on a comparable scale, perfecting the ability to recover and re-fly the Falcon 9 only a handful of times without overhaul will be enough to secure a dominant position in the commercial launch market for the foreseeable future no matter how competitors respond. It will have undeniably been earned.

All 9 Merlin 1-D's ready for the first Falcon 9 V1.1. Credit: SpaceX

It will also almost certainly be the final stop for the Merlin series of engines. What comes next was until late 2013, a matter of some debate. Throughout its history, SpaceX has periodically offered insights into its long term concepts for a series of boosters considerably larger and more powerful than the Falcon Heavy. Informally designated Falcon X or Falcon XX, these super heavy lift boosters, equaling or even surpassing the Saturn V, would necessarily be powered by much larger engines, likely on par with the mighty F-1. In fact, at several points Musk himself described the open cycle, regeneratively cooled Merlin 1-C as a working scale model of a much larger engine informally designated the Merlin 2 that the company planned to introduce. SpaceX propulsion engineer Tom Markusic lent further credence to the plan with a much publicized presentation at the 2011 AIAA Propulsion Development Conference in which he made the case for a massive open cycle keralox engine which was essentially a super-sized Merlin 1C, a development path which seemed to run counter to recent trends elsewhere in the launch industry.

It came as something of a surprise then, that when speaking to the British Aerospace Institute in late 2012, Musk observed that "the future would be methane," and that the promising but largely untried fuel source would be the basis of SpaceX engine development work going forward.

It is a decision which makes sense for several reasons. The first, and perhaps most important is directly tied to the goal of Mars colonization. As Robert Zubrin popularized in his 1996 book "The Case for Mars," the Red Planet's thin carbon dioxide atmosphere can be readily converted and broken down into methane and oxygen, using technology perfected over 100 years ago.

Known as the Sabatier reaction, it is a process which depends only on supplying a feed stock of liquid hydrogen and electricity to begin the conversion of Mars carbon dioxide atmosphere into rocket fuel. Zubrin's original plan called for transporting the hydrogen feedstock to Mars, but since it was introduced, the subsequent confirmation of water ice across wide swaths of the surface of Mars has suggested that the entire process can be performed using in-situ resources, with the hydrogen supplied through electrolysis of water. Given the desirable performance characteristics of a methane/LOX fuel combination as a "soft cryogenic" which is superior to keralox in terms of specific impulse but does not bring the more problematic aspects of hydrogen based systems, if your intention is a fully re-usable two way transportation system between the Earth and Mars, then methane is the undisputed fuel of choice.

There is another more prosaic reason which argues for methane as well, the price of the fuel. Quite simply, readily abundant methane is the lowest price rocket fuel available on Earth today, and is likely to remain so for the foreseeable future. While the price of fuel is inconsequential in terms of overall launch prices in an era where expendable launch vehicles cost over a hundred million dollars apiece, the equation quickly begins to change as reusable elements are introduced. For a fully, rapidly reusable system, the cost of fuel becomes a deciding factor, just as it is with other forms of transportation. At the time SpaceX began developing the first Merlin engine however, and despite its apparent advantages, there simply wasn't a significant body of experience relating to methane/lox engines from which to build, prompting Musk's team to adopt a keralox architecture for its first family of boosters. Three generations later, and now justifiably confident in its propulsion engineering capacity, SpaceX was ready to take the next step.

On October 23rd, 2013, news media in Mississippi reported that in a surprising move, SpaceX would soon be joining other aerospace companies in conducting major engine testing at NASA's Stennis Space Center. Equally surprising was the engine to be tested, an all methane, staged combustion 650,000 lb. vacuum thrust engine bearing the name "Raptor," a designation previously associated with a hydrogen based upper stage project. According to the guarded statements which soon followed, it is to be the first in an entirely new family of methane engines. The specific agreement called for SpaceX, together with funding from the State of Mississippi and NASA itself, to convert the relatively small E-2 facility which consists of a horizontal and a vertical test stand, each capable of supporting engines operating at up to 100,000 lbs. thrust, to a methane architecture. After beginning testing at the component level, for which the company will re-imburse NASA under a Space Act Agreement, SpaceX indicated it would like to test the entire engine at Stennis as well. With propulsion head Tom Mueller subsequently suggesting that the Raptor will be even more powerful than initially announced, capable of generating approximately 1 million lbs. of thrust, it will clearly require a much larger test stand.

Raptor was a remarkable announcement for several reasons. To begin with, it marked the first time SpaceX has elected to perform significant engine testing outside of McGregor, a major departure from the company's previous strategy. Second, and perhaps more significantly, like Falcon Heavy and Falcon R, it once again highlights SpaceX's absolute willingness to engage in challenging and demanding projects on its own initiative without waiting for NASA or the Department of Defense to agree to pick up the tab. It is difficult to overstate in terms of a national industrial base which is dependent on the goodwill of Russia for access to high performing hydrocarbon main engines, just how significant the Raptor program may become. It also raises the question once more, regarding the supine posture and lack of initiative on the part of the established aerospace base, now reduced to a single provider with the early 2013 merger of the remaining legacy propulsion companies Aerojet and Pratt and Whitney Rocketdyne to become Aerojet Rocketdyne, a company wholly owned by GenCorp.

Although SpaceX indicated component testing would begin sometime in 2014, the initial announcement did not provide any other insight into a development schedule, what the rest of the methane engine "family" might look like, or how the massive new engine would fit into existing or future booster plans. At the intended thrust level, six times that of the Merlin 1D Vacuum engine, unless derated, the new Raptor appeared to be overkill as a replacement for the Falcon second stage, and indeed was. Speaking during an interview in early 2014, SpaceX President Gwynne Shotwell revealed that the new Raptor is not meant for use on the existing Falcon family at all, but instead as the core building block of an enormous new booster designed for Mars. She was quick to add however, that for the immediate future the company is primarily focused on existing product lines and the new markets they enable. Even so, the willingness to publicly announce its intention to develop an engine larger than anything built in the United States since the Saturn V's F-1 served as further evidence that the hard realities of entering the launch business have done nothing to dim the passion for Mars.

In addition to ramping up its Raptor engine development program, SpaceX is also moving ahead on realizing another long standing ambition, that of developing separate launch facilities to service its different client bases, a goal which has led to a very public search for a new commercial launch facility to compliment SLC-40. It is an intriguing development made possible by the minimalistic approach the company takes to establishing its launch site infrastructure. It is also very good business.

The key to the company's goal of developing progressively better and more fully reusable launch systems lies in capturing the highest percentage of the total launch market possible and achieving a rapid flight rate in order to generate cash flow and increase flight proficiency. On the other hand, each of the three customer classes SpaceX needs to enter and dominate; NASA, DOD and Commercial, all have different requirements which are reflected in their respective launch facilities. The

Department of Defense is without a doubt the most restrictive, by maintaining a somewhat arbitrary insistence on vertical integration, as well as exceedingly tight and entirely understandable security procedures, places a higher demand, and thus a higher cost, on its launch providers. While the military often seeks to find ways to share costs with commercial providers, and would like to see its Cape Canaveral launch facilities handle commercial missions, it is somewhat problematic to bring foreign nationals onto the premises, a process which requires constant escorts and disrupts work flow. In other words, many of SpaceX's most coveted commercial customers are not exactly welcome at the SLC-40 launch facility, an aggravating situation if you happen to be the customer and it is your $250 million dollar satellite sitting on the pad. Combined with immediate presence of competitors, high range fees, outdated support equipment and all of the other challenging aspects of operating a commercial business off of a military installation, while it is a workable solution, it is certainly not preferable.

As a result, SpaceX has long been engaged in the search for a location to develop an entirely separate commercial launch facility which could operate independently of the U.S. Eastern Test Range at Cape Canaveral, and in doing so offer the best customer experience, as well as a cost structure which accurately reflects industry rather than military requirements. After considering sites in Georgia, Florida, Puerto Rico, Hawaii and Texas, SpaceX has seemingly settled on a site at Boca Chica beach in Cameron County Texas, only three miles from the Mexican border as the preferred location for a strictly commercial launch facility. Subsequent revelations that the company was quietly buying up adjacent property under the name "Dogleg Park," a none too subtle reference to the maneuver required to reach GSO from a non-equatorial launch site, suggested that while other sites might be of interest, Texas was in the lead.

With the help of enthusiastic marketing from local authorities, as well as changes to state law to allow temporary beach closures during launches, the Texas site offers a clear launch path to equatorial orbits, albeit one which much carefully thread the needle of Gulf oil rigs, as well as a latitude which is actually marginally lower than that of Cape Canaveral. From a commercial standpoint, the primary advantage of a separate launch facility with access over open water is the freedom to conduct its launch operations, while still under the authority of the FAA, but under its own guidelines and timeframes without undue interference from the constraints of a military base or the necessities of accommodating one of the more congested air spaces in the world.

From a political perspective, a Texas launch site, combined with ongoing development work at the McGregor test facility offers a high profile counterpoint to the Johnson Space Center's traditional contractor constituency, giving SpaceX additional standing to engage the always influential Texas Congressional delegation in actively considering its interests. Combined with its existing operations and planned improvements in Florida, as well as its new role at Stennis in Mississippi, SpaceX is slowly putting together a counterbalance to the large contractor coalition which has dominated the American space program since its earliest days. It remains to be seen what the long term effect might be, but there seems to be little doubt that combined with its undeniable success in launch and space operations, SpaceX is determined to earn its place at the legislative table, and a high profile commercial spaceport operating out of the Lone Star State is a good place to start.

Cameron County Texas is not the only new launch location the company has in mind however, and it may not even be the most important. SpaceX will soon become inextricably associated with the one facility which has been at the center of America's space history for the better part of the last five decades, the Kennedy Space Center's historic Pad 39A, home to all but one (Apollo 10) of the manned Apollo Moon missions, as well as the majority of Space Shuttle Launches, including the first flight of Columbia and the last 19 launches of the Shuttle program. Sadly, with the retirement of the Shuttle following its final flight in 2011, and with work already underway to modify the adjacent 39B to a "clean pad" configuration to support the replacement for cancelled Ares-1, the Space Launch System, NASA no longer has a use for its most famous pad.

Formal announcement of NASA/SpaceX agreement to lease historic Pad 39A for 20 years. Left to right are NASA Administrator Charles Bolden, SpaceX President Gwynne Shotwell and Kennedy Space Center Director Robert Cabana. 04/14/2014 Credit:Stewart Money

In early 2013, with no clear need for a second pad to support a system anticipated to launch at the tepid pace of once every two years, and facing the necessity of regular upkeep, NASA announced its intention to lease the 39A facility to a commercial provider if anyone was willing and a deal could be worked out. Initially the only interested party appeared to be SpaceX, but in an unanticipated development, long silent Blue Origin, which had still yet to release any details concerning its secretive development work beyond the necessary disclosures as part of the CCDev-2 program, put in a last minute bid as well. Somewhat curiously, Blue Origin's bid outlined a business concept of leasing the pad as a non-exclusive operator with the intention of making it open to multiple parties. While an interesting idea at least in theory, it appeared to ignore the reality that launch pads are not like airports, and unlike the generic fuel and cargo trucks, passenger ramps and food service fixtures which are common to commercial aircraft, it is a very different case where launch vehicles are concerned. Taken together with the sheer scale of 39A, and the rather notable lack of any other plausible launch vehicle in search of a home, upon closer scrutiny, the plan did not appear to make a lot of sense. Not surprisingly, there was more to the story.

NASA had wanted to make a decision regarding the lease of 39A by the end of the 2013 fiscal year on September 30th. Frustrating that effort, Blue Origin filed a protest with the GAO on September 3rd, while the bid process was still underway. The substance of the protest was that NASA had given preferential treatment to SpaceX in the course of discussions leading up to the formal bid announcement. As part of the protest, Blue Origin included a letter of support from United Launch Alliance. While ULA, like any other company, had been presented with the opportunity to turn in its own bid for leasing the historic pad, it elected not to do so, but then promptly joined Blue Origin in stating that it might possibly be interested in a future use of the pad under Origin's stewardship. At this point Alabama Congressman Robert Aderholt attempted to introduce an amendment in NASA's FY 2014 funding bill which would have denied NASA the ability to lease the pad to any company proposing exclusive use, a clear political shot aimed directly at SpaceX. When the amendment failed to pass, a coalition of Representatives and Senators whose districts rather pointedly happened to cover Blue Origin and United Launch Alliance sent letters expressing "deep concern" over the proposed lease addressed to the NASA Administrator. In an increasing sign of its growing influence however, that letter was soon followed by another, this one signed by the entire Florida house delegation, as well as another signed by influential Florida Senators Bill Nelson and Marco Rubio, supporting NASA's authority to decide for itself how to handle the facility. Within that context, it was difficult not to see the effort as less about opening opportunities for Blue Origin, as it was about yet another attempt by ULA to block opportunities for SpaceX, with the battle being fought as usual not in the open marketplace, but behind closed doors in Washington D.C. While it was only a matter of time until the media, intrigued by the notion of two internet billionaires fighting over the same piece of NASA property, picked up the story and ran with it, the public narrative completely missed the darker undertones, and the fact that the issue was about much more than a clash of super-sized egos.

The prospect that ULA might seriously consider sub-leasing the pad from Blue Origin was undercut by the fact that with its commitment to a vertical stacking process, it appeared highly unlikely that the company would commit the resources to build yet another vertical assembly structure at the NASA facility when it was comfortably ensconced at its own pad only a few miles away at Cape Canaveral Air Force Station. Furthermore, with SpaceX edging ever closer to breaking into the EELV launch business, it hardly seemed likely that ULA was facing a glut of launch opportunities in the near future. Even in anticipating the possible impact of a win in the Commercial Crew program, any gain in the total number of Atlas launches, at either one or two a year, were almost certainly going to be offset by corresponding SpaceX gains in the defense market. Annoyed by interference he described as a "phony blocking tactic," and highly skeptical of Blue Origin's own claim that it would make use of the facility by 2018 when it was showing no visible progress in developing its own booster, Elon Musk subsequently upped the ante, offering to make the historic pad available to any other willing users, while assessing Blue Origin's odds of successfully doing so as less than the prospect that SpaceX would "discover unicorns dancing in the flame duct." On Monday, April 14th as it was preparing for the CRS-3 launch to ISS, SpaceX and NASA announced the conclusion of negotiations over the historic pad, signing of a twenty year lease agreement which would likely see the first Falcon Heavy launch take place from the home of Saturn V and Space Shuttle in little over a year.

Despite the proximity of its own SLC-40 pad, SpaceX has a number of very sound reasons for seeking to lease 39A as well. On a basic level, the company wants to secure a dedicated NASA launch site for what it foresees as a growing list of NASA missions, including hopefully another round of cargo missions under the CRS program after the expiration of the original contract, and perhaps more significantly, any flights to be conducted as part of the Commercial Crew program. Finally, there is also the occasional NASA ordered launch under the National Launch Services II contract. In addition to the overall efficacy of using NASA facilities for NASA launches, one of the prime benefits is anticipated to be a reduction in the start-up costs for Commercial Crew, where the presence of a pre-existing Shuttle era crew escape system (a tower mounted zip line leading to an underground bunker), would spare the company the necessity of qualifying a new system to NASA standards. In this way, and in continuing to use SLC-40 for the unique demands of DOD launches, SpaceX has the opportunity to price each of its launch services according to the specific needs levied by the customer, thereby avoiding the potential for expensive NASA or DOD requirements to affect the cost structure of highly competitive commercial launches. There are however, a number of other benefits which makes the lease of 39A a good arrangement for both NASA and SpaceX.

Perhaps the most significant is the matter of image. Not only is KSC's 39A the best known launch pad in the world, it is also far and away the most visited. Each year more than a million people visit the Kennedy Space Center Visitors Complex, a number sure to increase with the added attraction of the Space Shuttle Atlantis exhibit. With many of those visitors coming from around the world, and drawn by the unique family vacation destination of Disney, Universal Studios and all that the Orlando area offers, the Kennedy Space Center stands as a monument not to escapist fantasy, but to real life human adventure and American accomplishment on the grandest stage possible. Significantly, a substantial number of those visitors take the bus tour to the Saturn V exhibit which is not on site at the Visitors Center, but within the expansive confines of the Kennedy Space Center itself. In the process, the tour buses pass by not only pad 39A, but actually stops at its dedicated viewing tower as well. With the infrastructure at Pad 39B substantially demolished to make room for a Space Launch System which will not host a crewed launch until 2021 at the earliest, it is all too easy to see the two historic pads not as symbols of American triumph, but of American decline. Sitting vacant and beginning to rust, it is not the impression either NASA or the United States really wants to send to its own citizens, as well as to the rest of the world. Compared to the alternative, the chance to lease the pad to an American company, launching an all-American booster on a regular basis to the International Space Station is a public relations godsend for NASA. From the moment work banners are hung up, and a Falcon rocket is on the pad at 39A, NASA will have a tailor made example of its concept of KSC as a 21st century spaceport. More importantly it will be an indelible symbol that the winds have shifted, and after the demoralizing end of the Shuttle program, the United States is on its way back to space, perhaps this time to stay.

Falcon 9 with landing gear being prepped for the CRS-3 mission. Credit: SpaceX

Although SpaceX couched its intentions towards 39A in terms of serving its customer NASA, there is another potential application which is rather intriguing. Built as it was to service first the Saturn V and then the Space Shuttle, the NASA facility contains some unique and very impressive infrastructure, including the shipping canal and turning basin used to unload Shuttle External Tanks and Saturn V stages built at NASA's Michoud facility on the outskirts of New Orleans and shipped to KSC by barge. At 12 feet in diameter, the current Falcon 9 is just narrow enough to be transported (in segments) legally and without great difficulty from Hawthorne California, to Cape Canaveral or to its prospective commercial site, via McGregor Texas. At whatever point SpaceX elects to introduce a new, larger diameter booster to supplant the Falcon Heavy, it will have little choice but to change its manufacturing and transportation arrangements. One possibility, and according to Elon Musk, perhaps the most likely, is to build the stage segments on site, and the availability of unique facilities at KSC offer not only that capacity as well, but also at least the potential of booster construction anywhere in the U.S. with available barge access, including presumably Michoud itself. With 39A's unique legacy, one cannot imagine a more fitting launch site for SpaceX's first super booster. That is provided of course, the pad is large enough to support such a rocket. Incredibly, there are some indications that as massive as it is, 39A may prove to be too small to accommodate the largest boosters SpaceX has in mind.

Until it reaches that point, there is an intriguing logistical angle to the 39A agreement. When combined with a commercial site in Texas, and the existing site at SLC-40, SpaceX would have as many as three Easterly oriented pads available at the same time. Should it have the need to do so, the company could potentially conduct three launches in support of a lunar or Mars mission within a matter of a few days, or even hours of each other. Under such a scenario, using a Falcon Heavy for all three missions would result in the capacity to place in excess of 150 tons into the same orbit at essentially the same time, effectively giving SpaceX the ability to launch more mass in support of a Lunar or Mars mission in 24 hours than any single version of SLS would be capable of achieving in the next twenty years, if ever. Furthermore, anticipating steady reductions in pad turnaround times, there would be little reason to believe SpaceX could not repeat the effort within a matter of weeks. That such a capacity could be readily available within a few years, and at no development cost to the taxpayer or the agency should perhaps provide NASA mission planners with a moment of reflection.

Whether it plans to utilize the Falcon Heavy to open a pathway to Mars, which based on more than a few cryptic comments as the Falcon 9 v1.1 came together suggests may be the case, or to wait on the introduction of an even more powerful booster in the future, the ultimate element in realizing SpaceX's vision of the future is implementing reusability, and substantially reducing the costs of space launch for all comers. Absent that development, and the subsequent emergence of entirely new commercial markets in low Earth orbit and beyond, the business case for any mega booster is highly suspect. It is for this reason that Musk so relentlessly preaches the gospel of a "rapid and fully reusable" space transportation architecture, and it is why the company has placed such a priority on the Grasshopper and Falcon-R programs that it stands a legitimate chance of successfully flying back a Falcon 9 first stage before the Falcon Heavy even makes its maiden launch.

When Elon Musk introduced the Falcon-R in September 2011, even he admitted the goal might not be achievable. When the Grasshopper made its first 1.8 meter hop a year later on September 21st, 2012, the prospect of stage recovery still appeared to be a long way off. What a difference a year makes. With a progressively higher series of hops conducted throughout the following year, SpaceX indicated after the eighth and final flight, a 744 meter (10 story) climb on October 7th, 2013 that it was already moving on to an entirely new round of Grasshopper flights. These would be conducted with a new test article based around a Falcon 9 v1.1 first stage equipped with flight weight retractable landing legs. Significantly, these would start in Texas, but then shift to Spaceport America in New Mexico.

The reason for the change of venue, made possible by a 3 year lease agreement signed on May 7, was that by moving testing to Spaceport America, the U.S. base of operations for Richard Branson's Virgin Galactic, SpaceX could take advantage of pre-existing FAA flight waivers allowing ascents to take place all the way up to the edge of space itself, unconstrained by the much more restricted 11,500 foot ceiling authorized at McGregor. Pursuing that opportunity, construction began on a simple 30 x 30 meter pad in July, with initial flights expected to begin shortly thereafter. Based on the surprising progress achieved on the first flight of the new Falcon 9 on September 29th, which saw the descending first stage, under power, come within range of cameras aboard the recovery vessel, it seems entirely likely that SpaceX will close the circle between the low altitude leaps begun in Texas and full flight recovery at Cape Canaveral sooner than anyone would have imagined.

As significant a moment as it may be, the immediate impact is likely to be more psychological than financial. Perhaps more than with any other issue attached to SpaceX, the questions raised by the march towards full and rapid reusability are substantial and open ended. With Musk having observed that 75% of the cost of the Falcon 9 resides in the vehicle's first stage, an unsurprising number given the 9:1 ratio of Merlin engines between the upper and lower stages, the introduction of a five mission lifespan for each Falcon-R first stage would appear to offer the potential for further blowing the floor out from under all existing launch vehicles, and dropping the cost per pound to LEO well under the magical threshold of $1,000. Unfortunately, it is not quite that easy.

The unfortunate downside of any reusable system, regardless of whether it is based on wings, parachutes or powered touchdowns, is the unavoidable impact on lift capacity, generally thought to be in the range of 40% for a fully recoverable two stage booster. A substantial figure, it would apparently reduce the Falcon 9's LEO payload capacity from close to 30,000 lbs. to just over 16,000 lbs. Both followers and most certainly competitors were shocked to learn in early 2014, that the advertised performance levels for the Falcon 9 v1.1 *already* include the margins necessary for first stage re-usability. A remarkable statement, not only does it mean that the company can pursue commercial contracts unhampered by reduced vehicle performance, it also implies that there is substantial untapped capacity in the fully expendable versions of both the Falcon 9 and Falcon Heavy.

Of course, not every flight requires a booster's maximum capacity. In the case of both the Falcon 9 and Falcon Heavy, the baseline booster is designed to the high side of its potential market, but with the potential to sacrifice performance to achieve re-usability and still be commercially relevant. This is in complete contrast with the United Launch Alliance approach, in which both Delta and Atlas boosters begin with a less capable core, progressively adding expensive strap-on elements to achieve higher performance.

For its part, SpaceX identified NASA's Commercial Resupply missions, which maximize the volume of the Dragon cargo capsule rather than the overall capacity of the booster as the ideal circumstances in which to progressively test its recovery capacity. Testing will also take place on commercial launches which do not tax the entire capacity of the Falcon 9. In each case, the fuel margin not required to reach the target destination becomes available to progressively test a return under power instead. SpaceX estimates even a fifteen percent margin is sufficient to conduct a controlled water landing over the ocean, with twice that amount required for a return to dry land. Anticipating a regular pace of CRS launches through the remainder of the initial contract expiring in 2016, as well as an equal or greater number through competitively competed follow-on contracts lasting the life of the station, SpaceX will not lack for opportunities. In terms of symbolism, just as with the Saturn V/MCT tie, there could not be a more appropriate location than 39A from which to build on the Shuttle's legacy as the world's first partially re-usable launch system.

So what might it all mean?

As of 2014, a fully expendable Falcon 9 v1.1 lists for $56.5 million, offering a maximum lift capacity of 28,991 lb. to a 28.5 degree low Earth orbit. In purely theoretical terms, the result is a cost of $1,948 per lb., a figure which symbolically at least breaks the $2,000 per lb. benchmark and makes it the lowest priced EELV class booster in the world. Momentarily leaving aside the fact that ISS occupies a 51.6 degree orbit, and that most Comsats are placed into far more demanding geostationary transfer orbits to which the same booster can only loft 10,692 lbs., how many flights would be required to break the $1,000 per lb. benchmark? The answer depends first and foremost on the ultimate lifespan of the Merlin 1-D engine, a figure SpaceX has not disclosed, and may not even know.

What is known, however, is that the Merlin 1D was designed and tested to exceed its single mission burn lifespan by a factor of four. Accepting this to be the case, a four mission lifespan for the Falcon 9-R first stage would mean that SpaceX could expect to re-coup the $42.4 million sales price (the costs are understandably confidential) of the first stage over its four flight lifespan, resulting in an amortized price of $10.6 million price per launch. Added to the price of the fully expendable second stage required for each of the four flights, and the total launch cost falls from $56.5 million to just over $25 million per flight. But there is still another factor which has to be taken into consideration; the cost to re-furbish each recovered stage and then integrate it with a second stage into a new launch vehicle.

Making what is likely to be a very conservative assumption, and assigning the entire recovery, refurbishment and re-integration amounts at 25% of price of a new first stage, the effective cost of the re-usable first stage doubles, to $21,187,500. With the addition of the second stage, the total cost per flight over a four mission lifespan becomes roughly $33.3 million. Still pretty good compared to $56.5 million, but at $1,250 per lb., it does not break the threshold.

This simplified example indicates why it may take quite some time to reach truly breakthrough pricing, enough to open entirely new fields of space commerce, with a re-usable vehicle. In terms of doing so, SpaceX will face a challenge on two fronts; extending the flight life of each Merlin 1D engine, and substantially reducing the costs of maintenance between flights. Fortunately, modest improvements offer the opportunity of significant gains. For example by extending the life of each first stage to 10 flights, and reducing the between flight maintenance to 10% of the first stage price, the barrier comes within reach. One other factor seems clear as well. The cost of the second stage, initially discounted, quickly becomes a major factor as the working life of the first stage surpasses five flights. At some point, it becomes more cost effective to recover the second stage, and achieve full reusability, than to wring further performance out of the first. On the other hand, in addition to the development costs, the performance hit is even more substantial. Every pound added to the second stage, whether for fuel, heat shielding or recovery thrusters, takes away an equal pound of payload capacity. It is going to be fascinating to see just where that threshold lies. For its part however, SpaceX has predicted that it could ultimately achieve a $5-7 million dollar launch cost for the Falcon 9-R, a figure which clearly requires second stage re-usability.

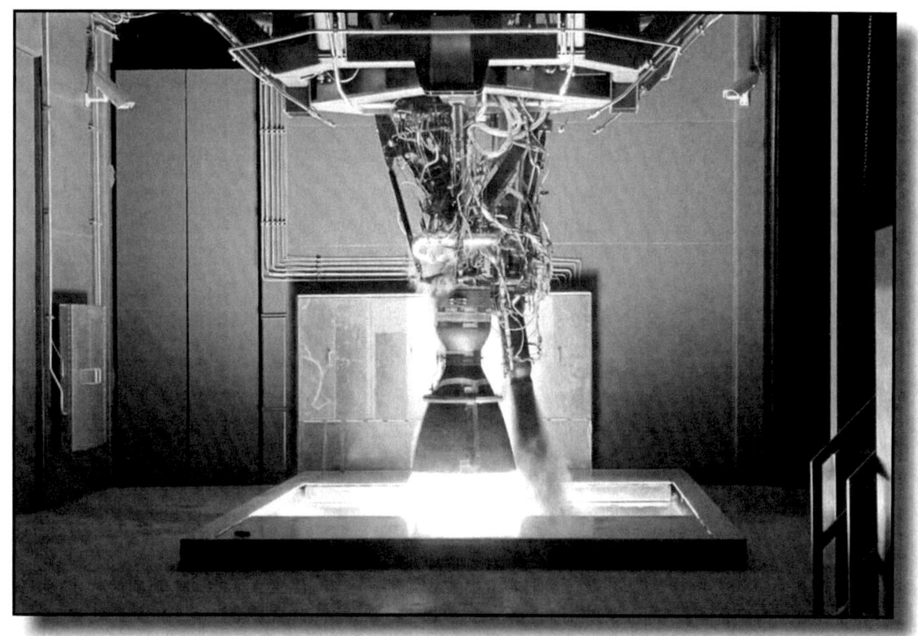

Merlin 1-D undergoing testing at McGregor. Credit: SpaceX

Although Elon Musk originally presented the re-usability efforts in regards to the Falcon 9, it has been clear from the outset that in terms of reducing the cost per pound to orbit, the largest gains are available to be made with the Falcon Heavy, where even as a fully expendable vehicle, the booster flirts with the $1,000 mark. When even modest first stage re-usability is taken into consideration, $500 per pound becomes a very achievable goal. The huge reduction in relative costs in going from the Falcon 9-R to Falcon Heavy reveals why Musk and SpaceX are so focused on developing a next generation 'super booster" even though the company clearly stands to capture a dominating share of every existing market available to it with its current products. While it is entirely feasible that the right public/private venture might be put together to conduct the first manned mission to Mars using near term SpaceX technology, just as we saw with the Apollo program, interest will fall quickly as the novelty wears off, and no nation is likely to support more than a handful of missions costing in excess of $1 billion per person. It is only by developing a much larger, fully re-usable launch system that SpaceX can drive the price to orbit low enough to facilitate the settlement of Mars by people who must buy their own ticket.

Interestingly, for those of us with no interest in permanently leaving Earth for an uncertain future on a cold, airless world, but who still might want to get away for a while, it seems altogether likely that a fully re-usable mega booster could open up the Earth/Moon system in a way which seems hardly believable just a few years out of the Shuttle era. The same elements which make it even conceivable to contemplate large scale emigrations to Mars would also make much simpler, cheaper and far shorter vacations to low Earth orbit affordable to hundreds of thousands, and perhaps even millions of people. For the trendsetters, possibly mortified by seeing the great unwashed invade the realm of orbital space, lunar tourism will almost certainly become the next big thing. All of this will take time however, which begs the question; is it likely that rather than waiting on new markets to emerge at some unspecified time in the future, that SpaceX would introduce its next generation booster with Mars colonization as the first, rather than the ultimate use?

While Elon Musk has offered a plan, or more realistically a vision, of Mars colonization based on a back of the envelope calculation of .0001% of the world's current population of 8 billion people who might be both willing and able to emigrate to Mars, there is a different way to look at the issue. The 80,000 number is based on a rudimentary assessment that if the cost of moving to Mars was roughly established at $500,000 dollars, the equivalent of the sale price of a median income house in California, then potentially at least, there is a considerable population of people which could have the means to travel provided they cashed in their savings and sold their houses and cars in order to do so. It is easy to become distracted by the seeming absurdity of what is an optimistically extrapolated s.w.a.g. of how many might ultimately go, and miss the far more relevant detail of how many might go at first.

Even notional details are very sketchy, but as Musk sees it now, the process could begin with as few as 10 people journeying to the Red Planet and initially living out of the landing module of the spacecraft which brought them to the surface. Later groups of settlers would arrive in slowly building waves, first by the 100's and then only later by the thousands.

Taking the original proposition at face value, and asking oneself "is this really on the horizon?" one way to look at it is by using the same example Elon Musk cited when introducing the Falcon Heavy, a launch vehicle he noted that was capable of placing the mass of a fully loaded Boeing 737 into orbit. As the most commonly flown airliner in the world, it is the one aircraft almost everyone who has ever flown is familiar with and can visualize the interior dimensions.

Though used to make a point regarding the Falcon Heavy's LEO lift capacity, and not its intended payload, it is still more than a little instructive particularly considering the few graphics SpaceX has released depicting a "hab" facility on Mars.

Consider Musk's analogy not in the context of a fully loaded, 135 passenger 737, but instead of a late night flight aboard a sparsely loaded plane with a total of 10 people aboard. With the inclusion of an expanded galley, lavatory and partitions, it is not all that difficult to imagine an equivalent volume comprising the living space for a crew of 10 colonists on a 6 month transit to Mars. By soft landing the entire structure, and deploying inflatable hab modules attached to the central hard shell, much like the far smaller arrangement built around linked Dragon capsules as envisioned by Mars One, the first colonists would arrive with credible resources and the beginnings of a home environment offering more cubic meters per person than a New York City "micro apartment." While lacking the endless entertainment and bustling humanity found on the other side of the apartment door, there is something to be said for exploring a new world, even if Mars suits are a permanent necessity. With the certainly of a constant stream of new arrivals enabled by the manageable costs of a fully re-usable launch system, and the legitimate prospect for a return to Earth at some point, Musk's vision becomes achievable, and perhaps even inevitable.

It would certainly not be without historical precedent. Contrary to popular depictions, it has not always been the poorest of the poor who have made the decision to leave home and strike out for new lands, but often, those somewhere in the middle, who have the means to move, accompanied by the belief that there is a very good reason for doing so. Is it really that surprising then, that if the most basic element of life on Mars, the 38% pull of gravity compared to that of Earth, proves to be acceptable to the human organism, that a small element of the existing population would elect to set off on what would undeniably be the adventure of a lifetime? If there is one clear message which comes down to us from the earliest eras, reaching all the way back to the day the first hominids left the African rift valley of Olduvai Gorge, it is that we, at least some of us, are incurable wanderers. Whenever there is a means to go, a few of us will. It is vanity of the highest order to presume that the world of the 21st century is so idyllic, that given the chance none of us will elect to go once again.

It is for certain a terrifying idea, the prospect of leaving, in most cases forever, the warm, wet planet which has given rise to human civilization, only to travel to a frigid and arid wasteland which makes the worst desert on Earth appear to be an oasis, but it is also a compulsion that some find overwhelming, even if still inexplicable. In the end, it does not matter if it is 80,000, or 800,000; presented with the opportunity, a measurable fraction will go.

Embarking on its own second decade, SpaceX is in the process of creating just that opportunity. When it arrives, potentially as soon as the beginning of its third decade, a few irrepressible souls will inevitably answer the call humanity has always answered and pave the way for those who follow. It is for this reason that despite all the cries to the contrary, and all the carefully reasoned arguments regarding the political underpinnings of the space program, we do not need a motivating national rationale or an enduring value proposition on which it must depend. We merely need the means to buy the ticket and book the passage. Human nature will take care of the rest. The moment it is introduced, as improbable as it may sound, Elon Musk's Mars Colonial Transport becomes simply a matter of scale, and humanity becomes a two planet species.

It is indeed time to boldly go.

The Dragon V2 on display at SpaceX Credit: Stewart Money

Epilogue

One of the challenges in telling what is very much an ongoing story is deciding where to establish an ending point, leaving future developments to the next writer, or at least to the next volume. *Here Be Dragons* nominally concludes with the SES-8 flight out of Cape Canaveral, and the beginning of commercial operations for the Falcon 9 v1.1. As the spring of 2014 progressed however, a series of surprising and even bizarre events unfolded, creating a near crisis in the U.S. civilian and defense space programs, and thrusting SpaceX even further into the spotlight. When viewed together, the events of a single 90 day period encapsulated many of the themes which emerged during the first era of SpaceX; failed government policy, emerging opportunity, entrenched political opposition, brilliant technological achievement juxtaposed with nagging technical glitches and most of all an unswerving focus on Mars. In short, events suggested the necessity for an epilogue.

Barely 30 days after SpaceX successfully launched its first commercial GTO mission with the lift-off of the SES-8 satellite from Cape Canaveral on December 3, 2013, the company was back on the pad with its second commercial mission to GTO, the launch of the Thaicom 6 satellite on January 6, 2014. This time, the third Falcon 9 v1.1 booster overall, and the second from SLC-40, lifted off precisely on time at 5:06 pm, and just as with the previous launch, the flight was flawless. By any measure, SpaceX had turned in a remarkable performance in achieving two flights in such a short window, but the company had also done something which was perhaps more significant. SpaceX had achieved the all-important three launch threshold to qualify it for consideration in the EELV program under the USAF New Entrants Criteria, and specified in the Cooperative Research and Development Agreement (CRADA) governing the process. With the Air Force indicating that the maiden flight of the new Falcon 9, the CASSIOPE launch on September 29, 2013 would count towards the three launch minimum despite the problem with an optional second stage re-ignition, as SpaceX began preparing for its next mission, the CRS-3 flight to ISS, it appeared that the company was at last ready to move into an era of routine operations and finally begin working down the backlog of commercial launches on its manifest. With the first openly competed EELV launch opportunities expected to come within a year, a successful 2014 flight campaign appeared to be the only substantial impediment to reaching the long sought goal.

The first sign that things were not as they appeared came in the form of a late January announcement that the Air Force and United Launch Alliance had concluded negotiations regarding the 36 core “block buy” the company had been seeking for years. The announcement itself was hardly a surprise, as the space press had reported much the same thing on December 18, but offering little in the way of details. The January announcement came with something else however, a claim, cited by an Air Force official, that the block buy had already resulted in $4.4 billion in savings. The extremely questionable claim was difficult to break down however, because neither the Air Force nor ULA was inclined to reveal the source of the numbers, nor in fact, the contract itself, which remained a secret. Furthermore, a review of the recent annual program budget clearly demonstrated that regardless of what ULA claimed, or how the facts were nuanced by the varying number of cores under procurement in any given year, the gross per flight cost was still outrageous; $365 million in FY 2013, $278 million in FY 2014 and back up to $460 million in FY 2015. Nevertheless, with friendly legislators practically tripping over themselves to tout the “savings” which in fact appeared to be almost entirely illusory, and based on comparisons against wildly inflated projections from 2011, SpaceX found itself in a difficult position. The discomfort however, was mollified by the awareness that competition was right around the corner, and SpaceX could now look forward to aggressively contesting the "up to 14 cores" set aside for that purpose.

Not everyone in Congress was as sanguine about the state of the EELV program however, and with Russia fomenting unrest in Eastern Ukraine, the issue of the Atlas V RD-180 engine, always in the background, was beginning to attract concern. As a result, both Elon Musk and ULA President Michael Gass were summoned to appear on March 5th before the Senate Committee on Appropriations, Subcommittee on Defense, chaired by Sen. Richard Durbin of Illinois. Titled simply "National Security Space Launch Program" the somewhat testy hearing began with a long winded exposition by Alabama Senator Richard Shelby proclaiming the purported $4.4 billion in savings while calling into the questioning the overall value of competition.

It offered a prelude for what was to come, with Elon Musk repeatedly drawing attention to the enormous price differential between the Falcon and ULA boosters, as well as ULA's incongruous dependence on the Russian RD-180 engine for the Atlas V, and Michael Gass attempting to explain how a subsidy was not a subsidy, how competition was good in other places but not the launch industry, and above all else, the priceless nature of EELV payloads. The one thing he could not explain however was precisely why his company had elected to keep those same payloads dependent on Russia.

With relations between the U.S. and Russia steadily souring as Russian troops entered and occupied Crimea, the SpaceX founder observed that it did not appear like a good time to be furthering that dependence. In a somewhat curious twist which highlighted the animosity between the two companies, both left the hearing with a "homework" assignment to submit questions to ask of each other, with written responses due in thirty days. Apparently still unsatisfied, Shelby announced that he would be composing an additional list of questions for both to answer as well. If only the same requirement could be levied on elected representatives.

Two days after the hearing, the entire context of the EELV program took an ominous turn with reports that of the 14 cores set aside for competition, seven were being removed from the list, and the remaining seven opportunities might be further reduced, perhaps to as few as one. The ostensible reason put forward was a slowdown in Air Force procurement of GPS III satellites which would see at least five of the fleet delayed until 2019. The GPS cutback was a curious policy, and one which appeared to be in direct conflict with other statements by Air Force officials that the existing GPS fleet was more fragile than it appeared. With many of the spacecraft launched around the same time, and now "old enough to vote," the constellation was increasingly susceptible to a host of age related maladies which could appear in clusters. In addition to the GPS related flights, two other launch opportunities were now being withdrawn from competition and assigned to ULA. The reason cited in their case was a lack of performance on the part of the Falcon 9, but the Air Force refused to provide any data to justify the claim. From the SpaceX perspective, what had been an acceptable, if not fully equitable, attempt to incorporate competition into the EELV market was quickly turning in to yet another in the long line of accommodations to the ULA monopoly which had driven launch prices to absurd levels. It quickly got much worse.

A letter from the Secretary of the Air Force to Senator Diane Feinstein in response to questions raised during a separate hearing on the EELV program revealed that one of the apparent reasons behind the Air Force change in plans was a still secret provision in the ULA/Air Force purchase agreement which effectively required the Air Force to purchase its full annual allotment of ULA cores before it could open any subsequent launches for competition. If the pace slowed for any reason, ULA would be conveniently protected and competition would be forestalled. This was exactly one of the primary concerns the GAO highlighted in its highly critical 2011 audit of the proposed block buy, and an outcome opponents had feared. It was difficult to reach any other conclusion other than the entire purpose of the block buy from the start was to forestall competition until almost the end of the decade, and the Air Force had gone along with it, despite the 2012 directive from undersecretary of defense for acquisition Frank Kendall authorizing the block buy but directing the Air Force to "obtain the benefits of competition as soon as possible."

That was not all. In addition to plowing ahead with the block buy only a few weeks before SpaceX successfully completed its third qualification flight, and thus would have been well positioned to compete for nearly all single core launches beginning with opportunities in 2015, the Air Force inexplicably advanced the purchase of a single Atlas V launch planned for 2017 more than a year ahead of schedule, making the sole source award to ULA only two days before announcing the block buy. As a result, the Air Force would possibly not be offering up any bids for open competition in FY 2015. For a growing group of bi-partisan Senators, it was too much, with Arizona Senator John McCain entering the fray once again by firing off letters to the Secretary of the Air Force Debra Lee James regarding the "apparently incomplete and incorrect nature" of her prior testimony regarding EELV procurement, and to the DOD Inspector General asking for an investigation.

McCain's letters were not the only shoes to drop that day. Having reached the limit of its patience, SpaceX filed suit against the Air Force and the block buy in the United States Court of Federal Claims on the morning of April 28, 2014, seeking an injunction against the award, and a decision in favor of its right to compete for EELV launches which had apparently been taken away. The SpaceX lawsuit came at what could only be called a most intriguing time. On March 17, the U.S. and European Union had imposed sanctions on 21 Russian and Ukrainian officials linked to Russia's annexation of the Crimea, and the ongoing unrest in Eastern Ukraine. One of those officials, and as it soon turned out, the most vocal of all, was Deputy Prime Minister Dmitry Rogozin, head of the Russian space and defense sectors. As one of the counts in its lawsuit, SpaceX pointed out that ongoing purchases of Russian RD-180 rocket engines for the Atlas V, which ULA was in the process of accelerating, might very well constitute a violation of the sanctions. Even if they did not, the overall issue of escalating tensions with Russia clearly called into question the sanity of a purchase agreement which was further ensuring that same dependence. Interestingly however, the SpaceX complaint did not ask for an injunction against the engines specifically, only against the block buy as it pertained to single core vehicles. Adding a new element to a rapidly evolving story, the Federal judge overseeing the case issued a temporary injunction against the engines anyway, intended to remain in place until the three Federal departments with oversight on the issue, Commerce, Justice and State, could provide letters stating that further purchases did not violate the sanctions.

Despite cries of impending havoc from the usual suspects, the ban was not long lasting. A few days after it was put in place, the injunction was lifted when the Department of Commerce demurred, even as State and Justice stated that they had no substantive proof that Rogozin benefitted directly from the sale of rocket engines to the U.S. On the surface it did not appear as if either department had looked very hard, or even at all, and on another level, it raised the question of why impose sanctions in the first place.

The temporary ban now lifted, ULA could proceed with purchasing even more RD-180 engines, and the debate could resume regarding the overall merits of the case. At least it appeared that way, until Dmitry Rogozin stepped in with an announcement of his own, which in a sign of the times, was delivered by way of Twitter. Apparently in response to the sanctions, and a somewhat confusing restriction on direct communications between NASA and the Russian space program imposed at the direction of the White House, Rogozin suggested that if any further sanctions were forthcoming, NASA might need a trampoline to reach the International Space Station. To drive home the point, he somewhat cheerfully included a picture of a trampoline emblazoned with the NASA logo. His next "tweet" was more explicit.

Going forward, the U.S. would only be allowed to purchase more Russian rocket engines if it could guarantee they would not be used for military purposes. Throwing in an entirely new wrinkle, Rogozin then suggested that Russia would not be participating in the International Space Station program past 2020. Suddenly, the future of ISS, recently buoyed by the announcement that NASA would seek to extend its operation until at least 2024, now appeared to be very much in doubt. Also now in question, was the business case for the Commercial Crew program. In the space of a few tweeted characters, U.S. civil and defense space policy was thrown into turmoil by a single angry Russian official. Adding to the confusion, although Rogozin subsequently repeated his statements in a speech in Russia, the absence of any formal notification, or official clarification of Russia's position left a string of questions, and sent American policymakers scrambling to see what it all might mean. The vacuum created the perfect opportunity for all parties to see what they wanted to see.

Heading in to the final phase of the Commercial Crew program, a suddenly self-righteous Congress was belatedly willing to come closer to fully funding the program along the lines the Administration requested, and which Administrator Bolden continued to assert was absolutely necessary to avoid yet another purchase of Soyuz seats for American astronauts. Old habits weren't completely going away however. On both the House and Senate side, the FY-2015 funding bill still short-changed the program by roughly $63 and $43 million respectively, even as the Senate sought to shower $320 million more on SLS and Orion than NASA had requested. It even took the additional step of specifying in its authorizing language the precise amounts those two programs would receive, thereby directing any funding cuts to come out of other accounts, namely Commercial Crew. No amount of national humiliation was too much it appeared, to explicitly recognize the folly of its own policy in delaying the program from the outset.

As for the Commercial Crew contestants, all three prospective providers were looking at new uncertainty in the long term business case, with Boeing and Sierra Nevada left to contemplate both the wisdom, and the viability, of depending on Russian engines for their entry in America's efforts to regain a domestic spaceflight capability. By contrast, the SpaceX decision to pursue an all domestic booster coupled with an aggressive development schedule for its CCiCap milestones, appeared to be paying off, potentially offering NASA the prospect of an all American launch solution a year ahead of the alternatives, and perhaps the margin to avoid the humiliation of writing another large check to Russia.

Regarding the RD-180 predicament, Rogozin's announcement also appeared to preclude any prospective use by Orbital Sciences Corporation of Aerojet modified NK-33 engines for potential defense related launches in the Antares launch vehicle as well. America's self-inflicted dependence on Russia for both civilian and military launches, one clear manifestation of the "failure of national leadership" identified by the Columbia Accident Investigation Board was finally coming home to roost, and there was no denying one inescapable fact, nearly every institution involved; NASA, the Air Force, Congress, two Presidents, as well as a host of defense contractors, had only themselves to blame. Whether by fate or fortune, SpaceX was left standing alone as the one American institution which had charted a course independent of Russia. Even Boeing, which might have chosen its own Delta IV rocket as the booster for its CST-100 capsule, had instead gone with the Atlas V.

For its part, United Launch Alliance, which might have elected to take the issue as a turning point, and a time to begin extricating itself from Russian dependence and a now compromised procurement strategy, instead chose to blame SpaceX for the entire affair, posting a statement on its website saying:

"If recent news reports are accurate, it affirms that SpaceX's irresponsible actions have created unnecessary distractions, threatened U.S. military satellite operations, and undermined our future relationship with the International Space Station."

For all the bluster regarding the "priceless" nature of the national security payloads it launched, and the unimpeachable record of success it had complied in doing so, the one thing ULA could not sidestep was the responsibility for putting a potential adversary squarely on the critical path for performing those launches. With the terms of the block buy now thrown in to further disarray by uncertainty over how many Atlas missions could be performed, depending on how the sanctions were to be enforced, or even if they were for real in the first place, Dmitry Rogozin seemed to be making SpaceX's case for them. On the other hand, he was also setting the stage for an entirely new round of arguments, as U.S. policymakers sought to devise a response to the possible lack of a rocket engine upon which they had partially placed the security of the United States.

The one certainty resulting from the situation was that it very predictably brought out calls for the federal government to subsidize an American investment in a domestic counterpart to the RD-180, an outcome which would no doubt be well received by ULA, as well as Aerojet Rocketdyne, the sole remaining independent liquid rocket engine manufacturer in the U.S. If there is anything to be learned from recent history, as witnessed by NASA's abortive work on the J-2X, it is that whether new or re-engineered, engines must be considered within the context of the overall market as well as the launch vehicle for which their use is intended. To that point, a hastily convened high level review concluded that precisely replicating the RD-180 in the U.S. would be problematic, which was an interesting finding considering the fact that Michael Gass had stated that his company had spent "hundreds of millions of dollars" to demonstrate doing so would not be a problem. Ignoring the obvious contradiction, the review panel suggested the U.S. should begin funding a new hydrocarbon engine program, even as it offered little in the way of specifics for how it might be integrated with an existing booster. Somewhat ironically, Air Force General William Shelton, commander of Air Force Space Command, told the press that the funds to pay for the engine's development, likely in the range of $1 billion, could not come out of the Air Force space budget, because it was already so stretched. Apparently that $4.4 billion in "savings" so recently created by the block buy had already disappeared into the ether. Seemingly disinclined to further pursue the billions in actual savings being offered by Elon Musk and SpaceX, (a course which would remove the necessity for a new engine in the first place) the USAF cast a furtive glance towards the civilian side of the space program. NASA wasted little time in suggesting that it had little interest in an EELV class hydrocarbon engine program, even as the Air Force went on to assess NASA's own technology as "a thing of the past," a point rather difficult to dispute.

However the EELV lawsuit is ultimately resolved, the role of SpaceX in leading the U.S. out of the RD-180 debacle cannot be dismissed. With nine consecutive successful Falcon 9 launches to its credit as of the CRS-3 mission, SpaceX is clearly on the cusp of entering the defense launch market unless the most blatant bias imaginable is levied against the company. As a result, the Falcon 9 will be an established provider years before any newly re-powered version of the Atlas V makes a test flight, much less enters service. Furthermore, with the Delta IV still remaining in the ULA stable, and the Delta Heavy assured of continued service as the nation's sole booster for heavy lift payloads until the Falcon Heavy has gone through its own round of qualification launches, for the first time in its history, the EELV program will have achieved each of the three goals it originally outlined in moving to a two vendor model in the first place. Perhaps poetically, the arguments once put forward to accommodate both Boeing and Lockheed Martin now turn squarely against the latter. DOD will have two fully independent U.S. providers offering competing service across the entire range of launch vehicles, even if only one of the two is fulfilling the original goal of offering steady innovation and lower costs through a strong commercial presence. Although the Delta IV is more expensive than the Atlas, and reportedly offers a rougher ride to orbit, its record, as ULA never forgets to point out, is exemplary.

Absent a major change in overall U.S. space policy, such as one which integrated NASA and DOD efforts, the rationale for a new hydrocarbon engine program is highly suspect for a defense launch market which will have both the Delta and Falcon families. It becomes even more challenging when taking full consideration of the costs of designing, testing and qualifying a new engine for what will necessarily be a new booster, and then developing all new heavy lift variants and their infrastructure, on both coasts. The case is further undercut by the likelihood that the resulting booster would have limited economic opportunity in a commercial market driven by SpaceX and its headlong pursuit of lower costs and full reusability. The latter point may be the most damning of all where an American RD-180 or its equivalent is concerned. With SpaceX edging ever closer to recovering a Falcon 9 first stage, the pendulum has already begun to swing in favor of reusability, and any new effort which does not incorporate that ability from the outset, and as well as the architecture to take advantage of that capacity, is extremely likely to be obsolete before it begins. Still, Congress did not let rational design considerations get in the way of the Space Launch System in 2011, and with many of the same cast of characters, both on Capitol Hill and in the aerospace sector involved, the nation could very well follow the same path again.

On another level, the rush to fund a new domestic engine for the EELV program, a move which would presumably be a windfall for ULA and Aerojet Rocketdyne, would seem to be painfully myopic, and to many, an undeserved reward for both companies' failure to invest in the domestic industrial base as well as their own future. It would also be a slap in the face not only to SpaceX, but to other emerging launch efforts such as Blue Origin, and Paul Allen's Stratolaunch project. On the other hand, a more open and competitive, COTS-like program focusing on technological risk reduction, component development and demonstration could prove a benefit to multiple parties, SpaceX, ULA and Aerojet Rocketdyne, without unduly prejudicing the outcome. Given the overwhelming success of the COTS program, as demonstrated by both SpaceX and Orbital Sciences, one might conclude that taking a similar approach for engine development would offer a compelling alternative, but just as NASA's own long range plans suggest, the aerospace establishment is still very effective at deflecting uncomfortable proposals.

With the EELV battle now being waged by lawyers, and any engine program years away, SpaceX had other, equally pressing items on its agenda. Following the nearly disastrous CRS-2 mission to ISS in March 2013, a combination of ISS scheduling and the introduction of Orbital Science's own Cygnus resupply vehicle provided a much needed break for a planned upgrade of the Dragon spacecraft, as well as the introduction and first three flights of the Falcon 9 v1.1. It also allowed for something else, taking the next step in the quest for reusability with the addition of flight weight landing legs to the booster slotted to launch the CRS-3 mission scheduled for early 2014.

Though externally identical to the Dragon spacecraft which performed the first three flights to ISS, the CRS-3 capsule would feature enhanced capability to deliver powered cargo, with a provision for four powered units in a pressurized section, as well as two powered units in the Dragon trunk. Complete with new flight computers and an upgraded power distribution system, the first Dragon to launch on the new Falcon 9 was specifically intended to support an accelerated use of ISS research capacity as NASA and its two commercial providers entered the era of routine, dependable transportation.

The one thing which remained less than fully dependable however, was the ability to keep a tightly defined launch schedule. Originally slotted to launch on March 16, the CRS-3 mission was initially delayed two days prior to liftoff when SpaceX personnel preparing the craft for flight discovered what appeared to be an oily residue on the surface of insulating blankets inside the Dragon's trunk. With the mission scheduled to launch a much anticipated laser communications payload to ISS, as well as four high definition video cameras to be mounted on the outside of the Station, all of which were riding to orbit inside the trunk assembly, neither NASA nor SpaceX wanted to take the chance that outgassing could harm the very sensitive hardware.

As it turned out, the residue was a natural result of the manufacturing process, and determined to not pose a threat. Following a resolution of the problem, one which allowed SpaceX to address several other smaller issues, the launch was re-set for March 30th. This time, it was not a SpaceX glitch, or even a NASA problem, but instead a breakdown in the Air Force's outdated and trouble-prone range tracking equipment which caused a delay. One day before an Atlas V launch scheduled for March 25, a radar tracking station south of Cape Canaveral caught on fire, triggering a two week delay for the entire range as repairs were made. After some debate as to which booster would launch first, the Atlas V mission was given the go ahead, performing the launch of NROL-67 on April 10, meaning SpaceX's launch date would now fall to Monday, April 14th. Even this date became suspect when a backup external controller on the International Space Station, known as the Multiplexer/Demultiplexer (MDM) failed. Responsible for controlling the deployment of the Station's constantly changing solar arrays, the loss of the backup MDM threatened a further delay as NASA mission managers conducted a Sunday morning meeting to determine whether the CRS-3 mission could launch without breaking safety protocols which called for the Station to rotate its solar arrays out of the exhaust plume of visiting vehicles.

With a contingency plan in place, and a last minute load of insulating repair tape intended for the MDM spacewalk flown to KSC and stowed aboard the Dragon, SpaceX received a green light for a CRS-3 launch scheduled for Monday, April 14th at 4:58 PM EDT. In contrast to its three previous missions to ISS, the upgraded Dragon spacecraft would now be more heavily laden, containing nearly 5,000 lbs. of supplies and experiments, including a set of "legs" for the "Robonaut" aboard the Station, replacements for failed spacesuit components, including the one which had nearly drowned Italian astronaut Luca Parmitano during a spacewalk the previous July. Also aboard were a variety of experiments, including the first protein crystal growth experiments awarded by CASIS. Launch day began in an unusually festive mood, with SpaceX announcing a 20 year lease agreement with NASA for Shuttle Launch Pad 39A in a brief press conference at the historic facility only a few hours before liftoff. Everything, including the weather, appeared set for a possibly history making day which was drawing nearly as much press attention for the recovery attempt as for the launch. Everything that is, except the Falcon rocket. Roughly thirty minutes from T-0, as the booster was pressurizing for flight, a helium valve in the pneumatic stage separation mechanism failed to maintain pressure, triggering a hold, which immediately turned into a scrub for the day.

Following a return to the hangar for diagnosis and a replacement of the failed component, the flight was re-scheduled for four days later, Friday the 18th, which happened to be Good Friday on the Christian calendar and the beginning of Easter Weekend. Pressing ahead, even though the weather forecast had degraded to a 50/50 estimate of another scrub, SpaceX's luck held, and the Falcon 9 lifted off on its flight to ISS at 3:25 PM EDT.

With a clean launch and trouble-free first stage burn, stage separation brought a new moment of truth, even as the second stage accelerated on its way to orbit. After a brief pause, the first stage re-ignited three of its nine Merlin 1-D engines, and once again successfully slowed itself for a controlled reentry into the atmosphere as it had done on the booster's maiden flight. This time however, the Falcon was equipped with the stabilizing effect of four flight-weight carbon finder landing legs, as well as a significantly upgraded cold nitrogen gas roll control system. As it entered the lower atmosphere, the Falcon first stage was still on target for a historic soft landing. Unfortunately, the Atlantic Ocean waiting below had something else in mind, offering a seething caldron of 20 foot waves and driving winds in the midst of what amounted to a furious late winter storm. With possible recovery vessels kept in port due to the dangerous conditions, the Falcon 9 was descending alone into a mariner's nightmare. Even NASA's photographic chase plane was grounded by icing conditions. Undeterred, Elon Musk sent his personal plane into the fray, hastily equipped with an antenna partially fashioned from a pizza pan in hopes of picking up telemetry from the descending booster.

Though it would take several days to reconstruct the narrative of what took place from the fragmentary data stream and garbled video captured by on-board cameras, the results were nothing short of amazing. Apparently oblivious to the raging weather, the Falcon 9 first stage had descended tail first and under complete control until it reached a steady hover just over the surface of the pitching ocean. At that point, with landing legs fully deployed, the center mounted Merlin engine terminated thrust, allowing the booster to fall intact, on its side, where flight computers continued to transmit data eight seconds after going horizontal. By the time recovery vessels made it to the area days later, there was little to be found but floating debris, including most of one leg, with the main body of the rocket almost certainly ruptured and sunk by the violent sea state within minutes or hours of having come to rest. Still, SpaceX had achieved truly stunning progress since the last recovery effort, leaving little doubt that the company was well on its way to scoring one of the red letter achievements in aerospace history, a first ever powered booster recovery, possibly as early as the next flight on the manifest, the upcoming Orbcomm OG2 launch.

As for the rest of the CRS-3 mission, after resolving one brief problem with a single isolation valve as the Dragon entered orbit, NASA and SpaceX enjoyed another successful trip to the Station, only to be plagued by another water immersion issue when the Dragon splashed down. After a 28 day stay, during the course of which much of U.S. space policy was cast into doubt back on Earth, the Dragon departed ISS on Sunday, May 16th, once again making a successful re-entry in the Pacific Ocean off the coast of Baja, Mexico. This time there was no electrical failure, but a problem was revealed nonetheless when an amount of seawater described as "substantial" was discovered on the floor of the pressure shell, as it was opened for the removal of rapid retrieval items. According to both NASA and SpaceX, no damage was incurred. At the same time, no explanation for the presence of sea water was immediately forthcoming either, leading to speculation that a combination of high seas, an extended recovery time and a more heavily laden capsule riding lower in the water may have led to sea water entering through an open valve. As usual, following the voyage back to Los Angeles, the Dragon was soon on its way to McGregor for complete unpacking and post flight analysis.

There was once again, a sordid political aftermath to a SpaceX launch. On April 29, yet another Alabama congressman, this time Mike Rogers, Chairman of the House Subcommittee on Strategic Forces sent a nearly unintelligible letter to both NASA and the Air Force, putting forward an allegation that SpaceX had missed its target orbit on the CRS-3 launch. The rest of the letter presented a jumbled re-hash of prior, publicly documented SpaceX incidents as if they had only recently come to light, asking for further investigations. It should perhaps come as no surprise that Roger's district happens to border on the Decatur, Alabama facility of United Launch Alliance. As to the specific charge of a missed orbit, and implied cover-up, SpaceX quickly put forward a strongly worded denial, pointing out that with the Air Force and NASA both receiving reams of launch data, and in near daily communication, the suggestion of concealment was patently absurd.

Perhaps it was the letter, and what appeared to be naked political smearing, or maybe it was an overall frustration with the EELV battles, but on May 22, Elon Musk launched a broadside which surprised even his most ardent supporters. Engaging in what one source called a "trial by twitter" the SpaceX founder issued a series of tweets in response to an on-line article detailing the fact that only a few months after retiring from the Air Force, a key official in the EELV program office, and one who had reportedly signed off on the ULA block buy, had taken a high level job with Aerojet Rocketdyne. As an indispensable supplier to ULA, building both the RS-68 and RL-10 rocket engines, and having a stake in the RD-180 engine imported by RD-Amross, the tie was too much for Musk, who tweeted that the official had previously sought a job with SpaceX. Going a step further, Musk suggested the likelihood of a revolving door, quid pro quo which might have played a role in the block buy, asked for an investigation. One way or another, the long festering EELV debate was now clearly personal as well.

Legal and political drama aside, back at Cape Canaveral, it was time to perform.

Following the CRS-3 liftoff, SpaceX wasted no time in getting another Falcon 9 rolled out to the pad, and initially it appeared that the company might actually return to space before the Dragon even left the Station. The next mission, and one which marked in earnest the start of a fast launch pace intended to carry throughout the year, was the long standing Orbcomm OG2 flight carrying six small M2M comsats. Significantly, the booster would be burdened with a relatively light load, and launching into a friendly orbit which would see the Falcon 9 first stage re-enter the atmosphere much closer to Cape Canaveral than took place on the 51.6 degree inclined CRS-3 launch, and in broad daylight as well. In short, the Orbcomm mission represented a nearly ideal opportunity for a new recovery effort, and one promising much better weather. The gremlins however, had yet to leave SLC-40. Scheduled for Saturday May 10th, the first attempt was scrubbed due to an undisclosed helium pressurization problem which cropped up during fueling for a static fire on May 9th. With a delay in resolving the helium leak pushing the Orbcomm launch back until at least mid-June, likely resulting in a more problematic night time first stage recovery effort, SpaceX was now facing an altogether more important issue, meeting the rest of the year's launch calendar. This was one criticism nearly impossible to deflect, and a point repeatedly made by ULA even as other arguments lost water with each new launch.

Throughout its history, SpaceX has invariably displayed an ability to delight supporters and infuriate detractors by focusing attention on future projects, even in the midst of more immediate challenges, and the tumultuous spring of 2014 was no exception. Following Dmitry Rogozin's "trampoline tweet" on April 29, Elon Musk responded with a tweet of his own saying, "Sounds like this might be a good time to unveil Dragon Mk 2 spaceship that @SpaceX has been working on with @NASA," adding "no trampoline needed." The next tweet set the date, "curtain drops May 29, flight hardware, not a mockup."

The reveal took place one month later at SpaceX's Hawthorne, Ca. headquarters. Scheduled to begin at 6:00 pm PT, like so many other SpaceX launches, it did not quite go off on time, but once it did, no-one was disappointed. With a crowd of government customers, VIPs and media crammed into a confined area at the cavernous entry to the factory floor, and the COTS 1 Dragon capsule suspended overhead, Elon Musk mounted the stage in front of a curtain standing between the crowd and the future of space transportation. If Musk's performance lacked the flash and Tony Stark flair some may have been expecting, (no AC/DC, no dancing girls) the genuine enthusiasm and unrestrained exuberance for the new spacecraft came shining through.

When the curtain finally dropped, standing before the crowd on its own extendable landing legs was a generational advance in the art of spacecraft form and function. In other words, pretty much what SpaceX had been promising all along. Although built around a similar pressure shell as the original Dragon design, the new version, informally designated "V2," stood both taller than its predecessor, and noticeably wider, with much of the change due to fairings covering the 8 SuperDraco engines. Arranged two by two into four nacelles offset around the base of the spacecraft, with each faring ending in a recessed exhaust ramp to direct the flow from the engines, and flanked by three conventional Draco thrusters on each side, the SuperDraco engines eliminated the consistent curvature of the original Dragon, giving it a decidedly more ship-like aspect.

At the top, the expendable nose cap is replaced by a hinged cap protecting the SpaceX manufactured docking adapter. Inside the capsule, the seat arrangement remained as originally depicted with four seats on the upper level, and three on the lower. The Tesla influence is not hard to detect either, with the seats trimmed in leather and embossed with the SpaceX logo, and a fold away control panel consisting of Tesla Model S display panels ruggedized for launch conditions and backed up by a smaller conventional control panel covering fundamental functions. It is a design clearly optimized to make the most of the small space which is required to contain at least seven astronauts, and one which includes five surprisingly large, oval shaped windows arrayed around the forward facing wall of the cabin, including one in the smartly hinged hatch itself.

The true measure of the change being wrought by the Dragon 2.0 comes not in the interior arrangement, or the appealing, almost sci-fi shape and contrasting black and white color scheme, but in the SuperDraco engines and the heat shield designed to enable re-flight. While SpaceX had published a video of a SuperDraco test firing as early as 2012, the flight weight version of the 16,000 lb. thrust engine used to achieve qualification for launch offered a surprise. It was according to Musk, the first ever rocket engine constructed by 3-D printing, a solution he described as a "Hail Mary" attempt to achieve both low production costs and maintain mass margins. Built out of Inconel, a high heat and corrosion tolerant steel alloy notorious for its difficult working characteristics, "3-D printing" is achieved using a laser sintering process. The procedure works by hitting minute amounts of powdered metal with carefully aimed laser pulses, allowing a complex shape which would otherwise require expensive tooling and time intensive procedures. Built layer by layer, the result was a small, lightweight engine with big engine features, including the narrow channels required to offer a regenerative cooling capacity. Every bit as crucial to the SpaceX plan going forward as the far larger Merlins, the SuperDraco would be called upon to work under a wider variety of conditions, including both instant full power, very deep throttling, and of course, multiple restarts.

With a pair of SuperDracos embedded in each fairing, the Dragon is designed to offer redundancy sufficient to accommodate the loss of two engines and still perform its mission, one which Elon Musk emphasized throughout the evening, included not just crew escape, but the ability to touch down anywhere on the planet "with the accuracy of a helicopter, the way a 21st Century spaceship should land." Compared to the rough and tumble, almost violent nature of even the best Soyuz landings, the SpaceX founder was not only likely taking a shot at Boeing, he was also offering benefactor NASA an honorable retort to the beating it had recently been taking at the hands of Russia.

The plan for demonstrating just that landing capability had actually already slipped out the week before in the form of an FAA Environmental Impact Statement (EIS) filed by SpaceX to cover a new testing program at the McGregor Rocket Development Facility. Much like the previous 2011 assessment which revealed details of the soon to be announced Grasshopper program, the latest EIS painted a picture of an aggressive, even thrilling flight program for a re-usable test article named DragonFly.

The planned DragonFly test regime calls for four different types of flight operations, beginning with helicopter aided parachute drops from 10,000 feet, with propulsive assist enabled by a five second burn of the SuperDraco engines. Next would be full propulsive landings, also following a parachute drop and consisting of five second burns. The remaining two regimes, described as "hopping" begin on the ground with DragonFly boosted to altitude by its SuperDraco thrusters and then brought back down either by a combination of parachutes and thrusters, or on the power of thrusters alone. In both cases, the engines would burn for 25 total seconds. The schedule was more than a little revealing; whereas SpaceX anticipated "up to" two annual flights taking place under each of the first three scenarios, the company foresaw as many as eighteen flights under the latter. Clearly, thruster assisted landing is much more than a novelty, or a long term goal, it is a fundamental tenet of Elon Musk's vision of a rapidly re-usable spacecraft, and may in fact become the first operational component of just such a system. Interestingly, some of the tests are scheduled to take place with a mock trunk assembly attached, and brought back to rest on its own set of legs.

In looking forward to actual orbital flights, the trunk assembly offered a number of surprises as well. The first is that it would be included at all, as previous comments suggested the passenger friendly version of Dragon might fly with no trunk, and no solar panels, relying on battery power alone for the short transit time to ISS. Now however, the trunk will take on added importance with the inclusion of four fins designed to help aerodynamically stabilize the craft, as well as lower the center of gravity in the event of an emergency separation and boost while still in the thicker part of the atmosphere. Once in space, the trunk, fins included, generates power through the use of solar panels embedded on its surface, providing augmentation to the on-board batteries. Here again was another distinction from both the Boeing CST-100 and the Sierra Nevada Dream Chaser, and a signal to the rest of the aerospace world, the new Dragon might be able to touch down anywhere on land, but it was being built for space, and for journeys much longer than the quick jaunts to ISS.

It was also being built for re-use, a prospect considerably improved by the fact that it would be returning for touch-down, rather than splash-down. To paraphrase Heinlein, the sea, as SpaceX was continuing to experience, can be a harsh mistress. So too can the atmosphere, particularly when you slam into it at 18,000 miles per hour, and in combatting that environment, Musk revealed the second major feature of the new Dragon. Over its five flight history, SpaceX engineers had continued to work on improving the PICA-X heat shielding which had successfully protected each incoming Dragon spacecraft to date. Describing the most recent CRS-3 Dragon as featuring a "version 2" improvement over the original shielding cladding the COTS 1 capsule hanging over his head, the new, V2 Dragon would have a third generation heat shield offering significantly reduced abrasion rates, and allowing for up to 10 flights before requiring refurbishment. Alternatively, it would also allow higher re-entry speeds with what Musk described as the best heat rejection capacity known to science. Though unspoken, the implication was clear; while the future might offer fleets of far larger and more capable vessels such as Mars Colonial Transport, the new Dragon was already engineered to ensure a safe landing capability on Mars. To that point, the SpaceX founder offered a bit of insight into his company's philosophy on continuing, iterative technological improvement such as that revealed in the next generation Dragon. It would continue unabated he said, "Until there is a city on Mars." Sitting on the stage behind him it seemed, wasn't just the SpaceX entry into NASA's Commercial Crew competition, it was a new Caravel, designed to sail resolutely past the Dragons which have kept the human race always 20 years away from Mars, and in doing so, lay the cornerstone for just such a city, and the civilization to come.

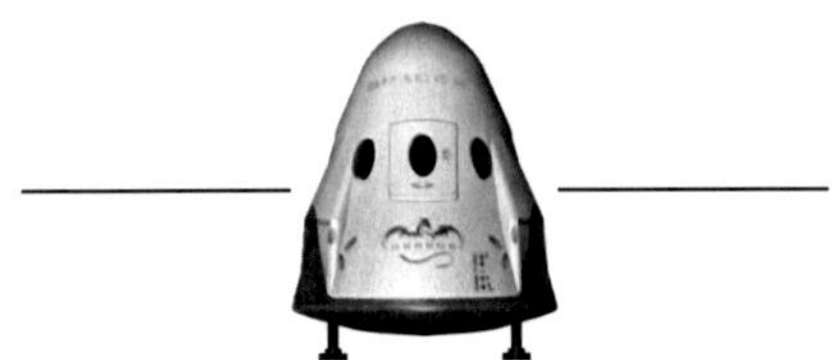

Notes

1 Elon Musk,"Opportunities in Space: Mars Oasis," http://ecorner.stanford.edu/author/elon_musk.
2 Musk,"Opportunities in Space: Mars Oasis."
3 Encyclopedia Astronautica,Saturn V, http://www.astronautix.com/lvs/saturnv.htm .($431 million in 1967 dollars)
4 National Security Decision Directive Number 42," National Space Policy," July 4th , 1982, http://www.hq.nasa.gov/office/pao/History/nsdd-42.html.
5 FAA, Office of Commercial Space Transportation, 2002 Year in Review, January 2003, http://www.faa.gov/about/office_org/headquarters_offices/ast/media/2002yir.pdf.
6 The White House, National Science and Technology Council, Fact Sheet National Space Policy, September 19, 1996, http://history.nasa.gov/appf2.pdf.
7 Department of Defense, Office of Force Transformation Trends, Operationally Responsive Space ExperimentTacSat-1,Oct 17, 2003, http://www.hsdl.org/?view&did=446601.
8 Futron Corporation, Space Transportation Costs: Trends in Price Per Pound to Orbit, 1990-2000, September 6, 2002,http://www.futron.com/upload/wysiwyg/Resources/Whitepapers/Space_Transportation_Costs_Trends_0902.pdf.
9 Stuart Brown,"Winging it to Space," Popular Science, May 1989, p-128.
10 Futron, September 6, 2002
11 Henry M. Holden,The Legacy of the DC-3, (Wind Canyon Books, 2002) p. 123
12 NASA,The Space Launch Initiative: Technology to pioneer the space frontier, MSFC Fact Sheet FS-2002-04-87-MSFC, Release Date 04/02, http://www.nasa.gov/centers/marshall/news/background/facts/slifactstext02.html.
13 California Secretary of State Office, http://www.sos.ca.gov.
14 Michal Belfiore, Behind the Scenes With the World's Most Ambitious Rocket Makers, Popular Mechanics, September 1, 2009, http://www.popularmechanics.com/science/space/rockets/4328638.
15 Encyclopedia Astronautica, Atlas, http://www.astronautix.com/lvs/atlas.htm.
16 GAO, Evolved Expendable Launch Vehicle, DOD Needs to Ensure New Acquisition Strategy Is Based on Sufficient Information, GAO-11-641, September 2011, p. 21
17 I-Shih Chang , Space Launch Vehicle Reliability, The Aerospace Corporation, Crosslink, Winter 2000/2001
18 Todd Halvorson , Mogul Eyes Space Coast for Rocket Missions, Florida Today, January 1, 2009, http://archive.today/BFxuo.
19 Orbital Sciences Corporation,Taurus Launch System Payload User's Guide, March 2006, Release 4.0, http://www.orbital.com/NewsInfo/Publications/taurus-user-guide.pdf.
20 Space is Becoming Big Business,Truax Engineering Multimedia Archive, Neverworld.net/Truax.
21 RIA Novosti, Russia May Supply Soviet-Era Engines for U.S. Space rockets, March 31, 2010,http://en.ria.ru/russia/20100331/158375997.html.
22 NASA, Fastrac Engine,A Boost for Low Cost Space Launch, NASA Fact Sheet FS-1999-02-002-MSFC, Release Date 02/99,http://www.nasa.gov/centers/marshall/news/background/facts/fastrac.html.
23 Jason Paur, Inside SpaceX's Texas Rocket-Testing Facility,Wired.com, October 10, 2012, http://www.wired.com/2012/10/spacex-texas-rocket-test.
24 Spacex.com, updates.
25 NASA, The X-33 History Project Home Page, http://www.hq.nasa.gov/office/pao/History/x-33/menu1.htm.
26 "Remember Beal Aerospace?" Space News, December 11, 2001http://www.spaceprojects.com/beal2001.htm.
27 Jeff Foust, The Falcon and the Showman, The Space Review, December 8, 2003,http://www.thespacereview.com/article/70/1.

28 M. Hurley, Tactical Microsatellite Experiment (TacSat-1), NRL Review, http://www.nrl.navy.mil/research/nrl-review/2004/space-research-and-satellite-technology/hurley.

29 National Snow and Ice Data Center, Defense Meteorological Satellite Program, Satellite F13, NSIDC.org, http://nsidc.org/data/docs/daac/f13_platform.gd.html

30 Encyclopedia Astronautica, Buran, www.astronautix.com/craft/buran.htm.

31 Space Launch Vehicles Broad Area Review Report, Nov. 1999, http://klabs.org/richcontent/Reports/Failure_Reports/Space_Launch_Vehicles_Broad_Area_Review.pdf.

32 Leonard David, SpaceX Private Rocket Shifts to Island Launch, Space .com, August 8, 2005, http://www.space.com/1422-spacex-private-rocket-shifts-island-launch.html.

33 Kimbal Musk, Kwajalein Atoll and Rockets, kwajrockets.blogspot.com.

34 NASA, Columbia Accident Investigation Board, August 2003,Chapter 11, Recommendations, http://www.nasa.gov/columbia/caib/html/start.html

35 NASA, President Bush Announces New Vision for Space Exploration Program, Remarks by President on U.S. Space Policy, NASA Headquarters, Washington, D.C. January 14, 2004, http://history.nasa.gov/Bush%20SEP.htm,

36 NASA, Report of the President's Commission on Implementation of United States Space Exploration Policy, A Journey to Inspire, Innovate and Discover, June 2004, http://www.nasa.gov/pdf/60736main_M2M_report_small.pdf.

37 NASA, Taking the Vision to the Next Step, October 5, 2004, WWW.NASA.gov/missions/solarsystem/vision_concepts.html.

38 Frank Sietzen, Jr. and Keith Cowing, NASA Studying Unmanned Solution to Complete Space Station as Return to Flight Costs Grow, Spaceref.com, July 24. 2005, http://www.spaceref.com/news/viewnews.html?id=1052.

39 NASA, Exploration Systems Architecture Study-Final Report, November 2005, http://www.nasa.gov/exploration/news/ESAS_report.html, Why NASA Chose to Utilize a Shuttle-Derived Crew Launch Vehicle(CLV) Instead of Human Rating an Evolved Expendable Launch Vehicle(EELV), February 8, 2008, http://www.nasa.gov/pdf/211102main_eelv_faq.pdf

40 NASA, Office of the General Counsel, Public Law 105-303, October 28, 1998, http://www.nasa.gov/offices/ogc/commercial/CommercialSpaceActof1998.html.

41 NASA, NASA Awards study contracts for Space Station contingency cargo launch services, MSFC Release 00-251, August 25, 2000.

42 HMX Presentation, Alternative Access to Space, summary March 1, 2003, http://www.hmx.com/AAS_Briefing_Edited.pdf.

43 Greg Lamm, "Rocket maker loses $227 million deal," Puget Sound Business Journal, July 7, 2004, http://www.bizjournals.com/seattle/stories/2004/07/05/story1.html?page=all.

44 Michael Griffin, Remarks to Space Transportation Association, Washington, D.C. June, 21, 2005, http://www.nasa.gov/pdf/119275main_Griffin_STA_21_June_2005.pdf.

45 NASA, Commercial Orbital Transportation Services Demonstrations, NASA Announcement Number COTS-01-05, as amended February 18, 2006, http://www.nasa.gov/pdf/225439main_COTS%20Final%20Announcement%20(Amend%201,%20%202-17-06).pdf

46 Valin Thorn, Commercial Crew and Cargo Overview, AIAA Aerospace Sciences Meeting, Janurary 11, 2007,http://www.nasa.gov/pdf/168735main_AIAA_2007_COTS.pdf

47 Business Editors."SPACEHAB Finalist as NASA's Commercial Space Station Logistics Supplier," The Houston Chronicle, May 5, 2006, http://www.chron.com/news/article/BW-SPACEHAB-Finalist-as-NASA-s-Commercial-Space-1480751.php.

48 "Andrews Space Reveals Cargo Vehicle Design Work," Spaceref, December 14, 2007, http://www.spaceref.com/news/viewpr.html?pid=24293

49 Brian Berger, "t/Space Offers an Option for Closing Shuttle, CEV Gap," Space.com, May 9, 2005, http://www.space.com/1043-space-offers-option-closing-shuttle-cev-gap.html.

50 Dave Klingler, "50 Years to orbit: Dream Chaser's crazy Cold War back story," September 6, 2012, arstechnica.com, http://arstechnica.com/science/2012/09/the-long-complicated-voyage-of-the-dream-chaser-may-yet-end-in-space.
51 NASA, NASA Selects Crew and Cargo Transportation to Orbit Partners, NASA Press Release, August 18, 2006, spaceref.com, http://www.spaceref.com/news/viewpr.html?pid=20628.
52 NASA Partners With Orbital Sciences for Space Transport Service, 02/19/2008
53 NASA, Space Act Agreement Between National Aeronautics and Space Administration and Space Exploration Technologies Corporation for Commercial Orbital Transport Services (COTS), August 18, 2006, http://www.nasa.gov/centers/johnson/pdf/189228main_setc_nnj06ta26a.pdf
54 NASA, Space Act Agreement Amendment Two Between National Aeronautics and Space Administration and Space Exploration Technologies Corporation for Commercial Orbital Transport Services (COTS), February 28, 2008, http://www.nasa.gov/centers/johnson/pdf/216459main_spacex_amend_2.pdf.
55 Braddock Gaskill, "Interview with Elon Musk, SpaceX Has Magical Goals for Falcon 9," NASAspaceflight.com, August 5, 2006, http://www.nasaspaceflight.com/2006/08/spacex-has-magical-goals-for-falcon-9/
56 SpaceX, Updates
57 GAO, NASA Ares I and Orion Projects Risks and Key Indicators to Measure Progress, Testimony Before Congress, April 3, 2008, http://www.gao.gov/new.items/d08186t.pdf.
58 Scott Pelley, "Tesla and SpaceX: Elon Musk's Industrial Empire," cbsnews.com, March 30, 2014, http://www.cbsnews.com/news/tesla-and-spacex-elon-musks-industrial-empire.
59 Eric Fleischauer, "ULA set to add 50 workers, up output," Decatur Daily.com, April 10, 2011, http://static.decaturdaily.com/stories/ULA-set-to-add-50-workers-up-output,77928
60 Michael Hiltzik, "787 Dreamliner Teaches Boeing a Costly Lesson on Outsourcing," The Los Angeles Times, February 15, 2011, http://articles.latimes.com/2011/feb/15/business/la-fi-hiltzik-20110215.
61 NASA Fact Sheet, Super Lightweight External Tank, April 2005, http://www.nasa.gov/centers/marshall/pdf/113020main_shuttle_lightweight.pdf
62 SpaceX Demo Flight 2 Flight Review Update, June 15, 2007, http://www.spaceref.com/news/viewsr.html?pid=24539.
63 "SpaceX Wins Orbcomm Contract to Launch 18-Satellite Constellation," Satellitetoday.com, September 4, 2009,.http://www.satellitetoday.com/telecom/2009/09/04/spacex-wins-orbcomm-contract-to-launch-18-satellite-constell ation
64 NASA, Workforce Transition Strategy, Initial Report, Space Shuttle and Constellation Workforce, March 2008, http://www.nasa.gov/pdf/220259main_Workforce_Transition_Strategy_rpt.pdf.
65 The Florida Senate, Review of Space Florida's Infrastructure Projects, Issue Brief 2010-307, September 2009, http://archive.flsenate.gov/data/Publications/2010/Senate/reports/interim_reports/pdf/2010-307cm.pdf.
66 NASAspaceflight.com, August 5, 2006.
67 SpaceX Press Release, Space Successfully Completes First Stage Firing, October 21, 2009, http://spaceref.com/onorbit/spacex-successfully-completes-first-stage-9-engine-rocket-firing.html.
68 Barack Obama Campaign Fact Sheet "Barack Obama's Plan for Lifetime Success Through Education," http://www.carleton.edu/departments/educ/vote/external%20documents/PreK-12EducationFactSheet.pdf.
69 Robert Block and Mark K. Matthews, "NASA chief Griffin buck's Obama's transition team," December 11, 2008, Orlando Sentinel, http://articles.orlandosentinel.com/2008-12-11/news/nasa11_1_nasa-employees-administrator-mike-griffin-obama-transi tion-team.
70 NASA, Review of Human Spaceflight Plans Committee, Seeking a Human Spaceflight Program Worthy of a Great Nation, http://www.nasa.gov/pdf/396093main_HSF_Cmte_FinalReport.pdf.

71 The White House, Office of the Press Secretary, Remarks by the President on Space Exploration in the 21st Century, April 15, 2010, http://www.nasa.gov/news/media/trans/obama_ksc_trans.html.

72 Ryan Grim and Jason Cherkis, "Elon Musk, SpaceX Founder, Battles Entrenched Rivals Over NASA Contracts," Huffington Post, February 20, 2013, http://www.huffingtonpost.com/2013/02/20/elon-musk-spacex_n_2727312.html.

73 Stephen Clark , "Falcon Countdown Rehearsal 'A Great Success'" , SpaceFlight Now, Feb 26, 2010, http://www.spaceflightnow.com/falcon9/001/100226wetdress.

74 Miles O' Brien, "All in a day's work," This Week in Space, Interview with Ken Bowersox, Jun 26, 2010.

75 Hutchison Statement on SpaceX Test Flight, June 4, 2010, Spaceref.com,http://www.spaceref.com/news/viewpr.html?pid=30980

76 Tony Romm, "Falcon 9 Launches Successfully," Politico.com, June 4, 2010, http://www.politico.com/news/stories/0610/38145.html

77 Need for Commercial Cargo to ISS, William Gerstenmaier, FAA Commercial Space Transportation Advisory Council , May 18, 2007, http://www.faa.gov/about/office_org/headquarters_offices/ast/17th_cst_presentations/media/NASA_FAA_Human_Spacefligh t_Bill_Gerstenmaier.pdf.

78 Leonard David, "Exclusive: Rules Set for $50 Million 'America's Space Prize'" , November 8, 2004, http://www.space.com/517-exclusive-rules-set-50-million-america-space-prize.html.

79 Rob Coppinger, "Picture: UK built SpaceX capsule revealed," Flightglobal.com, April 15, 2008, http://www.flightglobal.com/news/articles/picture-uk-built-spacex-capsule-revealed-222995/

80 Dennis R. Jenkins, Space Shuttle, The History of Developing the National Space Transporation System, (Walsworth Publishing Company, 1996) p. 132

81 Joint Statement of Space Exploration Technologies Corp. and Valador Inc., August 2, 2011, spaceref.com, http://www.spaceref.com/news/viewpr.html?pid=34247.

82 GAO, Commercial Launch Vehicles, NASA Taking Measures to Manage Delays and Reduce Risks, GAO-11-692T, May 26, 2011, http://www.gao.gov/products/GAO-11-692T.

83 GAO, May 26, 2011

84 NASA, Appendix A, Summary of Findings ISS AC-ASAP Review of SpaceX "Dragon" and Orbital "Cygnus" Logistic Vehicles, August 9, 2011, NASA-JSC, http://www.spacepolicyonline.com/pages/images/stories/Stafford_Dyer_CRS_Assessment_Aug_2011.pdf

85 Mark K. Matthews, "New NASA moon rocket could cost $38 billion," The Orlando Sentinel, August 5, 2011, http://articles.orlandosentinel.com/2011-08-05/news/os-nasa-next-moonshot-20110805_1_constellation-moon-program-nas a-supporters-internal-nasa-documents.

86 Public Law 109-155, NASA Authorization Act of 2005, January 4, 2005,http://www.gpo.gov/fdsys/pkg/BILLS-109s1281enr/pdf/BILLS-109s1281enr.pdf.

87 Aerospace Safety Advisory Panel, Annual Report for 2009, p.11, http://oiir.hq.nasa.gov/asap/documents/2009_ASAP_Annual_Report.pdf

88 NASA Fact Sheet, Commercial Crew Development Round 2, FS-2012-05-104-KSC, http://www.nasa.gov/pdf/657336main_06.06.12_CCP_CCDev2_508.pdf.

89 NASA Commercial Crew Program, Session 4, Key Driving Requirements Walkthrough, July 26, 2011

90 Anatoly Zak, History of the Almaz Space History Station, RussianSpaceWeb.com, http://www.russianspaceweb.com/almaz.html.

91 Space Act Agreement No. NNK11MS04S Between National Aeronautics and Space Administration and Space Exploration Technologies Corp. for Commercial Crew Development Round 2(CCDeV 2), http://commercialcrew.nasa.gov/page.cfm?ID=42&CFID=68401&CFTOKEN=67548540.

92 Commerce, Justice, Science and Related Agencies Appropriations Bill, Report 112, p. 70 Trouble HERE

93 Space Act Agreement –NNK12MS02S, Amendment One (01) Between National Aeronautics and Space Administration and Space Exploration Technologies Corp. for Commercial Crew Integrated Capability CCiCap), http://commercialcrew.nasa.gov/page.cfm?ID=38.

94 GAO, Report to Congressional Committees, NASA Assessments of Selected Large Scale Projects, April 2013, p-61, http://www.gao.gov/products/gao-13-276sp.
95 NASA, Request for Proposal (RFP) NNK14467515R, Commercial Crew Transportation Capability Contract (CCtCap), Kennedy Space Center, November 19, 2013, https://prod.nais.nasa.gov/eps/eps_data/158768-SOL-001-001.pdf.
96 FAA, Office of Commercial Space Transportation, 2006 Year in Review, https://www.faa.gov/about/office_org/headquarters_offices/ast/media/2006YIR.pdf
97 Anatoly Zak, "Disaster at Xichang," Air & Space Magazine, February 2013, http://www.airspacemag.com/history-of-flight/disaster-at-xichang-2873673.
98 Lt Gen Thomas S. Moorman, Jr., DoD Space Launch Modernization Plan, Briefing to Congressional Staffers, May 17, 1994, http://www.fas.org/spp/military/program/launch/dod1994.pdf.
99 Frank Sietzen, Jr, "USAF Planning additional EELV funding," Spacelift Washington, January 26, 2002, Spaceref.com, http://www.spaceref.com/news/viewnews.html?id=426
100 Peter B. Teets, Press Conference on the Evolved Expendable Launch Vehicle Program, DOD, July 24, 2003, http://www.defense.gov/Transcripts/Transcript.aspx?TranscriptID=2886
101 "Lockheed Martin Sues Boeing, Former Employees Over Rocket," KOMOnews.com, August 31, 2006, http://www.komonews.com/news/archive/4095326.html
102 Jason Bates, "EELV Launches Dished Evenly to Boeing and Lockheed Martin," Space.com, April 11, 2005, http://www.space.com/945-eelv-launches-dished-evenly-boeing-lockheed-martin.html,
103 NASA, Falcon 9 Launch Vehicle NAFCOM Cost Estimate, August 8, 2011, NASA Deputy Administrator for Policy, http://www.nasa.gov/pdf/586023main_8-3-11_NAFCOM.pdf.
104 GAO, September 2011.
105 Mike Wall, "NASA to Pay $70 Million a Seat to fly Astronauts on Russian Spacecraft," Space.com, April 30, 2013, http://www.space.com/20897-nasa-russia-astronaut-launches-2017.html.
106 Irene Klotz," Bigelow's Fares Show SpaceX Trumps Boeing on Price," Spacenews.com, January 31, 2013, http://www.spacenews.com/article/civil-space/33436bigelow's-fares-show-spacex-trumps-boeing-on-price
107 NASA, Office of Inspector General, Status of NASA's Development of the Multi-Purpose Crew Vehicle, Report No.IG-13-022 (Assignment No. A-12-002-00) August 15, 2013, p. 2-3.http://oig.nasa.gov/audits/reports/FY13/IG-13-022.pdf.
108 Augustine, 8.4
109 Augustine, 8.4
110 Inspiration Mars, Architecture Study Report Summary, Inspirationmars.org, http://www.inspirationmars.org/IM%20Architecture%20Study%20Report%20Summary.pdf
111 Dennis Tito, Written Statement Commercial Space Hearing, Subcommittee on Space, Committee on Science, Space and Technology, November 20, 2013,http://www.inspirationmars.org/Written_Testimony_DTito_Nov2013.pdf.
112 Robert Zubrin, "How We Can Fly to Mars in This Decade-And on the Cheap," The Wall street Journal, May, 14, 2011, http://online.wsj.com/news/articles/SB10001424052748703730804576317493923993056
113 Karcz et al., "Red Dragon: Low-Cost Access to the surface of Mars Using Commercial Capabilities, Lunar and Planetary Institute," http://www.lpi.usra.edu/meetings/marsconcepts2012/pdf/4315.pdf.
114 Mars One Roadmap, mars-one.com, http://www.mars-one.com/mission/roadmap
115 James R. French, Alan Stern, Max Vozoff, Taber McCallum and Charles Deiterich, Architecture for Lunar Return Using Existing Assets, The Golden Spike Company, AIAA Publication, http://goldenspikeco.wpengine.com/wp-content/uploads/2013/07/GSC_Architecture-for-Lunar-Return_JSR_20131.pdf

Index

D

E

F

G

H

I

T

U

V

W

X

Y

Z

Here Be Dragons: The Rise of SpaceX and the Journey to Mars
 ISBN 978-1926837-33-8
Second Edition

Printed and bound in the USA.
Apogee Prime is an imprint of Griffin Media Inc. Box 62034, Burlington, Ontario, Canada, L7R 4K2
www.apogeeprime.com